PARKINSON'S DISEASE

PARKINSON'S DISEASE

Molecular Mechanisms Underlying Pathology

Edited by

PATRIK VERSTREKEN
Center for the Biology of Disease
Flemish Institute for Biotechnology (VIB)
KU Leuven Department of Human Genetics
Leuven, Belgium

ACADEMIC PRESS

An imprint of Elsevier
elsevier.com

Academic Press is an imprint of Elsevier
125 London Wall, London EC2Y 5AS, United Kingdom
525 B Street, Suite 1800, San Diego, CA 92101-4495, United States
50 Hampshire Street, 5th Floor, Cambridge, MA 02139, United States
The Boulevard, Langford Lane, Kidlington, Oxford OX5 1GB, UK

Library of Congress Cataloging-in-Publication Data
A catalog record for this book is available from the Library of Congress

British Library Cataloguing-in-Publication Data
A catalogue record for this book is available from the British Library

ISBN: 978-0-12-803783-6

For information on all Academic Press publications
visit our website at https://www.elsevier.com

  Working together
to grow libraries in
developing countries

www.elsevier.com • www.bookaid.org

Publisher: Mara Conner
Acquisition Editor: Natalie Farra
Editorial Project Manager: Kristi Anderson
Production Project Manager: Karen East and Kirsty Halterman
Designer: Matthew Limbert

Typeset by Thomson Digital

Contents

3. Mitochondrial Fission and Fusion

V.L. HEWITT, A.J. WHITWORTH

4. Axonal Mitochondrial Transport

E. SHLEVKOV, T.L. SCHWARZ

5. Mitophagy

L. KING, H. PLUN-FAVREAU

List of Contributors

Leire Abalde-Atristain Neurodegeneration and Stem Cell Programs, Institute for Cell Engineering; Graduate Program in Cellular and Molecular Medicine, Baltimore, MD, United States

Liesbeth Aerts Center for the Biology of Disease, Flemish Institute for Biotechnology (VIB); Center for Human Genetics, Leuven Institute for Neurodegenerative Disorders and University Hospitals Leuven, Leuven, Belgium

Carolina Cebrián Department of Neurology, Columbia University Medical Center, New York, NY, United States

Ted M. Dawson Neurodegeneration and Stem Cell Programs, Institute for Cell Engineering; Graduate Program in Cellular and Molecular Medicine; Department of Neurology; Solomon H. Snyder Department of Neuroscience; Department of Pharmacology and Molecular Sciences, Johns Hopkins University School of Medicine, Baltimore, MD; Adrienne Helis Malvin Medical Research Foundation; Diana Helis Henry Medical Research Foundation, New Orleans, LA, United States

Valina L. Dawson Neurodegeneration and Stem Cell Programs, Institute for Cell Engineering; Department of Physiology; Graduate Program in Cellular and Molecular Medicine; Department of Neurology; Solomon H. Snyder Department of Neuroscience, Baltimore, MD; Adrienne Helis Malvin Medical Research Foundation; Diana Helis Henry Medical Research Foundation, New Orleans, LA, United States

Victoria L. Hewitt Medical Research Council Mitochondrial Biology Unit, Cambridge Biomedical Campus, Cambridge, United Kingdom

Hao Jia Neurodegeneration and Stem Cell Programs, Institute for Cell Engineering; Department of Physiology, Baltimore, MD, United States

Jungwoo Wren Kim Neurodegeneration and Stem Cell Programs, Institute for Cell Engineering; Department of Physiology, Baltimore, MD, United States

Louise King University College London Institute of Neurology, Queen Square, London, United Kingdom

Christine Klein Institute of Neurogenetics, University of Luebeck, Luebeck, Germany

Patrick A. Lewis School of Pharmacy, University of Reading, Reading; Department of Molecular Neuroscience, UCL Institute of Neurology, London, United Kingdom

Christina M. Lill Institute of Neurogenetics, University of Luebeck, Luebeck, Germany

Ian Martin Neurodegeneration and Stem Cell Programs, Institute for Cell Engineering; Department of Neurology, Baltimore, MD; Jungers Center for Neurosciences Research, Parkinson Center of Oregon, Department of Neurology, Oregon Health and Science University, Portland, OR, United States

Vanessa A. Morais Center for the Biology of Disease, Flemish Institute for Biotechnology (VIB); Center for Human Genetics, Leuven Institute for Neurodegenerative Disorders and University Hospitals Leuven, Leuven, Belgium; Institute of Molecular Medicine (IMM), Faculty of Medicine, University of Lisbon, Lisbon, Portugal

Maria Perez-Carrion Centre for Integrative Biology (CIBIO), Università degli Studi di Trento, Trento, Italy

Giovanni Piccoli Centre for Integrative Biology (CIBIO), Università degli Studi di Trento, Trento, Italy; Dulbecco Telethon Institute

Hélène Plun-Favreau University College London Institute of Neurology, Queen Square, London, United Kingdom

Thomas L. Schwarz F. M. Kirby Neurobiology Center, Boston Children's Hospital; Department of Neurobiology, Harvard Medical School, Harvard University, Boston, MA, United States

Evgeny Shlevkov F. M. Kirby Neurobiology Center, Boston Children's Hospital; Department of Neurobiology, Harvard Medical School, Harvard University, Boston, MA, United States

Sandra F. Soukup Center for the Biology of Disease, Flemish Institute for Biotechnology (VIB), KU Leuven Department of Human Genetics, Leuven, Belgium

David Sulzer Department of Neurology, Columbia University Medical Center; Departments of Psychiatry and Pharmacology, Columbia University Medical Center, New York State Psychiatric Institute, New York, NY, United States

Patrik Verstreken Center for the Biology of Disease, Flemish Institute for Biotechnology (VIB), KU Leuven Department of Human Genetics, Leuven, Belgium

Sven Vilain Center for the Biology of Disease, Flemish Institute for Biotechnology (VIB), KU Leuven Department of Human Genetics, Leuven, Belgium

Alexander J. Whitworth Medical Research Council Mitochondrial Biology Unit, Cambridge Biomedical Campus, Cambridge, United Kingdom

Preface

Parkinson's disease, first described as the "shaking palsy," was originally described in 1817 by James Parkinson. Now 100 years later, we have learned a great deal about the etiology of this very common disease, but a cure still does not exist. In this book, the leading scientists in the field of Parkinson's disease discuss key aspects of molecular and cellular dysfunction associated with the disease and we highlight potential therapeutic avenues that may be explored.

Parkinson's disease is the second most common neurodegenerative disorder and millions of people around the world have been diagnosed with this disease. A number of "typical" features are associated with Parkinson's that include difficulties to initiate movements and a loss of automatism in moving, shaking, rigid muscles, posture changes as well as difficulties to smell, sleep, swallow, constipation etc. Pathologically, several of these dysfunctions are ascribed to the loss of dopaminergic neurons in the substantia nigra pars compacta but the disease is much more systemic and other neurons suffer as well. While several books on the clinical and pathological aspects of Parkinson's disease have been published, compendia on critical molecular and cellular defects associated with Parkinson's are much scarcer, yet understanding the underlying molecular and neuronal dysfunction will be important when developing therapeutic interventions.

While only decades ago it was thought that Parkinson's disease was a purely sporadic disease, caused by environmental causes, scientists have discovered numerous genetic factors, causative genes, and risk loci in the genome, that are strong and important contributors to the disease. The genetic and genomic era in Parkinson's disease has brought (and will bring in the future) many important breakthroughs that are being discussed in this volume. In many instances, the discoveries made regarding these genetic factors are also important to understand how environmental factors contribute to the disease. This book therefore takes off with a classification of the genetic factors involved in Parkinson's disease and continues with a discussion on the different molecular and cellular pathways that have been implicated in the disease. This knowledge will be critical to eventually understand how Parkinson's disease manifests itself at the level of neuronal circuits and the brain.

This book is build up around different key pathways and cellular defects that have been connected to Parkinson's disease and not around individual genes or environmental stress factors causative to the disease. It integrates the data obtained with different in vivo and in vitro systems, from cultured cells and yeast to nematodes, flies, mice, rats, and where applicable humans. The time is right for this book as we are at the brink of taking the molecular and cellular discoveries to the next level for the benefit of patients and society.

Patrik Verstreken
Professor, KU Leuven
Group Leader, VIB

The Neurogenetics of Parkinson's Disease and Putative Links to Other Neurodegenerative Disorders

C.M. Lill, C. Klein

Institute of Neurogenetics, University of Luebeck, Luebeck, Germany

OUTLINE

Parkinson's Disease. http://dx.doi.org/10.1016/B978-0-12-803783-6.00001-8

1 INTRODUCTION

Two decades after the discovery of the first gene causing a monogenic form of Parkinson's disease (PD), that is, *alpha-synuclein (SNCA)*,[1] the etiology of classical PD remains an imperfectly understood complex puzzle of genes and the environment. Although only a minority (ie, ~5%) of cases is due to well-defined genetic causes, important clues about the more common, "idiopathic" PD (iPD) can be garnered from these monogenic model diseases, which will be discussed in more detail in the first part of this introductory chapter. The idiopathic form of PD constitutes the majority (>90%) of all PD cases and typically shows no or a less pronounced family history than monogenic PD. Importantly, iPD cannot be attributed to a single genetic mutation but rather the combination and interaction of dozens to hundreds of genetic risk variants and few environmental factors (such as pesticide exposure, history of head injury, and possibly coffee consumption and smoking history)[2].

Postencephalitic and MPTP-induced parkinsonism lack of convincing concordance rates among monozygotic and dizygotic twins—except for those with an early age of onset[3]—and the identification of environmental risk factors[2] had initially all supported the hypothesis of an exogenous cause of PD. Thus, the identification of monogenic forms of PD has revolutionized this previously held view of a largely nongenetic etiology for this progressive movement disorder. The genetics have clearly established the existence of several distinct entities of PD, and has greatly advanced our understanding of both monogenic and iPD.

Major findings in this context include but are not limited to the discovery of SNCA as the main component of Lewy bodies in both *SNCA* mutation-linked and iPD[4] and the intriguing observation that *SNCA* mutations can not only be causative of PD but also that variants in the very same gene confer risk to iPD.[5–7]

Detailed multimodal analyses of individuals with monogenic forms of PD have provided unique opportunities to pursue the mechanisms of neuronal degeneration in PD highlighting the "Bermuda triangle" of PD pathogenetic mechanisms with (1) impaired protein turnover, (2) mitochondrial dysfunction, and (3) disturbances in synaptic and endosomal vesicle and protein trafficking and recycling[8] in postmitotic neuronal cells (Fig. 1.1).

An improved understanding of monogenic PD and of the genetic contribution to iPD is highly imperative, as it is conceivable that at least a subset of PD may be causally treatable. In this context, neurogenetics provides a unique opportunity to identify and study individuals at risk of this neurodegenerative disorder in its earliest stages, which likely are the ones most amenable to neurorestaurative or even preventive treatment. Although causative PD mutations are rare, all monogenic forms combined and considered across different ethnic populations constitute a significant proportion (~5%) of all PD.

Althogether, the numbers of people with PD in the most populous nations worldwide have been estimated at ~4 million in 2005 and are

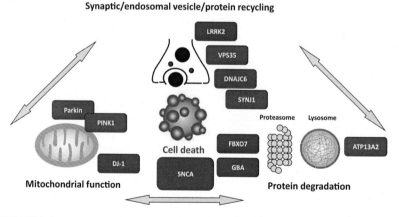

FIGURE 1.1 "Bermuda triangle" of disease mechanisms implicated in monogenic (and idiopathic) Parkinson's disease highlighting the role of confirmed genes for monogenic PD and parkinsonism, as well as for the *GBA* gene in the context of protein degradation, mitochondrial function, and synaptic and endosomal vesicle and protein recycling.

expected to more than double to ~9 million by the year 2030,[9]. These figures clearly highlight the need for continued research efforts to tackle monogenic forms and iPD, which will constitute an increasing health problem and socioeconomic burden in the upcoming years.

Although the research community, neurologists, as well as PD patients have been rightfully excited about the rapid advances in gene discovery and improved understanding of the mechanisms underlying PD, success in translational research is in danger to fall behind the high expectations that have been raised by these new discoveries. In this chapter, we will review important research questions and data gaps and highlight some of the burning issues related to the genetics of PD, including the definition and classification of genetic PD, promises and pitfalls of recent genetic insights and genetic testing, the role of reduced penetrance in disease manifestation, genetic susceptibility, and lessons learned from other neurodegenerative diseases that may relate to PD.

2 DEFINITION OF (GENETIC) PARKINSON'S DISEASE

As the discovery of different monogenic forms of PD has challenged the previous concept of a single clinical and nosological entity, it is important to reconsider the definition of PD and to clarify the nomenclature and categorization used in the remainder of this and the other chapters of this book. The following paragraph is based on and contains verbatim quotes from recommendations of the Task Force on the "Definition of Parkinson's Disease," which published their landmark consensus paper in *Movement Disorders*,[10] the official journal of the International Parkinson and Movement Disorder Society (MDS), in 2014: "Most clinicians would endorse the diagnostic gold standard (of PD) as a combined clinical and pathological syndrome, consisting of the following: (1) a motor clinical syndrome, with levodopa-responsive parkinsonism, typical clinical characteristics, and an absence of markers suggestive of other disease. (2) Pathologic confirmation of alpha-synuclein deposition and dopamine neuronal loss in the substantia nigra pars compacta (SNpc). Only at this point is the diagnosis termed "definite." If typical synuclein pathology is not found, the clinical diagnosis is considered incorrect. Likewise, the pathology is "incidental" in the absence of clinical symptoms or attributed to another disease if parkinsonism did not dominate the clinical picture (eg, diffuse Lewy body disease or primary autonomic failure). Therefore, a motor clinical syndrome is the entry point, and pathology is the arbiter of diagnosis."[10]

However, clinico-pathological findings in monogenic forms of PD have called this definition into question in several ways. For example, although the majority of mutation carriers in the *Leucine-rich repeat kinase 2* (*LRRK2*) gene show typical SNCA-positive Lewy bodies,[11] there are reports on variable *postmortem* findings, even among members of the same LRRK2 family who showed different pathologies, with and without SNCA deposition.[12] Even more strikingly, while most carriers of Parkin (*PARK2*) mutations meet typical clinical criteria of PD,[13] SNCA deposition and Lewy bodies are frequently lacking.[14] To account for these emerging challenges in definition, the MDS Task Force proposed a separate "clinico-genetic" category irrespective of the occurrence of SNCA deposition: "This category would refer specifically to highly penetrant mutations in which the majority of affected meet clinical PD criteria, regardless of whether autopsy specimens of patients with this mutation find SNCA pathology. In research studies, this diagnostic subcategory could be included or not according to the context. For example, an autopsy study validating clinical diagnostic criteria might exclude such patients, a randomized trial of symptomatic dopaminergic therapy might include them, and a neuroprotective trial may elect to include or exclude, depending upon the mechanism of the agent."[10]

Of further note, the umbrella term "parkinsonism" is often used in conjunction with monogenic forms of PD and refers to the typical clinical hallmarks of PD, that is, bradykinesia, resting tremor, rigidity, and postural instability, of which usually at least two have to be present. However, parkinsonism is observed in multiple clinical contexts beyond PD and is, for example, found in patients with multiple system atrophy or progressive supranuclear palsy (PSP), is a common feature in the dystonia–parkinsonisms, and may be encountered as a clinical syndrome in patients with other neurologic conditions, such as stroke or neuroinflammatory disease.

3 CLASSIFICATION AND NOMENCLATURE OF MONOGENIC PARKINSON'S DISEASE

The discovery of monogenic forms of PD did not only pose challenges to the definition of PD but also to its classification and the nomenclature of genetic forms of PD, as the old naming system became increasingly faulty and obsolete. This system of locus (ie, the "PARK" locus system) was originally established to specify chromosomal regions that had been linked to a familial form of PD where the gene was yet unknown. According to this scheme, a number suffix was assigned to each PARK locus according to the chronological description of these loci in the literature (eg, "PARK1," "PARK2," etc.). This system has been adopted by clinicians and researchers to provide names for the condition and is often used synonymously for the chromosomal region.[15] However, as techniques of genetic research and our knowledge have evolved—especially in the light of next-generation sequencing (NGS)—a number of problems have arisen with this system including (1) the inability to distinguish disease-causing genes from genetic risk factors, (2) an inconsistent relationship between list membership and PD phenotypes including those with very atypical features, (3) missing locus symbols for some established monogenic PD genes, (4) more than one symbol being assigned for the same disorder (eg, PARK1 and PARK4 both designating *SNCA*), and (5) unconfirmed reports of a putative PD gene or locus. This state of affairs led to the foundation of the MDS Task Force for Nomenclature of Genetic Movement Disorders in 2012, which very recently published their international consensus recommendations for a new system for naming of genetically determined PD and other movement disorders.[16]

The newly proposed system takes into account the two key notions derived from the neurogenetics of PD: First, there are multiple clinically different forms of PD caused by the same genotype and second, multiple PD genes may cause a similar clinical picture. The Task Force recommendations for the inclusion of genes into the list of confirmed PD genes are as follows: (1) genes should only be included when genetic testing is possible. Accordingly, a disorder should only be listed once the causative gene is identified. The exception to this recommendation is when a founder haplotype is diagnostic, as in the case of X-linked dystonia–parkinsonism. (2) Previously used number suffixes should be replaced by the gene name, that is, the PARK designation should be followed by the name of the disease-causing gene (eg, PARK-*SNCA* [currently PARK1 and PARK4]). (3) Only disease-causing genes should be considered in this naming system, whereas genetic risk loci should not be included. For the latter, the PD-Gene website (http://www.pdgene.org) provides a genome-wide catalog of genetic association results in PD and highlights established as well as putative PD genetic risk factors.[6,7] (4) To avoid inaccuracies and redundancies that currently permeate the lists of locus symbols, the threshold of evidence should be raised before assigning locus symbols according to the guidelines of the US National Human Genome Research Institute: (1) presence of the variant in multiple unrelated affected individuals. (2) Evidence for segregation. (3) The variant should be conserved across different species. (4) The variant should be predicted to alter the normal biochemical effect of the gene product, if possible as supported by functional evidence in human tissue or well-established cellular or animal models or by other biochemical or histological abnormalities.[17] In the following, these criteria will be applied to monogenic PD.

4 MONOGENIC FORMS OF PARKINSON'S DISEASE AND PARKINSONISM

A total of 23 genes and loci have currently been assigned a "PARK" designation (Table 1.1).

TABLE 1.1   The Current List of Locus Symbols for Hereditary PD and Parkinsonism

Symbol	Gene locus	Gene	Inheritance	Status and remarks
PARK1	4q21-22	SNCA	AD	Confirmed
PARK2	6q25.2-q27	PARK2 (Parkin)	AR	Confirmed
PARK3	2p13	Unknown	AD	Unconfirmed; Causative mutation and gene not identified since locus description in 1998
PARK4	4q21-q23	SNCA	AD	Erroneous locus (identical to PARK1)
PARK5	4p13	UCHL1	AD	Unconfirmed (could not be replicated by independent studies
PARK6	1p35-p36	PINK1	AR	Confirmed
PARK7	1p36	PARK7 (DJ-1)	AR	Confirmed
PARK8	12q12	LRRK2	AD	Confirmed; Variations in LRRK2 gene include risk-conferring variants and disease-causing mutations.
PARK9	1p36	ATP13A2	AR	Confirmed
PARK10	1p32	Unknown	Risk factor	Unconfirmed. This locus did not show robust association signals in the most recent GWAS.
PARK11	2q36-27	Unknown	AD	Initially described mutations in GIGYF2 later also found in controls; replication studies could not confirm GIGYF2 as causative of PD
PARK12	Xq21-q25	Unknown	Risk factor	Unconfirmed. This locus did not show robust association signals in the most recent GWAS.
PARK13	2p12	HTRA2	AD or risk factor	Could not be confirmed by independent studies
PARK14	22q13.1	PLA2G6	AR	Confirmed. The majority of cases do not include parkinsonism but present as infantile neuroaxonal dystrophy

(*Continued*)

TABLE 1.1   The Current List of Locus Symbols for Hereditary PD and
Parkinsonism (*cont.*)

Symbol	Gene locus	Gene	Inheritance	Status and remarks
PARK15	22q12-q13	FBX07	AR	Confirmed
PARK16	1q32	Unknown	Risk factor	Unconfirmed. This locus did not show robust association signals in the most recent GWAS.
PARK17	4p16	VPS35	AD	Confirmed
PARK18	6p21.3	EIF4G1	AD	Unconfirmed. Initially described mutation in EIF4G1 later also found in controls
PARK19	1p31.3	DNAJC6	AR	Confirmed
PARK20	21q22.11	SYNJ1	AR	Confirmed
PARK21	3q22.1	DNAJC13	AD	Could not be confirmed by independent studies
PARK22	7p11.2	CHCHD2	AD	Not unequivocally confirmed
PARK23	15q22.2	VPS13C	AR	Could not be unequivocally confirmed by independent studies. First published in 2016 in three unrelated patients with rapidly progressive, early-onset parkinsonism

AD, autosomal dominant; AR, autosomal recessive.
Adapted from Ref. [16].

For eight of these 23 genes and loci, the relationship is not yet unequivo-cally confirmed (PARK3, 5, 11, 13, 18, 21, 22, 23) and three fall into the "risk factor" category (PARK10, 12, 16), of which none could be confirmed as risk locus by the most recent and largest genome-wide association study (GWAS; see later).[6,7] In addition, the PARK18 locus has been assigned to two different loci by two different groups, first to a (meanwhile estab-lished) PD risk locus (*HLA-DRB1*)[18] and second to a proposed autosomal dominant PD gene (*EIF4G1*).[19] This latter finding could, however, not be confirmed in independent replication studies and is most likely not a causative PD gene.[20] For PARK14, mutations in the linked gene, *PLA2G6*, present with a disease other than PD (infantile neuroaxonal dystrophy) in the overwhelming majority of mutation carriers. PARK1 and PARK4 are identical, both referring to the *SNCA* gene. In summary, this leaves 10 con-firmed monogenic genes for PD and parkinsonism (Table 1.2). Mainly in

TABLE 1.2   The Proposed New List of Hereditary PD and Parkinsonism

Designation and reference	GeneReviews and OMIM reference	Clinical clues	Inheritance	Previous locus symbol
1. CLASSICAL PD				
PARK-*SNCA*[1,21]	GeneReviews http://www. ncbi.nlm. nih.gov/ books/ NBK1223/ OMIM 168601	Missense mutations cause classical parkinsonism. Duplication or triplication mutations in this gene cause early onset parkinsonism with prominent dementia	AD	PARK1
PARK-*LRRK2*[22,23]	GeneReviews http://www. ncbi.nlm. nih.gov/ books/ NBK1208/ OMIM 607060	Clinically typical PD	AD	PARK8
PARK-*VPS35*[24,25]	GeneReviews http://www. ncbi.nlm. nih.gov/ books/ NBK1223/ OMIM 614203	Clinically typical PD	AD	PARK17
2. EARLY-ONSET PD				
PARK-*Parkin*[26]	GeneReviews http://www. ncbi.nlm. nih.gov/ books/ NBK1155/ OMIM 600116	Often presents with dystonia, typically in a leg	AR	PARK2
PARK-*PINK1*[27]	GeneReviews http://www. ncbi.nlm. nih.gov/ books/ NBK1223/ OMIM 605909	Often presents with psychiatric features	AR	PARK6

(Continued)

TABLE 1.2   The Proposed New List of Hereditary PD and Parkinsonism (*cont.*)

Designation and reference	GeneReviews and OMIM reference	Clinical clues	Inheritance	Previous locus symbol
PARK-*DJ1*[28]	GeneReviews http://www. ncbi.nlm. nih.gov/ books/ NBK1223/ OMIM 606324		AR	PARK7
3. PARKINSONISM				
PARK-*ATP13A2*[29]	GeneReviews http://www. ncbi.nlm. nih.gov/ books/ NBK1223/ OMIM 606693	Kufor–Rakeb syndrome with parkinsonism and dystonia; additional features: supranuclear gaze palsy, spasticity/ pyramidal signs, dementia, facial-faucial-finger mini-myoclonus, dysphagia, dysarthria, olfactory dysfunction	AR	PARK9
PARK-*FBXO7*[30]	GeneReviews http://www. ncbi.nlm. nih.gov/ books/ NBK1223/ OMIM: 260300	Early-onset parkinsonism with pyramidal signs	AR	PARK15
PARK-*DNAJC6*[31]	GeneReviews: n/a OMIM 615528	May present with mental retardation and seizures	AR	PARK19
PARK-*SYNJ1*[32,33]	GeneReviews: n/a OMIM 615530	May have seizures, cognitive decline, abnormal eye movements, and dystonia	AR	PARK20

AD, autosomal dominant; AR, autosomal recessive; n/a, not available.
Adapted from Ref. [16].

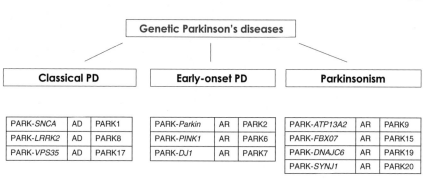

FIGURE 1.2 Clinico-genetic categorization of Parkinson's diseases according to Marras et al.[16]

an effort to guide clinicians but also of possible etiologic relevance, these have been divided into three categories: (1) those that have been attributed to a clinical picture closely resembling that of iPD; (2) those that present with PD but of early onset, and (3) phenotypically complex forms that have parkinsonism as a key clinical feature but in addition present with atypical, multisystem features (Fig. 1.2).

The six monogenic forms presenting with classical or early-onset PD will be described in more detail in the subsequent sections; GeneReview references and OMIM numbers are provided for the four forms characterized by atypical parkinsonism (listed in the third category in Table 1.2).

5 PARK-SNCA

Although mutations in *SNCA* are an extremely rare cause of PD, *SNCA* is likely the most intensely investigated PD gene, not only with respect to causative mutations[1,21] but also to risk variants, as well as function of the gene and the encoded protein. Unifying features of all *SNCA* mutations comprises an overall earlier age of disease onset than that seen in iPD, a faster decline of motor symptoms that are mostly levodopa-responsive, however, with a less sustained alleviation of symptoms than in iPD and with early occurrence of motor fluctuations, and presence of prominent nonmotor features.[34] Compared to the other PD-causing genes, the observation of mutation (type)-specific clinical expression appears to be rather unique to *SNCA*. *SNCA* triplication carriers have an about 10-year earlier onset than duplication carriers. In accordance with a dosage effect, *SNCA* triplication carriers also have a more severe phenotype and faster disease progression than duplication carriers with an about 8-year shorter duration from symptom onset to death. Importantly, other PD-related genes, such as *LRRK2* and *GBA* (acid beta-glucosidase), have also been linked to alterations of *SNCA* levels. Accumulation of SNCA can lead to inhibition of wild-type GBA by interfering with endoplasmic reticulum-to-Golgi trafficking of GBA, which, in turn, leads to decreased GBA activity and increasing accumulation of SNCA.[35] α-Synuclein-induced lysosomal dysfunction has recently been shown to occur through disruptions in protein trafficking.[36] While cell-to-cell transmission of α-synuclein has been demonstrated in both cell culture and animal models (eg, Ref. [37]), the exact sequence and molecular mechanisms of propagation of PD's neuropathology throughout the human brain remain elusive.[38]

6 PARK-LRRK2

Mutations in *LRRK2* are the most common pathogenic changes linked to autosomal dominant PD.[22,23] They account for 3–41% of familial cases and are also found at a lower rate in apparently sporadic cases.[39] The phenotype of LRRK2 p.G2019S mutations is indistinguishable from that of iPD, although tremor is more common and leg tremor may be a useful diagnostic clue[40]. *LRRK2* is a large gene that consists of 51 exons encoding the 2527-amino acid cytoplasmic LRRK2 protein. There are at least seven recurrent, confirmed pathogenic mutations (p.N1437H, p.R1441C, p.R1441G, p.R1441H, p.Y1699C, p.G2019S, p.I2020T). The p.G2019S mutation is by far the most prevalent; due to a founder effect, the p.R1441G is frequent in Basques and p.I2020T in Japanese patients. LRRK2 has a guanosine-5-triphosphate (GTP)-regulated serine/threonine kinase activity with pathogenic LRRK2 variants increasing autophosphorylation or kinase activity, raising potential not only for a mechanistic understanding of the effect of LRRK2 mutations but also for the development of biomarkers[41,42] and of LRRK2 kinase inhibitors to be employed as neuroprotective agents in PD.

7 PARK-VPS35

Two independent studies utilized whole-exome sequencing in a Swiss and an Austrian kindred to identify the same p.D620N (c.1858G > A) mutation in the *vacuolar protein sorting 35 homolog* (*VPS35*) gene as the cause of autosomal dominant PD.[24,25] This mutation was subsequently found in several additional families but is an overall rare cause of PD with a frequency lower than 0.1%.[43] The p.D620N mutation cosegregates with a phenotype similar to iPD and has incomplete, age-associated penetrance. VPS35 is a component of the retromer complex and is involved in retrograde transport from the endosomes to the trans-Golgi network. VPS35 localizes to dendritic spines and is involved in the trafficking of excitatory

AMPA-type glutamate receptors. Fundamental neuronal processes, including excitatory synaptic transmission and synaptic recycling are altered by VPS35 overexpression.[44]

8 PARK-PARKIN

Parkin mutations are the major cause of autosomal recessive and early-onset PD,[26] accounting for up to 77% of isolated PD with an age of onset <20 years[45] and for 10–20% of early-onset PD patients in general. The disease typically starts in the third or fourth decade of life (but can have childhood-onset, especially in Asian patients), is slowly progressive, responds well to dopaminergic treatment,[46] is commonly complicated by dystonia but very rarely by dementia.[13] The clinical phenotypes of *Parkin*-, *PINK1*-, and *DJ-1*-linked PD are similar; however, despite large data gaps, published phenotypic information on *Parkin* is the most comprehensive.[13] A large number and broad spectrum of Parkin mutations have been identified, including alterations in all 12 exons across various ethnic groups. Mutations comprise exchanges of single nucleotides, small deletions, and exonic deletions and duplications.[47] Parkin codes for a 465-amino-acid protein which functions as an E3 ubiquitin ligase in the process of ubiquitination and with a critical role in maintaining mitochondrial function and integrity[48,49] (see also Chapters 2–5).

9 PARK-PINK1

Mutations in the *PTEN-induced putative kinase 1 (PINK1)* gene are the second most common cause of autosomal recessive early-onset PD.[27] The frequency of *PINK1* mutations in patients is in the range of 1–9% with considerable variation across different ethnic groups.[50] *PINK1* mutation carriers have a similar phenotype to Parkin mutation carriers with the possible exception of a greater incidence of psychiatric and cognitive symptoms. Most of the affected have a disease onset in the fourth decade of life and a typical parkinsonian phenotype with asymmetric onset, slow progression, and an excellent response to levodopa.[51] In contrast to *Parkin*, the majority of the reported *PINK1* mutations are either missense or nonsense mutations, whereas only very few families with whole-exon deletions have been reported. PINK1 is a 581-amino-acid ubiquitously expressed protein kinase with most of the PINK1 mutations affecting the kinase domain. Parkin and PINK1 act in a common pathway of mitochondrial quality control, which will be described in detail in Chapter 5 of this book.

10 PARK-DJ-1

Mutations in *DJ-1* were first identified in two consanguineous families of Dutch and Italian origin, respectively,[28] and are a rare cause of autosomal recessive PD (1–2% of early-onset PD cases).[52] The seven exons of the DJ-1 gene encode a 189-amino-acid ubiquitous highly conserved protein, which functions as a cellular sensor of oxidative stress. Loss of DJ-1 has recently been shown to lead to an impaired antioxidant response by altered glutamine and serine metabolism and is accompanied by a constitutive proinflammatory activation of microglia.[53]

11 RECENT NOVEL PARKINSON'S DISEASE CANDIDATE GENES

In addition to the established monogenic genes reviewed in detail previously, a number of additional causative PD genes have been proposed recently. This includes *DNAJC13* (*DnaJ heat shock protein family (Hsp40) member C13*; assigned "PARK21," Table 1.1)[54] for a dominant form of PD, *CHCHD2* (*coiled-coil-helix-coiled-coil-helix-domain-containing protein 2*; assigned "PARK22," Table 1.1),[55] *VPS13C* (*vacuolar protein sorting protein 13C*; assigned "PARK23," Table 1.1) for an early onset, recessive form of PD with early cognitive decline,[56] and *RAB39B* (member RAS oncogene family 39B) for X-linked early-onset PD with intellectual disability.[57] Importantly, independent validation data for any of these genes are either missing (for *VPS13C*) or not unequivocally supportive (eg, for *CHCHD2*;[58-61] for *RAB39B*:[62,63]). Thus, additional data are needed to determine whether these genes can be considered as established causative PD genes.

12 SYSTEMATIC AND REGULARLY UPDATED OVERVIEW OF PARKINSON'S DISEASE MUTATIONS

As demonstrated earlier, the number of reported genes and mutations in PD and other movement disorders has substantially grown over the past few years and this trend will likely even increase, which is in part due to the growing application of whole-exome and whole-genome NGS technologies. However, not all reported putative new disease genes have been or will be confirmed, and the large number of publications makes it increasingly difficult to "separate the wheat from the chaff." Therefore, the MDS has launched the International Parkinson and Movement Disorder Society Genetic Mutation Database (MDSGene, available at http://www.mdsgene.org), which systematically assesses the pathogenicity of reported mutations and links them to movement disorder phenotypes and other demographic and clinical information. MDSGene will be launched in June 2016 with an initial set of nine movement disorder genes including *SNCA*, *VPS35*, *Parkin*, *PINK1*, and *DJ-1*. The content will be completed with all major movement disorder phenotypes and reported causative mutations until 2018 and content will be updated regularly thereafter.

13 REDUCED PENETRANCE AND VARIABLE EXPRESSIVITY OF GENETIC PARKINSON'S DISEASE

Another unexpected outcome of the major NGS-based efforts over the past few years was the identification of a surprisingly large number of carriers of an allegedly pathogenic mutation who eventually did not develop the disease in question. This phenomenon is commonly referred to as "reduced penetrance." Reduced penetrance is an evolving concept and in the current literature is less often used in the sense of all-or-none phenomenon than originally defined. Specifically, in conjunction with disease traits, such as PD, penetrance is typically age-dependent and may border on "variable expressivity" in cases of late or subtle disease manifestations or endophenotypes.

One of the remarkable discoveries of the 1000 Genomes Project in 2010 was that each person carries an average of 50–100 variants previously implicated in inherited diseases.[64] The recently completed third phase of the 1000 Genomes Project reported that the average genome of an individual differs from the reference human genome at 4.1–5 million sites.[64] When focusing on the variants most likely to affect gene function, an average genome was found to contain ~150 sites with protein-truncating variants, 10,000–12,000 sites with protein-altering variants, ~500,000 variant sites overlapping known regulatory gene regions. Furthermore, on average, every genome may contain up to 30 variants also implicated in rare diseases (using ClinVar as reference; http://www.ncbi.nlm.nih.gov/clinvar/). These findings are supported by data of the Exome Aggregation Consortium (ExAC, http://exac.broadinstitute.org) revealing an unexpectedly high frequency of protein-altering mutations in a collection of >60,000 unrelated individuals. Large-scale NGS efforts, such as the 1000 Genomes Project and ExAC, now allow at an unprecedented scale to assess penetrance of putatively pathogenic mutations and gene variants from the reverse perspective—based on large numbers of presumably nondiseased individuals. Notably, reduced penetrance has not only been described in autosomal dominant (eg, for p.G2019S, the most common mutation in *LRRK2*)[65] but also in recessive PD. While homozygous or compound-heterozygous mutations in *Parkin* and *PINK1* result in definite symptom expression, a growing body of evidence suggests that heterozygous mutations in these genes may predispose to PD in a

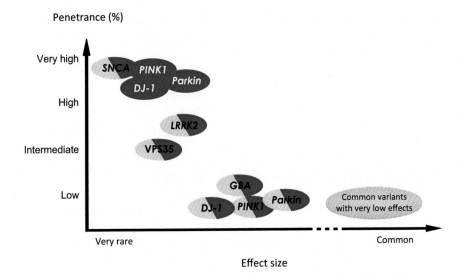

Penetrance (%)

FIGURE 1.3   Schematic distribution of mutations and genetic risk variants according to their penetrance in Parkinson's disease and their minor allele frequency in the general population.

dominant manner with highly reduced penetrance,[66] hypothetically, if a "second hit" is present. An alternative scenario would be the presence of protective factors counteracting the risk-conferring effects of the heterozygous mutation. This would provide support for the notion that "causative mutations with reduced penetrance" and "genetic risk factors" are different expressions on a continuum of effect sizes rather than representing a clearly dichotomous situation (Fig. 1.3). Finally, PD penetrance may also be influenced by ethnicity, as demonstrated for *LRRK2* showing markedly different penetrance rates of the autosomally transmitted p.G2019S mutation in Tunisian versus Norwegian carriers.[65]

14 GENETIC ASSOCIATION STUDIES

Similar to other genetically complex diseases, genetic risk factors of iPD are DNA sequence variants that also occur in the healthy population and individually exert only moderate effects on the risk to develop the disease (Fig. 1.3). The effect of these variants on disease risk can be measured in the context of genetic association studies using a case-unrelated control or a family-based design. Case–control studies allow comparing the frequency of a genetic variant in PD patients and unaffected controls by calculating the "odds ratio" (OR). For instance, an OR of 2 and 1.5 corresponds to a 2- and 1.5-fold increase in PD risk, respectively, for carriers of the allele of interest compared to the reference allele. An OR of 1 indicates no effect whereas an OR of less than 1 indicates a protective effect. Genetic association analyses can also be performed using a family-based design in which the segregation of the putative risk allele with the disease is assessed. However such family-based analyses are statistically less powerful for the detection of moderate genetic risk effects in comparison to studying unrelated cases and controls. Thus, larger sample sizes are needed, which are often beyond reach for family-based settings.

The spectrum of disease risk variants encompasses "common variants," that is, polymorphisms, as well as "rare variants." Frequency thresholds between both types of variants are arbitrary, for example, one commonly applied cutoff for the definition of a common variant or polymorphism is a frequency of at least 1% for the variant's minor allele (minor allele frequency [MAF]) in the general population. The most frequently studied type of polymorphism, especially in the context of GWAS (see later), is the "single-nucleotide polymorphism" (SNP), which represents a common single basepair exchange in the DNA sequence. A typical observation for genetic variants associated with complex traits is that they typically exert only modest effects, whereas rare variants have been proposed to invoke larger effects, that is, ORs >2. However, it should be noted that there are also a few common variants in genetically complex disease that exert larger effects and it is also plausible that rare variants may confer only modest effects. The latter scenario, however, would be very challenging if not impossible to prove as very large sample sizes (ie, ≫100,000 individuals) are needed to detect such effects with sufficient statistical confidence. Furthermore, their contribution to disease on a population-wide level would, by definition, only be very small.

15 CANDIDATE GENE STUDIES IN PARKINSON'S DISEASE

Prior to the commencement of the "GWAS era" in 2005, polymorphisms were typically assessed for their effect on disease susceptibility using a candidate gene approach. Candidate gene studies investigate varying numbers of polymorphisms in selected genes based on a priori knowledge or hypotheses about the pathophysiology of the disease of interest. In PD, for instance, candidates may comprise genes also causing the monogenic form of PD as well as their interactors, genes/variants implicated in other neurodegenerative diseases, and genes involved in pathways known to be involved in PD pathogenesis, for example, mitochondrial and lysosomal pathways.

Candidate gene studies unequivocally identified genetic variants in *SNCA*, *LRRK2*, *MAPT* (microtubule-associated protein tau), and *GBA* as

PD susceptibility factors. Already prior to having been studied by GWAS, all four loci showed genome-wide significant (ie, $p < 5 \times 10^{-8}$, see later) association with PD susceptibility upon meta-analysis.[6] Notably, both *SNCA* and *LRRK2* harbor causative PD mutations as well as polymorphisms that increase PD risk but are not sufficient to cause the disease. While the co-occurrence of rare causative mutations and risk-increasing variants in the same gene appears biologically plausible, it is not a commonly found scenario in complex genetics. For instance, in Alzheimer's disease, which is the most common neurodegenerative disease and shows neuropathological features overlapping with PD, the three known causative genes, that is, *APP* (amyloid precursor protein), *PS1* (presenilin 1), and *PS2* (presenilin 2) harbor causative mutations but do not appear to contain common risk variants.

SNCA was initially investigated[67] as a candidate gene following the identification of monogenic structural and point *SNCA* mutations[1,21,68] as well as the identification of its corresponding protein SNCA in pathognomonic Lewy bodies in iPD.[4] The most extensively studied[5,6] *SNCA* variant from the candidate gene era is the multiallelic Rep1 polymorphism (D4S3481), a microsatellite marker located approximately 10 kb upstream of the transcription start site. However, its relationship to subsequently identified GWAS-based SNP signals in the *SNCA* promoter region (see later) has never been systematically investigated. Preliminary data suggest that the association signal of Rep1 is not independent of the SNP-based GWAS signal at the 5′ end of *SNCA*.[69] Still, the question whether Rep1 has a functional role or is merely a proxy marker of the association signal in the *SNCA* promoter region (see later), currently remains unclarified.

Similarly as with *SNCA*, *LRRK2* was screened as a candidate gene for iPD (eg, Ref. [70–73]) due to its role in the monogenic disease form. The strongest *LRRK2* risk variant p.G2385R (rs3477838348), which was identified early in *LRRK2* candidate gene studies,[70,73] is present only in Asian populations (MAF ~3%) and shows ORs of ~1.7–2.2.[6,74] Subsequently, a comprehensive multiethnic study on *LRRK2* risk variants in PD did not find any risk association for rare *LRRK2* variants in PD but reported on several nonoverlapping, common PD risk polymorphisms in Caucasian (p.K1423K [rs11175964], p.M1646T [rs35303786]) and Asian samples (p.A419V [rs34594498], p.N551K [rs7308720], p.R1398H [rs7133914], p.G2385R), all of which seemed to exert independent effects. However, the statistical support for the two putative risk variants reported for the Caucasian populations is very limited in the most recent GWAS analysis[7] (data available at www.pdgene.org), and no secondary signals were observed in the *LRRK2* region[7]. Thus, further data are needed to clarify which variants, apart from the top GWAS variant in Caucasians (Table 1.3) and p.G2385R in Asians, are associated with PD in *LRRK2*.

TABLE 1.3 Overview of the 26 Genetic Risk Variants Showing Consistent Association With Parkinson's Disease in Genome-Wide Association Studies

SNP	Location (hg38)	Nearest gene	Alleles	Risk allele freq	OR	P value
rs35749011	1:155,162,810	SLC50A1 (GBA)	A/G	0.017	1.824	1.37×10^{-29}
rs114138760*	1:154,925,709	PMVK (GBA)	C/G	0.012	1.574	3.80×10^{-7}
rs823118	1:205,754,444	NUCKS1	T/C	0.559	1.122	1.66×10^{-16}
rs10797576	1:232,528,865	SIPA1L2	T/C	0.140	1.131	4.87×10^{-10}
rs6430538	2:134,782,397	ACMSD	C/T	0.570	1.143	9.13×10^{-20}
rs1474055	2:168,253,884	STK39	T/C	0.128	1.214	1.15×10^{-20}
rs12637471	3:183,044,649	MCCC1	G/A	0.807	1.188	2.14×10^{-21}
rs34311866	4:958,159	TMEM175 (GAK)	G/A	0.191	1.272	1.02×10^{-43}
rs34884217*	4:950,422	TMEM175 (GAK)	A/C	0.913	1.247	1.10×10^{-6}
rs11724635	4:15,735,728	BST1	A/C	0.553	1.126	9.44×10^{-18}
rs6812193	4:76,277,833	FAM47E	C/T	0.636	1.103	2.95×10^{-11}
rs356182	4:89,704,960	SNCA	C/T	0.367	1.316	4.16×10^{-73}
rs7681154*	4: 89,842,802	SNCA	C/A	0.498	1.189	7.09×10^{-19}
rs9275326	6:32,698,883	HLA-DQB1	C/T	0.906	1.211	1.19×10^{-12}
rs13201101*	6:32,375,827	C6orf10	T/C	0.053	1.192	3.84×10^{-6}
rs199347	7:23,254,127	GPNMB	A/G	0.590	1.110	1.18×10^{-12}
rs117896735	7:119,777,065	INPP5F	A/G	0.014	1.624	4.34×10^{-13}

(Continued)

TABLE 1.3   Overview of the 26 26 Genetic Risk Variants Showing Consistent Association With Parkinson's Disease in Genome-Wide Association Studies (*cont.*)

SNP	Location (hg38)	Nearest gene	Alleles	Risk allele freq	OR	P value
rs329648	11:133,895,472	*MIR4697HG*	T/C	0.354	1.105	9.83×10^{-12}
rs76904798	12: 40,220,882	*LRRK2*	T/C	0.143	1.155	5.24×10^{-14}
rs11060180	12:122,819,039	*CCDC62*	A/G	0.558	1.105	6.02×10^{-12}
rs11158026	14:54,882,151	*GCH1*	C/T	0.665	1.106	5.85×10^{-11}
rs2414739	15:61,701,935	*VPS13C*	A/G	0.734	1.113	1.23×10^{-11}
rs14235	16:31,110,472	*BCKDK*	A/G	0.381	1.103	2.43×10^{-12}
rs17649553	17:45,917,282	*MAPT*	G/A	0.774	1.300	2.37×10^{-48}
rs12456492	18:43,093,415	*RIT2*	G/A	0.307	1.106	7.74×10^{-12}
rs8118008	20:3,187,770	*DDRGK1*	A/G	0.657	1.111	3.04×10^{-11}

This table was adapted from Table 1.1 and Supplementary Table 1.3 of the original study.[7] It lists the 22 most significant SNPs per locus (defined in 1 Mb boundaries) that showed genome-wide significant ($p < 5 \times 10^{-8}$) association with Parkinson's disease (PD) status.[7] Furthermore, it displays four SNPs (labeled with a star [*]) that showed significant association (ie, $p < 1 \times 10^{-5}$ following Bonferroni correction) with PD risk upon conditioning on the most significant SNP in the same genetic region (ie, corresponding to the SNP listed in this table in the preceding line). Note that the nearest gene assigned to each SNP here (as determined according to RefGene as available on the UCSC genome browser [https://genome.ucsc.edu/]) does not necessarily represent the functional element underlying the genetic association. The genes in brackets refer to the more commonly used gene names for the respective locus. Full names of all official gene names listed here can be found in the EntrezGene database (http://www.ncbi.nlm.nih.gov/gene/). Alleles = the first allele represents the risk allele. hg38 = human genome build 38, Freq = frequency, OR = odds ratio.

Overlapping neuropathologies (and overlapping clinical features of parkinsonism) across several neurodegenerative diseases, that is, Alzheimer's disease, tauopathies such as PSP and "frontotemporal dementia and parkinsonism linked to chromosome 17" (FTDP-17), as well as PD, prompted the investigation of the *MAPT* gene in PD,[75–78] which encodes the microtubule associated protein tau. In all these diseases hyperphosphorylated tau aggregates within neurons of the central nervous system. The *MAPT* gene is located in an interval on chromosome 17 that in Caucasian populations is characterized by a large inversion giving rise to two extended haplotypes, H1 and H2.[79,80] Of these, H1 has been identified to confer risk for PD.[6,76–78] Interestingly, in East Asian populations, virtually all individuals are homozygous for the H1 haplotype,[6] thus, due to absence of the "allele contrast" (ie, H1 vs H2), no association of *MAPT* in Asian populations has been reported.[81] Genetic variants in *MAPT* have also been reported to increase the risk of PSP[82] and corticobasal degeneration.[83] Despite its role in the genetically complex form of PD, *MAPT* does not appear to be involved in causing the monogenic PD. Instead *MAPT* mutations have been established as a genetic cause of FTDP-17.[84]

The examination and identification of *GBA* as a risk gene for PD was based on the clinical observation that patients with type 1 Gaucher's disease and their relatives show more commonly signs of parkinsonism than expected (eg, Ref. [85,86]). Gaucher's disease is an autosomal recessive lysosomal storage disorder with a deficiency of the enzyme glucocerebrosidase (encoded by the gene *GBA*). The lack of enzymatic activity leads to accumulation of the glycolipid glucocerebroside in several tissues such as liver, spleen, bone marrow, and brain. The clinical observation of a link between Gaucher's disease and PD led to the discovery that several rare mutations in *GBA*, which cause Gaucher's disease in an autosomal recessive manner, confer risk of PD also in an heterozyous state.[6,87] The most frequently investigated and observed variant in PD is p.N370S, which is also the most common mutation in type 1 Gaucher's disease, and increases the risk of PD approximately ~threefold per risk allele.[6,87]

With four candidate genes, that is, *SNCA*, *LRRK2*, *MAPT*, and *GBA*, identified unequivocally as PD risk genes, candidate gene studies in PD have been comparatively successful. Still, similar to most other genetically complex diseases, the vast majority of the over 800 studies in PD reporting association results on hundreds of other candidate genes[6] provided false-positive results as can be judged retrospectively after the emergence of GWAS using increasing samples sizes (see later). This is due to a number of different reasons: (1) usage of underpowered sample collections, (2) application of inappropriate significance thresholds, (3) bias due to undetected and—due to the lack of informative markers—undetectable ethnic admixture, (4) insufficient insights into disease mechanisms and thus unsuitable selection of candidate genes, and (5) limited knowledge of the genetic variation landscape present in the human genome. Most if not all of these limitations have been meanwhile overcome with the availability and usage of genome-wide SNP arrays for GWAS.

16 GENOME-WIDE ASSOCIATION STUDIES

In contrast to hypothesis-driven candidate gene studies, GWAS allow to apply a hypothesis-free approach by testing the association between the phenotype of interest and several hundred thousands to millions of SNPs across the entire genome in one single experiment. This is achieved by high throughput parallel genotyping using specifically designed SNP arrays. Their design and subsequent optimization only became feasible as a consequence of landmark genomics projects such as the Human Genome Project,[88,89] the HapMap,[90,91] and most recently the 1000 Genomes Projects.[92] In this context, it is important to understand that GWAS have been designed to cover most of the common variation of the human genome without the need to genotype every single SNP. Therefore, many SNP arrays "only" genotype a number of "tag" SNPs in genetic regions with dozens to hundreds of variably correlated SNPs (ie, SNPs in "linkage disequilibrium" [LD]). Therefore, the best-associated GWAS-SNP in a region of interest does not necessarily represent the "pathophysiological culprit." Indeed, for none of the PD GWAS signals, the functional variants have been unequivocally identified to date. Furthermore, the investigation of hundred thousands of variants at the same time in a GWAS also led to the establishment of a stringent type-1 error threshold (ie, of the now widely accepted genome-wide significance threshold at $\alpha = 5 \times 10^{-8}$), which has subsequently also been adopted for smaller scale, for example, candidate gene, studies as a "universal" threshold to establish a genetic risk locus.[93]

Since 2005, several GWAS have been published for PD.[6,7,94–101] While the first studies were unsuccessful in establishing novel PD susceptibility loci (eg, Ref. [94–97]), technical, computational, and statistical advances in the field of genomics research and, possibly most importantly, increasing sample sizes led to a growing number of PD risk loci that showed consistent evidence of association. Along these lines, the most recent and largest GWAS, which was conducted on ~19,000 PD patients and 101,000 controls of Caucasian ancestry, established consistent, highly significant association of 26 independent genetic loci with PD risk[7] (Table 1.3; for details on all GWAS results see the PDGene database [http://www.pdgene.org][6]). This includes 22 primary genome-wide significant ($p < 5 \times 10^{-8}$) SNPs and 4 secondary SNPs (with $p < 1 \times 10^{-5}$ following Bonferroni correction).[7] The latter represent independently associated variants in regions that already show another (ie, the primary) signal, and have been identified for the *GBA*, *TMEM175/GAK* (transmembrane protein 175/ cyclin G associated kinase), *SNCA*, and *HLA* (human leukozyte antigen) loci.

As expected, the effect size estimates of the 26 established PD risk loci are only moderate with ORs ranging from 1.10 to 1.82 (median OR of 1.15). The strongest effect sizes (ORs >1.5) were estimated for relatively infrequent variants (ie, those with MAF ~0.01) located in the previously identified candidate gene locus *GBA*, as well as in the novel GWAS locus *INPP5F* (inositol polyphosphate-5-phosphatase F; Table 1.3). The two GWAS variants rs35749011 and rs114138760[7] are located 72 and 308 kb downstream of *GBA*. Whether they simply represent markers of functional, rarer PD risk variants within the *GBA* gene, such as those identified in the candidate-gene era, or pinpoint other independent signals currently remains elusive. This also highlights one of the main limitations of GWAS utilizing common SNP arrays: while the newer products have excellent resolution to assess common variants in the human genome, they currently have only low coverage of rare variants and are, by design, unable to detect novel variants. Thus, they cannot replace resequencing approaches for the functional fine mapping of association signals. Among the more common risk variants (MAF >0.05), the previously identified candidate genes *SNCA* and *MAPT* appear to exert the largest risk effects with ORs of ~1.30 (Table 1.3).[7] In the *SNCA* locus, GWAS revealed two independent signals.[7,99] The most significant primary SNP (rs356182) is located ~19 kb downstream of *SNCA*, while the secondary signal (rs7681154) is located in the promoter region. Another noteworthy result from recent GWAS is the identification of the *HLA* locus as a PD susceptibility locus.[7,18] This finding provided the first genetic evidence for the long-held hypothesis that the involvement of the immune system in PD does not only represent an epiphenomenon of the neurodegenerative processes but is also a primary contributor to its pathogenesis.[102] Interestingly, apart from the well-established role of the *HLA* locus in autoimmune diseases,[103–105] recent GWAS have also pinpointed its role in other neurological and psychiatric diseases such as Alzheimer's disease[106] and schizophrenia.[107] However, the *HLA* locus is presumably the most complex region in the human genome due to its high density of (often immunologically relevant) genes, its polymorphic nature, and extensive LD patterns.[108] Therefore, the clarification of the pathophysiological molecular mechanisms underlying this association in PD as well as in Alzheimer's disease or schizophrenia will be particularly challenging. Finally, the most recent PD GWAS also led to the identification of a PD risk variant in intron 1 of *GCH1* (GTP cyclohydrolase 1), which is known as a causative gene for levodopa-responsive dystonia–parkinsonism (Table 1.4) with an intriguing relationship between phenotypic expression and age at onset. While parkinsonism is overall rare in *GCH1* mutation carriers with an early age of onset (<15 years) and, if present, tends to appear only decades after the onset of dystonic features, more

TABLE 1.4   Genetic Parkinsonism With Atypical or Complex Phenotypes

Designation and reference[a]	GeneReviews and OMIM reference	Clinical clues	Inheritance	Previous locus symbol
NBIA/DYT/ PARK[b] - *PLA2G6*	GeneReviews http://www.ncbi.nlm.nih. gov/books/NBK1675/ OMIM 612953	*PLA2G6*-associated neurodegeneration (PLAN): Dystonia, parkinsonism, cognitive decline, pyramidal signs, psychiatric symptoms (adult phenotype), ataxia (childhood phenotype). Iron accumulation: GP, SN in some; adults may have striatal involvement; about half of INAD and the majority of adult-onset cases lack brain iron accumulation on MRI	AR	NBIA2, PARK14
DYT/PARK-*ATP1A3*[c]	GeneReviews http://www.ncbi.nlm.nih. gov/books/NBK1115/ OMIM 128235	Rapid onset dystonia–parkinsonism	AD	DYT12
DYT/PARK-*TAF1*	GeneReviews http://www.ncbi.nlm.nih. gov/books/NBK1155/ OMIM 314250	Dystonia and parkinsonism	X-linked	DYT3
DYT/PARK-*GCH1*	GeneReviews http://www.ncbi.nlm.nih. gov/books/NBK1508/ OMIM 128230	GTP cyclohydrolase I deficiency (mild form): childhood-onset dopa-responsive dystonia, adult-onset dystonia–parkinsonism. Additional clinical manifestations: diurnal fluctuation, pyramidal signs	AD	*DYT5a*
DYT/PARK-*GCH1*	GeneReviews http://www.ncbi.nlm.nih. gov/books/NBK1155/ OMIM 605407	GTP cyclohydrolase I deficiency (severe form)[26,27]: dystonia, parkinsonism. Additional clinical manifestations: developmental delay, truncal hypotonia, spasticity, oculogyric crises, seizures, with or without hyperphenylalaninemia	AR	None

DYT/PARK-*TH*	GeneReviews: http://www.ncbi.nlm.nih.gov/books/NBK1155/ OMIM 605407	Tyrosine hydroxylase deficiency:	AR	DYT5b
		Mild form: dopa-responsive infantile to early childhood onset dystonia.	AR	None
		Severe form: infantile-onset dystonia and parkinsonism, truncal hypotonia, global developmental delay	AR	None
		Very severe form: infantile-onset dystonia and parkinsonism, oculogyric crises, severe global developmental delay, truncal hypotonia, limb spasticity, autonomic dysfunction		
DYT/PARK-*SPR*	GeneReviews http://www.ncbi.nlm.nih.gov/books/NBK304122/ OMIM 612716	Sepiapterin reductase deficiency: dystonia, parkinsonism. Additional clinical features: motor and speech delay, truncal hypotonia, limb hypertonia and hyperreflexia, oculogyric crises, psychiatric symptoms, autonomic dysfunction, diurnal fluctuation and sleep benefit, no hyperphenylalaninemia	AR	None
DYT/PARK-*QDPR*	GeneReviews: n/a OMIM 612676	Dihydropteridine reductase deficiency: dystonia, parkinsonism. Additional clinical features: developmental delay, truncal hypotonia, seizures, autonomic dysfunction, hyperphenylalaninemia	AR	None

(Continued)

TABLE 1.4　Genetic Parkinsonism With Atypical or Complex Phenotypes (cont.)

Designation and reference[a]	GeneReviews and OMIM reference	Clinical clues	Inheritance	Previous locus symbol
DYT/PARK-PTS	GeneReviews: n/a OMIM 612719	6-pyruvoyl-tetrahydropterin synthase deficiency: dystonia, parkinsonism. Additional clinical features: neonatal irritability, truncal hypotonia, developmental delay, seizures, oculogyric crises, autonomic dysfunction, hyper-phenylalaninemia	AR	None
DYT/PARK-SLC6A3	GeneReviews: n/a OMIM 126455	Dopamine transporter deficiency syndrome: dystonia and parkinsonism (typically infantile-onset, atypical cases with juvenile-onset exist), occasionally chorea in infancy. Additional clinical features: mild developmental delay, truncal hypotonia, ocular flutter/oculogyric crises, saccade initiation failure, bulbar dysfunction	AR	None
DYT/PARK-SLC30A10	GeneReviews: n/a OMIM 611146	Hypermanganesemia with dystonia, polycythemia, and liver cirrhosis dystonia, parkinsonism. Additional clinical features: hypermanganesemia, polycythemia, chronic liver disease, dysarthria	AR	None
DYT/PARK-GLB1	GeneReviews: n/a OMIM 603921	GM1 gangliosidosis (type III, chronic/adult form): dystonia, parkinsonism. Additional clinical features: pyramidal signs, dysarthria, cognitive deficits (often mild initially), skeletal abnormalities and short statue, corneal clouding, vacuolated cells, cardiomyopathy, progressive disease	AR	None

				NBIA5
NBIA/PARK-WDR45	GeneReviews n/a OMIM 300894	Beta-propeller protein-associated neurodegeneration (BPAN, previously SENDA syndrome): Iron accumulation: SN > GP Halo of hyperintensity surrounding linear hypointensity in SN on T1 scans. Additional clinical features: developmental delay/intellectual disability, progressive cognitive decline, seizures, spasticity, Rett-like stereotypies, autistic-features, neuropsychiatric symptoms, sleep disorders, bowel/bladder incontinence, infantile epileptic encephalopathy	X-linked	
NBIA/DYT/PARK-CP	GeneReviews http://www.ncbi.nlm.nih.gov/books/NBK1493/ OMIM 604290	Aceruloplasminemia: dystonia, ataxia, chorea, parkinsonism, tremors. Iron accumulation: more homogeneous involvement of primarily, caudate, putamen, thalamus, dentate. Additional clinical features: Cognitive impairment, psychiatric symptoms, diabetes mellitus, retinal degeneration, anemia, liver iron storage	AR	

AD, autosomal dominant; AR, autosomal recessive; n/a, not available, DYT, dystonia; NBIA, neurodegeneration with brain iron accumulation.

aReferences provided for conditions with isolated "PARK" designation.

bMutations in this gene more commonly cause infantile neuroaxonal dystrophy (INAD): developmental delay/regression, hypotonia, spasticity/pyramidal signs, optic nerve atrophy, sensorimotor neuropathy, seizures.

cMutations in this gene also cause alternating hemiplegia of childhood, CAPOS (cerebellar ataxia, pes cavus, optic atrophy, and sensorineural hearing loss) syndrome, as well as CAOS syndrome (episodic cerebellar ataxia, areflexia, optic atrophy, and sensorineural hearing loss).

Adapted from Ref. [16].

than half of the patients with a later disease onset (>15 years) develop parkinsonism. In these patients, parkinsonism may even be the presenting sign and the clinical picture may be clinically indistinguishable from iPD in individual cases.[109] In line with this, *GCH1* mutations have also been described in members of dystonia–parkinsonism families who showed typical signs and symptoms of clinical PD without dystonia as well as in a cross-sectional PD case–control dataset.[110] Further studies including neuropathologically confirmed case studies are warranted to assess whether *GCH1* plays a role in PD-specific pathogenesis.[111]

Despite the successes of the GWAS approach in identifying genetic risk variants in PD, little is known about their effects beyond disease association. For instance, it currently remains largely unknown whether some or many of the identified loci also have an impact on other clinical characteristics in PD, such as the progression of motor symptoms, the manifestation of nonmotor symptoms, the response to therapy, or the occurrence of motor fluctuations. Notable exceptions were recently highlighted in a number of studies investigating the putative impact of GWAS risk SNPs on age at disease onset.[112–116] The expected direction of effect of PD risk variants on onset age is similar to what has been observed for *SNCA* multiplications as well as autosomal-recessive mutations in *Parkin*, *PINK1*, and *DJ-1*. Thus, the risk allele would be expected to lower the onset age, possibly by accelerating the neurodegenerative process. Indeed, the available studies seem to converge on age at onset-lowering effects for risk alleles in the risk loci *GBA*,[113,116] *TMEM175/GAK*,[113,114] and possibly *SNCA*.[112,116] The effect of the GWAS SNP in the *GBA* locus is in line with the onset-lowering impact of the functional *GBA* variants identified by candidate gene studies (see Section 15 and, eg, Ref. [87]). However, the cumulative impact of all established GWAS risk loci on the age at onset of PD appears to be negligible and explains less than 1% of its variance.[113–115] This suggests that other genetic and/or environmental and lifestyle factors remain to be identified that modify age at onset in iPD.

17 GENE–ENVIRONMENT INTERACTIONS IN PARKINSON'S DISEASE

As mentioned previously, genetically complex diseases result from the interplay between an individual's genetic make-up and environmental or lifestyle influences that he or she is exposed to. "Interaction" in a statistical sense (also known as "effect modification") implies that a genetic variant's effect on disease risk is altered by other (environmental/lifestyle) factors. The establishment of gene–environment (G×E) interaction effects has proven to be difficult in most complex diseases, with iPD representing no exception. This is at least in part due to the fact that most G×E studies to date have followed a candidate gene approach, which is affected by the same limitations as the candidate gene approach for main effects (see Section 15). However, a number of additional difficulties arise in the context of G×E studies, such as substantially reduced power in interaction analyses as compared to main analyses, potential measurement errors upon quantification of environmental and lifestyle exposures, and biased (eg, when using nonpopulation-based ascertainment schemes) selection of individuals. Obtaining sufficiently sized population-based samples with DNA specimens and good quality environmental data is challenging, especially for diseases such as PD with a comparatively low incidence/prevalence. Genome-wide interaction analyses (GWIA) in PD, which would at least overcome the limitations of a candidate gene approach, are still rare. To date, only two studies have been published using the same dataset, which has been derived from a clinic-based convenience sample of ~1500 cases and 900 controls: the first GWIA reported on a genome-wide significant interaction of coffee consumption and SNP rs4998386 in *GRIN2A*,[117] which, however, failed independent validation in several independent population-based datasets.[118] The only other GWIA in the same dataset reported a suggestive result for smoking history that currently lacks any replication data.[119] Thus, this type of research is still in its infancy in PD, and large, population-based and carefully ascertained datasets are needed to investigate gene–environment interactions in PD.

18 LESSONS LEARNED FROM OTHER MONOGENIC NEURODEGENERATIVE DISEASES

A relatively large and growing number of inherited diseases also present with symptoms and signs of parkinsonism and may thus add further clues to our understanding of isolated monogenic and possibly also sporadic PD. While Table 1.4 lists forms of genetic parkinsonism with atypical or complex phenotypes, Table 1.5 compiles disorders that usually present with additional phenotypes but can have predominant parkinsonism. These disorders will not be described in detail; however, GeneReview and OMIM numbers are provided for further reference.

19 TRANSLATIONAL GENETICS: FROM ANIMAL MODELS TO PERSONALIZED TREATMENTS

Several animal models have been employed to study genetic forms of PD.[120] While a comprehensive review of animal models in PD research is beyond the scope of the present chapter, we will highlight an example from Drosophila PD research[121] illustrating how the investigation of animal models may lead to the development of personalized treatments. In flies,

TABLE 1.5   Disorders That Usually Present With Additional Phenotypes but Can Have Predominant Parkinsonism

Designation and reference#	GeneReviews and OMIM reference	Clinical clues	Inheritance	Previous locus symbol
SCA-*ATXN2*	GeneReviews http://www.ncbi.nlm.nih. gov/books/NBK1275/ OMIM 183090	Marked nonataxia features, can have predominant parkinsonism or chorea; neuronopathy, dementia, myoclonus	AD	SCA2
HSP-*KIAA1840*	GeneReviews http://www.ncbi.nlm.nih. gov/books/NBK1210/ OMIM 640360	Pure or complex; may cause Kjellin syndrome; TCC, mental retardation, sensory neuropathy, amyotrophy, dysarthria, nystagmus, ataxia, maculopathy, white matter lesions. Occasional parkinsonism	AR	SPG11
HSP-*ZFYVE26*	GeneReviews http://www.ncbi.nlm.nih. gov/books/NBK1509/ OMIM 270700	Complex; Kjellin syndrome. TCC, WMLs, mental retardation, dysarthria, pigmentary maculopathy, peripheral neuropathy, distal amyotrophy. Occasional parkinsonism	AR	SPG15
POLG	GeneReviews http://www.ncbi.nlm.nih. gov/books/NBK26471/ OMIM 174763	Multiple syndromes often with progressive external ophthalmoparesis. Variable other neurological manifestations. Rare prominent parkinsonism	AD or AR	None
NBIA/ CHOREA-*FTL*	GeneReviews http://www.ncbi.nlm.nih. gov/books/NBK1141/ OMIM 606159	Neuroferritinopathy: Dystonia, chorea, parkinsonism. Iron accumulation: GP, caudate, putamen, SN, red nucleus; cystic BG changes—pallidal necrosis. Additional clinical features: oromandibular dyskinesia, dysphagia, cognitive impairment, behavioral symptoms, low serum ferritin	AD	NBIA3

HSP/NBIA-*FA2H*	GeneReviews http://www.ncbi.nlm.nih.gov/books/NBK1509/ OMIM 612319	Fatty acid hydroxylase-associated neurodegeneration (FAHN): Iron accumulation: GP (more subtle than other NBIAs). Additional clinical features: spastic tetraparesis, cognitive decline, cerebellar and brainstem atrophy, dysarthria, dysphagia, optic nerve atrophy, seizures	AR	SPG35
NBIA/DYT-*PANK2*	GeneReviews http://www.ncbi.nlm.nih.gov/books/NBK121988/ OMIM 234200	Pantothenate kinase-associated neurodegeneration (PKAN): Iron accumulation: GP—eye of the tiger sign. Additional clinical features: spasticity, dysarthria, cognitive decline, gaze palsy, psychiatric symptoms, pigmentary retinopathy	AR	NBIA1
HSP/NBIA-*C19orf12*	GeneReviews http://www.ncbi.nlm.nih.gov/books/NBK185329/ OMIM 614298	Mitochondrial membrane protein-associated neurodegeneration (MPAN): Iron accumulation: GP—hyperintense streaking of medial medullary lamina between GPi and GPe; SN. Additional clinical features: progressive spastic paresis, dysarthria, dysphagia, cognitive decline/dementia, motor axonal neuropathy, optic nerve atrophy, psychiatric symptoms, bowel/bladder incontinence	AR	NBIA4/SPG43

AD, autosomal dominant; AR, autosomal recessive; n/a, not available; SCA, spinocerebellar ataxia; HSP, hereditary spastic paraplegia; DYT, dystonia; NBIA, neurodegeneration with brain iron accumulation.
Adapted from Ref. [16].

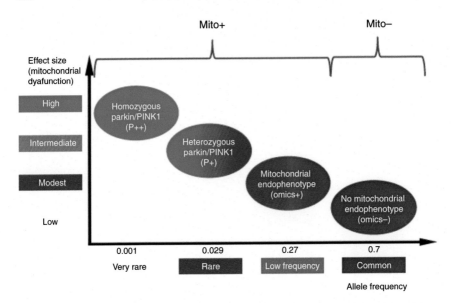

FIGURE 1.4 **Stratification of patients according to different levels of "mitochondrial genetic load" for personalized clinical trials and treatment approaches.** Patients with two mutated *Parkin* or *PINK1* alleles show the highest level of genetically determined mitochondrial dysfunction, followed by those with heterozygous mutations in these genes. Further included in the "Mito+" category are PD patients with susceptibility alleles for mitochondrial dysfunction. These three groups of patients will be contrasted to PD patients without genetic risk for mitochondrial dysfunction and are expected to show a more favorable response to mitochondrial enhancers than patients within the "Mito-" group.

phenotypes caused by mutations in *PINK1* are very well characterized and show defects of flying ability and mitochondrial abnormalities, including reduced ATP levels, impaired enzymatic activity of Complex I, disrupted membrane potential, and abnormal mitochondrial morphology.[122-125] A dominant modifier screen of PINK1 performed in *Drosophila* identified a total of 10 suppressors of the *PINK1* mutant phenotype including vitamin K_2, which functions as an electron carrier molecule in mitochondria leading to improved ATP production.[126] This leads to the hypothesis that genetically stratified subgroups of PD patients may show clinical improvement upon vitamin K2 therapy (Fig. 1.4). This hypothesis is being explored further in the first therapeutical clinical trial in the field of PD with genotype-specific patient selection while preparing this volume.

20 MOST IMPORTANT QUESTIONS FOR FUTURE PERSPECTIVES OF PD NEUROGENETICS

Detailed and systematic reporting of phenotypic data of patients with monogenic PD is currently lagging behind the advances in PD genetics and poses an important unmet need for successful translational efforts.

An improved pathogenetic understanding of the known PD genes and proteins is a key factor for the development of causative treatment strategies, as are the elucidation and possible exploitation mechanisms of endogenous disease protection as evidenced by reduced penetrance of many PD gene mutations. It is imperative that as yet unconfirmed, as well as newly detected PD genes undergo careful independent validation before uncritically being added to the literature and, even more importantly, diagnostic PD gene panels.

Although the mechanisms by which the vast majority of the >3000 GWAS loci currently known to be associated with common diseases—including 26 well-established risk loci for PD[6,7]—exert their pathogenic effects are still mostly unknown, recent technological advances now provide unprecedented opportunities to elucidate these effects. Large public databases, such as the ENCODE project data[127,128] or genome-wide 3D proximity maps of the human genome[129] (eg, 3D/virtual 4C genome browser [http://promoter.bx.psu.edu/hi-c/virtual4c.php] and HiC databases [http://hic.umassmed.edu]), provide unique resources of extensive functional genomic data that will help to close this knowledge gap. In addition, recent developments and improvements of new technologies (eg, chromatin conformation capture assays) now also allow identifying specific interactions of selected genomic regions with other loci opening up new avenues of research to investigate the pathogenic functions of noncoding variants in or close to disease-associated genes. Finally, even if we do not succeed to comprehensively elucidate the pathophysiologic mechanisms underlying monogenic and iPD, the causative PD genes and risk loci as well as the cellular and molecular pathways they are implicated in may represent prime targets for the development of novel neuroprotective or neurorestaurative therapeutic approaches.

References

1. Polymeropoulos MH, Lavedan C, Leroy E, Ide SE, Dehejia A, Dutra A, et al. Mutation in the alpha-synuclein gene identified in families with Parkinson's disease. *Science*. 1997;276(5321):2045–2047.
2. Bellou V, Belbasis L, Tzoulaki I, Evangelou E, Ioannidis JPA. Environmental risk factors and Parkinson's disease: an umbrella review of meta-analyses. *Parkinsonism Relat Disord*. 2016;23:1–9.
3. Tanner CM, Ottman R, Goldman SM, Ellenberg J, Chan P, Mayeux R, et al. Parkinson disease in twins: an etiologic study. *JAMA*. 1999;281(4):341–346.
4. Spillantini MG, Schmidt ML, Lee VM, Trojanowski JQ, Jakes R, Goedert M. Alpha-synuclein in Lewy bodies. *Nature*. 1997;388(6645):839–840.
5. Maraganore DM, de Andrade M, Elbaz A, Farrer MJ, Ioannidis JP, Krüger R, et al. Collaborative analysis of alpha-synuclein gene promoter variability and Parkinson disease. *JAMA*. 2006;296(6):661–670.
6. Lill CM, Roehr JT, McQueen MB, Kavvoura FK, Bagade S, Schjeide B-MM, et al. Comprehensive research synopsis and systematic meta-analyses in Parkinson's disease genetics: the PDGene database. *PLoS Genet*. 2012;8(3):e1002548.

7. Nalls MA, Pankratz N, Lill CM, Do CB, Hernandez DG, Saad M, et al. Large-scale meta-analysis of genome-wide association data identifies six new risk loci for Parkinson's disease. *Nat Genet*. 2014;46(9):989–993.

8. Volta M, Milnerwood AJ, Farrer MJ. Insights from late-onset familial parkinsonism on the pathogenesis of idiopathic Parkinson's disease. *Lancet Neurol*. 2015;14(10):1054–1064.

9. Dorsey ER, Constantinescu R, Thompson JP, Biglan KM, Holloway RG, Kieburtz K, et al. Projected number of people with Parkinson disease in the most populous nations, 2005 through 2030. *Neurology*. 2007;68(5):384–386.

10. Berg D, Postuma RB, Bloem B, Chan P, Dubois B, Gasser T, et al. Time to redefine PD? Introductory statement of the MDS Task Force on the definition of Parkinson's disease. *Mov Disord*. 2014;29(4):454–462.

11. Kalia LV, Lang AE, Hazrati L-N, Fujioka S, Wszolek ZK, Dickson DW, et al. Clinical correlations with Lewy body pathology in LRRK2-related Parkinson disease. *JAMA Neurol*. 2015;72(1):100–105.

12. Wider C, Dickson DW, Wszolek ZK. Leucine-rich repeat kinase 2 gene-associated disease: redefining genotype–phenotype correlation. *Neurodegener Dis*. 2010;7(1–3):175–179.

13. Grünewald A, Kasten M, Ziegler A, Klein C. Next-generation phenotyping using the Parkin example: time to catch up with genetics. *JAMA Neurol*. 2013;70(9):1186–1191.

14. Pramstaller PP, Schlossmacher MG, Jacques TS, Scaravilli F, Eskelson C, Pepivani I, et al. Lewy body Parkinson's disease in a large pedigree with 77 Parkin mutation carriers. *Ann Neurol*. 2005;58(3):411–422.

15. Marras C, Lohmann K, Lang A, Klein C. Fixing the broken system of genetic locus symbols: Parkinson disease and dystonia as examples. *Neurology*. 2012;78(13):1016–1024.

16. Marras C, Lang A, van de Warrenburg BP, Sue C, Tabrizi SJ, Bertram L, et al. Nomenclature of Genetic Movement Disorders: Recommendations of the International Parkinson and Movement Disorder Society Task Force. *Mov Disord*. 2016;31(4):436–457.

17. MacArthur DG, Manolio TA, Dimmock DP, Rehm HL, Shendure J, Abecasis GR, et al. Guidelines for investigating causality of sequence variants in human disease. *Nature*. 2014;508(7497):469–476.

18. Hamza TH, Zabetian CP, Tenesa A, Laederach A, Montimurro J, Yearout D, et al. Common genetic variation in the HLA region is associated with late-onset sporadic Parkinson's disease. *Nat Genet*. 2010;42(9):781–785.

19. Chartier-Harlin M-C, Dachsel JC, Vilariño-Güell C, Lincoln SJ, Leprêtre F, Hulihan MM, et al. Translation initiator EIF4G1 mutations in familial Parkinson disease. *Am J Hum Genet*. 2011;89(3):398–406.

20. Nichols N, Bras JM, Hernandez DG, Jansen IE, Lesage S, Lubbe S, et al. EIF4G1 mutations do not cause Parkinson's disease. *Neurobiol Aging*. 2015;36(8):2444.e1-4.

21. Singleton AB, Farrer M, Johnson J, Singleton A, Hague S, Kachergus J, et al. alpha-synuclein locus triplication causes Parkinson's disease. *Science*. 2003;302(5646):841.

22. Paisán-Ruíz C, Jain S, Evans EW, Gilks WP, Simón J, van der Brug M, et al. Cloning of the gene containing mutations that cause PARK8-linked Parkinson's disease. *Neuron*. 2004;44(4):595–600.

23. Zimprich A, Biskup S, Leitner P, Lichtner P, Farrer M, Lincoln S, et al. Mutations in LRRK2 cause autosomal-dominant parkinsonism with pleomorphic pathology. *Neuron*. 2004;44(4):601–607.

24. Vilariño-Güell C, Wider C, Ross OA, Dachsel JC, Kachergus JM, Lincoln SJ, et al. VPS35 mutations in Parkinson disease. *Am J Hum Genet*. 2011;89(1):162–167.

25. Zimprich A, Benet-Pagès A, Struhal W, Graf E, Eck SH, Offman MN, et al. A mutation in VPS35, encoding a subunit of the retromer complex, causes late-onset Parkinson disease. *Am J Hum Genet*. 2011;89(1):168–175.

26. Kitada T, Asakawa S, Hattori N, Matsumine H, Yamamura Y, Minoshima S, et al. Mutations in the Parkin gene cause autosomal recessive juvenile parkinsonism. *Nature*. 1998;392(6676):605–608.

27. Valente EM, Abou-Sleiman PM, Caputo V, Muqit MMK, Harvey K, Gispert S, et al. Hereditary early-onset Parkinson's disease caused by mutations in PINK1. *Science.* 2004;304(5674):1158–1160.

28. Bonifati V, Rizzu P, van Baren MJ, Schaap O, Breedveld GJ, Krieger E, et al. Mutations in the DJ-1 gene associated with autosomal recessive early-onset parkinsonism. *Science.* 2003;299(5604):256–259.

29. Ramirez A, Heimbach A, Gründemann J, Stiller B, Hampshire D, Cid LP, et al. Hereditary parkinsonism with dementia is caused by mutations in ATP13A2, encoding a lysosomal type 5 P-type ATPase. *Nat Genet.* 2006;38(10):1184–1191.

30. Shojaee S, Sina F, Banihosseini SS, Kazemi MH, Kalhor R, Shahidi G-A, et al. Genome-wide linkage analysis of a Parkinsonian-pyramidal syndrome pedigree by 500 K SNP arrays. *Am J Hum Genet.* 2008;82(6):1375–1384.

31. Edvardson S, Cinnamon Y, Ta-Shma A, Shaag A, Yim Y-I, Zenvirt S, et al. A deleterious mutation in DNAJC6 encoding the neuronal-specific clathrin-uncoating co-chaperone auxilin, is associated with juvenile parkinsonism. *PLoS ONE.* 2012;7(5):e36458.

32. Quadri M, Fang M, Picillo M, Olgiati S, Breedveld GJ, Graafland J, et al. Mutation in the SYNJ1 gene associated with autosomal recessive, early-onset parkinsonism. *Hum Mutat.* 2013;34(9):1208–1215.

33. Krebs CE, Karkheiran S, Powell JC, Cao M, Makarov V, Darvish H, et al. The Sac1 domain of SYNJ1 identified mutated in a family with early-onset progressive parkinsonism with generalized seizures. *Hum Mutat.* 2013;34(9):1200–1207.

34. Kasten M, Klein C. The many faces of alpha-synuclein mutations. *Mov Disord.* 2013;28(6):697–701.

35. Mazzulli JR, Xu Y-H, Sun Y, Knight AL, McLean PJ, Caldwell GA, et al. Gaucher disease glucocerebrosidase and α-synuclein form a bidirectional pathogenic loop in synucleinopathies. *Cell.* 2011;146(1):37–52.

36. Mazzulli JR, Zunke F, Isacson O, Studer L, Krainc D. α-Synuclein-induced lysosomal dysfunction occurs through disruptions in protein trafficking in human midbrain synucleinopathy models. *Proc Natl Acad Sci USA.* 2016;113(7):1931–1936.

37. McCann H, Cartwright H, Halliday GM. Neuropathology of α-synuclein propagation and braak hypothesis. *Mov Disord.* 2016;31(2):152–160.

38. Walsh DM, Selkoe DJ. A critical appraisal of the pathogenic protein spread hypothesis of neurodegeneration. *Nat Rev Neurosci.* 2016;17(4):251–260.

39. Brice A. Genetics of Parkinson's disease: LRRK2 on the rise. *Brain.* 2005;128(Pt 12): 2760–2762.

40. Healy DG, Falchi M, O'Sullivan SS, Bonifati V, Durr A, Bressman S, et al. Phenotype, genotype, and worldwide genetic penetrance of LRRK2-associated Parkinson's disease: a case-control study. *Lancet Neurol.* 2008;7(7):583–590.

41. Fraser KB, Moehle MS, Alcalay RN, West AB. LRRK2 Cohort Consortium. Urinary LRRK2 phosphorylation predicts parkinsonian phenotypes in G2019S LRRK2 carriers. *Neurology.* 2016;86(11):994–999.

42. Grünewald A, Klein C. Urinary LRRK2 phosphorylation: Penetrating the thicket of Parkinson disease?. *Neurology.* 2016;86(11):984–985.

43. Kumar KR, Weissbach A, Heldmann M, Kasten M, Tunc S, Sue CM, et al. Frequency of the D620N mutation in VPS35 in Parkinson disease. *Arch Neurol.* 2012;69(10):1360–1364.

44. Munsie LN, Milnerwood AJ, Seibler P, Beccano-Kelly DA, Tatarnikov I, Khinda J, et al. Retromer-dependent neurotransmitter receptor trafficking to synapses is altered by the Parkinson's disease VPS35 mutation p.D620N. *Hum Mol Genet.* 2015;24(6):1691–1703.

45. Lücking CB, Dürr A, Bonifati V, Vaughan J, De Michele G, Gasser T, et al. Association between early-onset Parkinson's disease and mutations in the Parkin gene. *N Engl J Med.* 2000;342(21):1560–1567.

46. Lohmann E, Periquet M, Bonifati V, Wood NW, De Michele G, Bonnet A-M, et al. How much phenotypic variation can be attributed to Parkin genotype?. *Ann Neurol.* 2003;54(2):176–185.

47. Hedrich K, Eskelson C, Wilmot B, Marder K, Harris J, Garrels J, et al. Distribution, type, and origin of Parkin mutations: review and case studies. *Mov Disord.* 2004;19(10):1146–1157.

48. Rakovic A, Shurkewitsch K, Seibler P, Grünewald A, Zanon A, Hagenah J, et al. Phosphatase and tensin homolog (PTEN)-induced putative kinase 1 (PINK1)-dependent ubiquitination of endogenous Parkin attenuates mitophagy: study in human primary fibroblasts and induced pluripotent stem cell-derived neurons. *J Biol Chem.* 2013;288(4):2223–2237.

49. Pickrell AM, Youle RJ. The roles of PINK1, Parkin, and mitochondrial fidelity in Parkinson's disease. *Neuron.* 2015;85(2):257–273.

50. Klein C, Djarmati A, Hedrich K, Schäfer N, Scaglione C, Marchese R, et al. PINK1, Parkin, and DJ-1 mutations in Italian patients with early-onset parkinsonism. *Eur J Hum Genet.* 2005;13(9):1086–1093.

51. Kasten M, Weichert C, Lohmann K, Klein C. Clinical and demographic characteristics of PINK1 mutation carriers—a meta-analysis. *Mov Disord.* 2010;25(7):952–954.

52. Pankratz N, Pauciulo MW, Elsaesser VE, Marek DK, Halter CA, Wojcieszek J, et al. Mutations in DJ-1 are rare in familial Parkinson disease. *Neurosci Lett.* 2006;408(3):209–213.

53. Meiser J, Delcambre S, Wegner A, Jäger C, Ghelfi J, d'Herouel AF, et al. Loss of DJ-1 impairs antioxidant response by altered glutamine and serine metabolism. *Neurobiol Dis.* 2016;89:112–125.

54. Vilariño-Güell C, Rajput A, Milnerwood AJ, Shah B, Szu-Tu C, Trinh J, et al. DNAJC13 mutations in Parkinson disease. *Hum Mol Genet.* 2014;23(7):1794–1801.

55. Funayama M, Ohe K, Amo T, Furuya N, Yamaguchi J, Saiki S, et al. CHCHD2 mutations in autosomal dominant late-onset Parkinson's disease: a genome-wide linkage and sequencing study. *Lancet Neurol.* 2015;14(3):274–282.

56. Lesage S, Drouet V, Majounie E, Deramecourt V, Jacoupy M, Nicolas A, et al. Loss of VPS13C function in autosomal-recessive parkinsonism causes mitochondrial dysfunction and increases PINK1/Parkin-dependent mitophagy. *Am J Hum Genet.* 2016;98(3):500–513.

57. Wilson GR, Sim JCH, McLean C, Giannandrea M, Galea CA, Riseley JR, et al. Mutations in RAB39B cause X-linked intellectual disability and early-onset Parkinson disease with α-synuclein pathology. *Am J Hum Genet.* 2014;95(6):729–735.

58. Liu Z, Guo J, Li K, Qin L, Kang J, Shu L, et al. Mutation analysis of CHCHD2 gene in Chinese familial Parkinson's disease. *Neurobiol Aging.* 2015;36(11):3117.e7-8.

59. Zhang M, Xi Z, Fang S, Ghani M, Sato C, Moreno D, et al. Mutation analysis of CHCHD2 in Canadian patients with familial Parkinson's disease. *Neurobiol Aging.* 2016;38:217.e7-8.

60. Fan T-S, Lin H-I, Lin C-H, Wu R-M. Lack of CHCHD2 mutations in Parkinson's disease in a Taiwanese population. *Neurobiol Aging.* 2016;38:218.e1-2.

61. Koschmidder E, Weissbach A, Brüggemann N, Kasten M, Klein C, Lohmann K. A nonsense mutation in CHCHD2 in a patient with Parkinson disease. *Neurology.* 2016;86(6):577–579.

62. Löchte T, Brüggemann N, Vollstedt E-J, Krause P, Domingo A, Rosales R, et al. RAB39B mutations are a rare finding in Parkinson disease patients. *Parkinsonism Relat Disord.* 2016;23:116–117.

63. Yuan L, Deng X, Song Z, Yang Z, Ni B, Chen Y, et al. Genetic analysis of the RAB39B gene in Chinese Han patients with Parkinson's disease. *Neurobiol Aging.* 2015;36(10):2907.e11-2.

64. Auton A, Brooks LD, Durbin RM, Garrison EP, Kang HM, 1000 Genomes Project Consortium et al. A global reference for human genetic variation. *Nature.* 2015;526(7571):68–74.

65. Hentati F, Trinh J, Thompson C, Nosova E, Farrer MJ, Aasly JO. LRRK2 parkinsonism in Tunisia and Norway: a comparative analysis of disease penetrance. *Neurology.* 2014;83(6):568–569.

66. Klein C, Lohmann-Hedrich K, Rogaeva E, Schlossmacher MG, Lang AE. Deciphering the role of heterozygous mutations in genes associated with parkinsonism. *Lancet Neurol.* 2007;6(7):652–662.

67. Krüger R, Vieira-Saecker AM, Kuhn W, Berg D, Müller T, Kühnl N, et al. Increased susceptibility to sporadic Parkinson's disease by a certain combined alpha-synuclein/apolipoprotein E genotype. *Ann Neurol.* 1999;45(5):611–617.

68. Krüger R, Kuhn W, Müller T, Woitalla D, Graeber M, Kösel S, et al. Ala30Pro mutation in the gene encoding alpha-synuclein in Parkinson's disease. *Nat Genet.* 1998;18(2): 106–108.

69. Pihlstrøm L, Lill CM, Hansen J, Bjørnarå KA, Dizdar N, Fardell C, et al. Allelic heterogeneity and risk stratification at the SNCA locus in Parkinson's disease. Under review.

70. Skipper L, Li Y, Bonnard C, Pavanni R, Yih Y, Chua E, et al. Comprehensive evaluation of common genetic variation within LRRK2 reveals evidence for association with sporadic Parkinson's disease. *Hum Mol Genet.* 2005;14(23):3549–3556.

71. Paisán-Ruíz C, Lang AE, Kawarai T, Sato C, Salehi-Rad S, Fisman GK, et al. LRRK2 gene in Parkinson disease: mutation analysis and case control association study. *Neurology.* 2005;65(5):696–700.

72. Biskup S, Mueller JC, Sharma M, Lichtner P, Zimprich A, Berg D, et al. Common variants of LRRK2 are not associated with sporadic Parkinson's disease. *Ann Neurol.* 2005;58(6):905–908.

73. Di Fonzo A, Wu-Chou Y-H, Lu C-S, van Doeselaar M, Simons EJ, Rohé CF, et al. A common missense variant in the LRRK2 gene, Gly2385Arg, associated with Parkinson's disease risk in Taiwan. *Neurogenetics.* 2006;7(3):133–138.

74. Ross OA, Soto-Ortolaza AI, Heckman MG, Aasly JO, Abahuni N, Annesi G, et al. Association of LRRK2 exonic variants with susceptibility to Parkinson's disease: a case-control study. *Lancet Neurol.* 2011;10(10):898–908.

75. Morris HR, Janssen JC, Bandmann O, Daniel SE, Rossor MN, Lees AJ, et al. The tau gene A0 polymorphism in progressive supranuclear palsy and related neurodegenerative diseases. *J Neurol Neurosurg Psychiatr.* 1999;66(5):665–667.

76. Golbe LI, Lazzarini AM, Spychala JR, Johnson WG, Stenroos ES, Mark MH, et al. The tau A0 allele in Parkinson's disease. *Mov Disord.* 2001;16(3):442–447.

77. Maraganore DM, Hernandez DG, Singleton AB, Farrer MJ, McDonnell SK, Hutton ML, et al. Case-Control study of the extended tau gene haplotype in Parkinson's disease. *Ann Neurol.* 2001;50(5):658–661.

78. Martin ER, Scott WK, Nance MA, Watts RL, Hubble JP, Koller WC, et al. Association of single-nucleotide polymorphisms of the tau gene with late-onset Parkinson disease. *JAMA.* 2001;286(18):2245–2250.

79. Stefansson H, Helgason A, Thorleifsson G, Steinthorsdottir V, Masson G, Barnard J, et al. A common inversion under selection in Europeans. *Nat Genet.* 2005;37(2):129–137.

80. Lill CM, Bertram L. Towards unveiling the genetics of neurodegenerative diseases. *Semin Neurol.* 2011;31(5):531–541.

81. Satake W, Nakabayashi Y, Mizuta I, Hirota Y, Ito C, Kubo M, et al. Genome-wide association study identifies common variants at four loci as genetic risk factors for Parkinson's disease. *Nat Genet.* 2009;41(12):1303–1307.

82. Höglinger GU, Melhem NM, Dickson DW, Sleiman PMA, Wang L-S, Klei L, et al. Identification of common variants influencing risk of the tauopathy progressive supranuclear palsy. *Nat Genet.* 2011;43(7):699–705.

83. Kouri N, Ross OA, Dombroski B, Younkin CS, Serie DJ, Soto-Ortolaza A, et al. Genomewide association study of corticobasal degeneration identifies risk variants shared with progressive supranuclear palsy. *Nat Commun.* 2015;6:7247.

84. Hutton M, Lendon CL, Rizzu P, Baker M, Froelich S, Houlden H, et al. Association of missense and 5'-splice-site mutations in tau with the inherited dementia FTDP-17. *Nature.* 1998;393(6686):702–705.

85. Neudorfer O, Giladi N, Elstein D, Abrahamov A, Turezkite T, Aghai E, et al. Occurrence of Parkinson's syndrome in type I Gaucher disease. *QJM.* 1996;89(9):691–694.

86. Bembi B, Zambito Marsala S, Sidransky E, Ciana G, Carrozzi M, Zorzon M, et al. Gaucher's disease with Parkinson's disease: clinical and pathological aspects. *Neurology*. 2003;61(1):99–101.

87. Sidransky E, Nalls MA, Aasly JO, Aharon-Peretz J, Annesi G, Barbosa ER, et al. Multicenter analysis of glucocerebrosidase mutations in Parkinson's disease. *N Engl J Med*. 2009;361(17):1651–1661.

88. Venter JC, Adams MD, Myers EW, Li PW, Mural RJ, Sutton GG, et al. The sequence of the human genome. *Science*. 2001;291(5507):1304–1351.

89. Lander ES, Linton LM, Birren B, Nusbaum C, Zody MC, Baldwin J, et al. Initial sequencing and analysis of the human genome. *Nature*. 2001;409(6822):860–921.

90. International HapMap ConsortiumA haplotype map of the human genome. *Nature*. 2005;437(7063):1299–1320.

91. International HapMap 3 Consortium, Altshuler DM, Gibbs RA, Peltonen L, Altshuler DM, Gibbs RA, et al. Integrating common and rare genetic variation in diverse human populations. *Nature*. 2010;467(7311):52–58.

92. 1000 Genomes Project Consortium. A map of human genome variation from population-scale sequencing. *Nature*. 2010;467(7319):1061–1073.

93. Sawcer S. Bayes factors in complex genetics. *Eur J Hum Genet*. 2010;18(7):746–750.

94. Fung H-C, Scholz S, Matarin M, Simón-Sánchez J, Hernandez D, Britton A, et al. Genome-wide genotyping in Parkinson's disease and neurologically normal controls: first stage analysis and public release of data. *Lancet Neurol*. 2006;5(11):911–916.

95. Maraganore DM, de Andrade M, Lesnick TG, Strain KJ, Farrer MJ, Rocca WA, et al. High-resolution whole-genome association study of Parkinson disease. *Am J Hum Genet*. 2005;77(5):685–693.

96. Simón-Sánchez J, Schulte C, Bras JM, Sharma M, Gibbs JR, Berg D, et al. Genome-wide association study reveals genetic risk underlying Parkinson's disease. *Nat Genet*. 2009;41(12):1308–1312.

97. Pankratz N, Wilk JB, Latourelle JC, DeStefano AL, Halter C, Pugh EW, et al. Genome-wide association study for susceptibility genes contributing to familial Parkinson disease. *Hum Genet*. 2009;124(6):593–605.

98. Simón-Sánchez J, van Hilten JJ, van de Warrenburg B, Post B, Berendse HW, Arepalli S, et al. Genome-wide association study confirms extant PD risk loci among the Dutch. *Eur J Hum Genet*. 2011;19(6):655–661.

99. Spencer CCA, Plagnol V, Strange A, Gardner M, Paisan-Ruiz C, Band G, et al. Dissection of the genetics of Parkinson's disease identifies an additional association 5′ of SNCA and multiple associated haplotypes at 17q21. *Hum Mol Genet*. 2011;20(2):345–353.

100. Nalls MA, Plagnol V, Hernandez DG, Sharma M, Sheerin U-M, Saad M, et al. Imputation of sequence variants for identification of genetic risks for Parkinson's disease: a meta-analysis of genome-wide association studies. *Lancet*. 2011;377(9766): 641–649.

101. Pankratz N, Beecham GW, DeStefano AL, Dawson TM, Doheny KF, Factor SA, et al. Meta-analysis of Parkinson's disease: identification of a novel locus, RIT2. *Ann Neurol*. 2012;71(3):370–384.

102. Deleidi M, Gasser T. The role of inflammation in sporadic and familial Parkinson's disease. *Cell Mol Life Sci*. 2013;70(22):4259–4273.

103. Moutsianas L, Jostins L, Beecham AH, Dilthey AT, Xifara DK, Ban M, et al. Class II HLA interactions modulate genetic risk for multiple sclerosis. *Nat Genet*. 2015;47(10): 1107–1113.

104. Lenz TL, Deutsch AJ, Han B, Hu X, Okada Y, Eyre S, et al. Widespread non-additive and interaction effects within HLA loci modulate the risk of autoimmune diseases. *Nat Genet*. 2015;47(9):1085–1090.

105. Lill CM. Recent advances and future challenges in the genetics of multiple sclerosis. *Front Neurol*. 2014;5:130.

106. Lambert JC, Ibrahim-Verbaas CA, Harold D, Naj AC, Sims R, Bellenguez C, et al. Meta-analysis of 74,046 individuals identifies 11 new susceptibility loci for Alzheimer's disease. *Nat Genet*. 2013;45(12):1452–1458.

107. Shi J, Levinson DF, Duan J, Sanders AR, Zheng Y, Pe'er I, et al. Common variants on chromosome 6p22.1 are associated with schizophrenia. *Nature*. 2009;460(7256):753–757.

108. Mungall AJ, Palmer SA, Sims SK, Edwards CA, Ashurst JL, Wilming L, et al. The DNA sequence and analysis of human chromosome 6. *Nature*. 2003;425(6960):805–811.

109. Tadic V, Kasten M, Brüggemann N, Stiller S, Hagenah J, Klein C. Dopa-responsive dystonia revisited: diagnostic delay, residual signs, and nonmotor signs. *Arch Neurol*. 2012;69(12):1558–1562.

110. Mencacci NE, Isaias IU, Reich MM, Ganos C, Plagnol V, Polke JM, et al. Parkinson's disease in GTP cyclohydrolase 1 mutation carriers. *Brain*. 2014;137(Pt 9):2480–2492.

111. Weissbach A, Klein C. Hereditary dystonia and parkinsonism: two sides of the same coin?. *Brain*. 2014;137(Pt 9):2402–2404.

112. Brockmann K, Schulte C, Hauser A-K, Lichtner P, Huber H, Maetzler W, et al. SNCA: major genetic modifier of age at onset of Parkinson's disease. *Mov Disord*. 2013;28(9):1217–1221.

113. Lill CM, Hansen J, Olsen JH, Binder H, Ritz B, Bertram L. Impact of Parkinson's disease risk loci on age at onset. *Mov Disord*. 2015;30(6):847–850.

114. Nalls MA, Escott-Price V, Williams NM, Lubbe S, Keller MF, Morris HR, et al. Genetic risk and age in Parkinson's disease: continuum not stratum. *Mov Disord*. 2015;30(6):850–854.

115. Pihlstrøm L, Toft M. Cumulative genetic risk and age at onset in Parkinson's disease. *Mov Disord*. 2015;30(12):1712–1713.

116. Davis AA, Andruska KM, Benitez BA, Racette BA, Perlmutter JS, Cruchaga C. Variants in GBA, SNCA, and MAPT influence Parkinson disease risk, age at onset, and progression. *Neurobiol Aging*. 2016;37:209.e1–209.e7.

117. Hamza TH, Chen H, Hill-Burns EM, Rhodes SL, Montimurro J, Kay DM, et al. Genome-wide gene-environment study identifies glutamate receptor gene GRIN2A as a Parkinson's disease modifier gene via interaction with coffee. *PLoS Genet*. 2011;7(8):e1002237.

118. Ahmed I, Lee P-C, Lill CM, Searles Nielsen S, Artaud F, Gallagher LG, et al. Lack of replication of the GRIN2A-by-coffee interaction in Parkinson disease. *PLoS Genet*. 2014;10(11):e1004788.

119. Hill-Burns EM, Singh N, Ganguly P, Hamza TH, Montimurro J, Kay DM, et al. A genetic basis for the variable effect of smoking/nicotine on Parkinson's disease. *Pharmacogenomics J*. 2013;13(6):530–537.

120. Valadas JS, Vos M, Verstreken P. Therapeutic strategies in Parkinson's disease: what we have learned from animal models. *Ann NY Acad Sci*. 2015;1338:16–37.

121. Vanhauwaert R, Verstreken P. Flies with Parkinson's disease. *Exp Neurol*. 2015;274 (Pt A):42–51.

122. Clark IE, Dodson MW, Jiang C, Cao JH, Huh JR, Seol JH, et al. Drosophila pink1 is required for mitochondrial function and interacts genetically with Parkin. *Nature*. 2006;441(7097):1162–1166.

123. Park J, Lee SB, Lee S, Kim Y, Song S, Kim S, et al. Mitochondrial dysfunction in Drosophila PINK1 mutants is complemented by Parkin. *Nature*. 2006;441(7097):1157–1161.

124. Morais VA, Verstreken P, Roethig A, Smet J, Snellinx A, Vanbrabant M, et al. Parkinson's disease mutations in PINK1 result in decreased Complex I activity and deficient synaptic function. *EMBO Mol Med*. 2009;1(2):99–111.

125. Vilain S, Esposito G, Haddad D, Schaap O, Dobreva MP, Vos M, et al. The yeast complex I equivalent NADH dehydrogenase rescues pink1 mutants. *PLoS Genet*. 2012;8(1):e1002456.

126. Vos M, Esposito G, Edirisinghe JN, Vilain S, Haddad DM, Slabbaert JR, et al. Vitamin K2 is a mitochondrial electron carrier that rescues pink1 deficiency. *Science*. 2012;336(6086):1306–1310.

127. ENCODE. Project ConsortiumAn integrated encyclopedia of DNA elements in the human genome. *Nature*. 2012;489(7414):57–74.
128. Diehl AG, Boyle AP. Deciphering ENCODE. *Trends Genet*. 2016;32(4):238–249.
129. Lieberman-Aiden E, van Berkum NL, Williams L, Imakaev M, Ragoczy T, Telling A, et al. Comprehensive mapping of long-range interactions reveals folding principles of the human genome. *Science*. 2009;326(5950):289–293.

Electron Transport Chain

L. Aerts*,**, V.A. Morais*,**,†

*Center for the Biology of Disease, Flemish Institute for Biotechnology (VIB), Lisbon, Belgium; **Center for Human Genetics, Leuven Institute for Neurodegenerative Disorders and University Hospitals Leuven, Leuven, Belgium; †Institute of Molecular Medicine (IMM), Faculty of Medicine, University of Lisbon, Lisbon, Portugal

Parkinson's Disease. http://dx.doi.org/10.1016/B978-0-12-803783-6.00002-X

1 INTRODUCTION

Cellular processes require energy. Usually, this energy is provided in the form of adenosine triphosphate (ATP). This universal energy carrier is produced primarily via oxidative phosphorylation in the mitochondria, which is why these organelles are often referred to as the power plants of the cell. An alternative but less efficient way to produce ATP is via glycolysis in the cytoplasm. Both metabolic routes rely on the dietary supply of glucose and other carbohydrates such as fructose and galactose. Via glycolysis, 1 molecule of glucose yields maximally 2 molecules of ATP, while oxidative phosphorylation has a maximum yield of 36 molecules of ATP per molecule of glucose and is responsible for the bulk of cellular energy production (Fig. 2.1). Both processes play an important role in metabolism, not only with regard to ATP production, but also for the production of nucleic acids, amino acids, and lipids.

2 GLYCOLYSIS IN THE CYTOSOL

Energy-rich food contains a large amount of sugars or carbohydrates, as these compounds are the main drivers of cellular energy production. Glycolysis describes the metabolic conversion of glucose into pyruvate, a 10-step process producing two high-energy compounds, ATP and NADH.[1] In addition to glucose, other sugars such as galactose and fructose can be converted to enter the glycolytic pathway at intermediate entry points. If there is an abundance of sugars, the product of the first step in glycolysis,

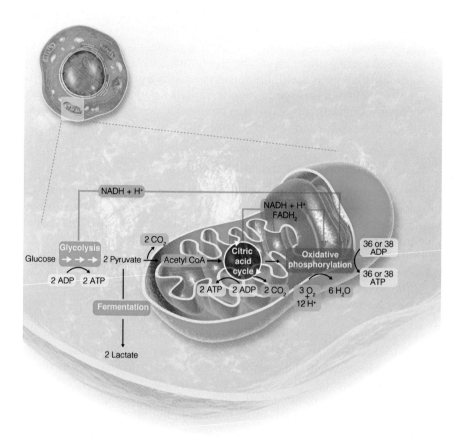

FIGURE 2.1 **Cellular ATP production pathways.** Adenosine triphosphate (ATP) is a high-energy compound that provides energy for a wide range of cellular processes. ATP is produced through the breakdown of glucose via two pathways. Via glycolysis in the cytoplasm, one molecule of glucose is converted to two molecules of pyruvate, NADH, and ATP. Pyruvate can either be fermented or translocated to the mitochondria, where it is converted to Acetyl-CoA and metabolized to CO_2 and H_2O in the Krebs cycle or citric acid cycle. The Krebs cycle provides substrates for the electron transport chain (ETC) and oxidative phosphorylation. If glucose is metabolized via the latter pathway, it yields an additional 2 molecules of ATP in the Krebs cycle and 32 more by oxidative phosphorylation, resulting in a total of 36 molecules of ATP from the breakdown of 1 molecule of glucose. Therefore, mitochondrial energy production is responsible for the bulk of the production of cellular energy. The ETC is at the core of this process, since it reduces O_2 and generates the proton motive force that drives the production of 32 molecules of ATP. The processes in the Krebs cycle, the ETC, and oxidative phosphorylation thus act in concert during cellular respiration, consuming O_2 and producing CO_2.

glucose-6-phosphate, is converted to glycogen for storage in the liver. Other intermediates are also used as building blocks for cellular carbohydrates in several anabolic pathways. Since glycolysis is not dependent on O_2, it can take place under anaerobic conditions.

Although glycolysis is a highly conserved pathway that does not require molecular oxygen, its energetic yield is limited. To increase the efficiency of energy production, pyruvate must be translocated to the mitochondria where it can be converted to acetyl-CoA.

2.1 ATP Production in the Mitochondria

Mitochondria are highly compartmentalized organelles comprised of two phospholipid bilayers: the outer and inner mitochondrial membrane. The inner mitochondrial membrane surface is enlarged via extensive invaginations called cristae and separates the intermembrane space from the matrix. These spatial arrangements enable the coordination of three different processes that lead to the production of ATP via respiration: the Krebs cycle, the electron transport chain (ETC), and oxidative phosphorylation.

The Krebs or citric acid cycle is an eight-step cyclic reaction catalyzing the complete oxidation of acetyl-CoA to CO_2, which takes place in the mitochondrial matrix.[2] The citric acid cycle is conserved in all aerobic organisms (but takes place in the cytosol of prokaryotes) and acquires its name from citric acid, which is consumed and regenerated during the cyclic reaction. Acetyl-CoA is derived from the breakdown of certain amino acids, from fats by fatty acid oxidation, or from carbohydrates through glycolysis. Its stepwise oxidation enables the production of NADH and $FADH_2$, two important high-energy molecules that feed into the ETC.

The ETC is a series of four multisubunit mitochondrial complexes that participate in the transport of electrons, coupled with the translocation of protons across the inner mitochondrial membrane. This generates an electrochemical gradient or proton motive force across the membrane, which drives the conversion of ADP to ATP by the F_1F_0-ATP synthase. The mammalian mitochondrial ETC includes three proton-pumping enzymes, namely complex I (NADH:ubiquinone oxidoreductase), complex III (cytochrome $bc1$ oxidase), and complex IV (cytochrome c oxidase), and a smaller complex, complex II (succinate:ubiquinone oxidoreductase), which constitutes an additional entry point for electrons into the chain (Fig. 2.2).[3] Electrons are transported within each complex through several redox centers, and from one complex to another via two mobile, membrane-embedded carriers: ubiquinone and cytochrome c. At complex I, NADH is oxidized and two electrons are transferred to ubiquinone (also referred to as CoQ_{10}, short for coenzyme Q_{10}), which is reduced to ubiquinol. Four matrix protons are translocated to the intermembrane space

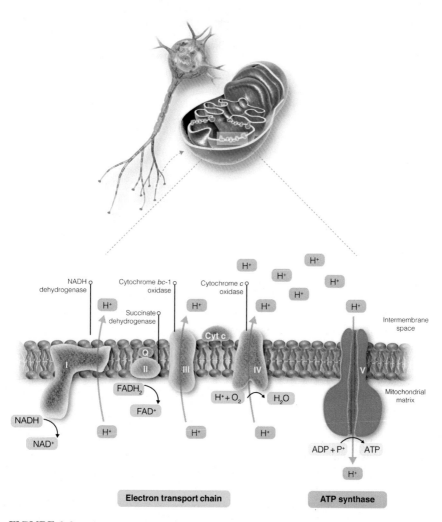

FIGURE 2.2   The electron transport chain. The electron transport chain consists of four multisubunit complexes, located in the inner mitochondrial membrane. Electron transport between the complexes occurs via two mobile carriers, ubiquinone (Q) and cytochrome c (Cyt c). Complex I or NADH:ubiquinone oxidoreductase (also known as NADH dehydrogenase), the largest of all four complexes, channels the transfer of two electrons from NADH to ubiquinone. At the level of complex III or cytochrome bc_1 oxidase, these electrons are transferred to two molecules of cytochrome c, which are subsequently used to reduce O_2 to two molecules of H_2O at complex IV (cytochrome c oxidase). Complex II or succinate:ubiquinone oxidoreductase (also known as succinate dehydrogenase) reduces ubiquinone as well, by converting succinate into fumarate. The transfer of electrons in complexes I, III, and IV is accompanied by the translocation of a total of 10 protons from the matrix to the intermembrane space. This generates an electrochemical potential across the inner mitochondrial membrane, which drives the conformational change of F_1F_0-ATP synthase and the generation of ATP.

(IMS) during this redox reaction. Although complex II does not pump any protons, it also reduces ubiquinone by converting succinate to fumarate. Ubiquinol can freely diffuse through the membrane and donates two electrons at the level of complex III, leading to the reduction of two cytochrome c molecules. Complex IV utilizes four electrons from four molecules of cytochrome c for the reduction of O_2, resulting in the production of two H_2O molecules. This process is also known as O_2 consumption. Overall, the oxidation of one molecule of NADH leads to the transport of 10 protons across the inner mitochondrial membrane. This proton motor force couples the ETC to oxidative phosphorylation by the F_1F_0-ATP synthase, sometimes referred to as the 5th ETC complex, even though it does not contribute to electron transport. The F_0 subunit of F_1F_0-ATP synthase serves as an ion channel through which protons are transported back into the matrix. This transport drives conformational changes in the F_1 subunit, catalyzing the formation of ATP from ADP and inorganic phosphate. While this is the "default" mode of operation, the F_1F_0-ATP synthase can also work in the reverse direction, pumping protons from the matrix into the IMS at the expense of ATP.[4] In this way, the mitochondrial membrane potential is conserved, even in the absence of proton translocation by the ETC complexes, for example during ischemia in the absence of O_2.[5,6] In other words, the F_1F_0-ATP synthase can compensate temporarily for defects in the ETC to conserve mitochondrial function, at the expense of ATP produced in the cytosol.

ATP production requires the coupling of the ETC to oxidative phosphorylation, but in some conditions, this coupling is impeded. Uncoupling proteins dissipate the mitochondrial membrane potential before it can be used by F_1F_0-ATP synthase. As such, the energy is converted into heat instead of ATP, a process that is physiologically important in brown adipocytes of hibernating animals.[7] However, a number of exogenously applied pharmacological agents also act as uncouplers and are very harmful to overall mitochondrial function.

Besides oxidative phosphorylation, the proton motive force also drives mitochondrial Ca^{2+} uptake, ATP/ADP exchange, Na^+ and K^+ fluxes, and the import of several nuclear-encoded proteins across the inner mitochondrial membrane. Calcium homeostasis is of vital importance in neurons, as their excitation and synaptic activity are dependent on the calcium gradient across the cell membrane. Mitochondria aid in the regulation of cytoplasmic calcium fluxes by buffering Ca^{2+} levels in the matrix.[8] Indeed, the role of mitochondria in eukaryotic cells extends well beyond energy production, and also includes the regulation of apoptosis. Upon induction by death stimuli (eg, DNA damage, UV radiation, or serum starvation), the mitochondrial membranes are permeabilized, for example, by opening of the mitochondrial permeability transition pore. In addition to membrane depolarization and a halt in ATP production, several apoptosis-inducing factors are released, which bind apoptosis-inhibiting proteins in the cytoplasm and set in motion caspase-dependent cell death mechanisms.[9,10] One of the apoptosis signaling molecules that is released from the intermembrane space is the ETC compound cytochrome c.[11] As these mitochondrial processes are strongly interconnected, and they all rely on the maintenance of the mitochondrial membrane potential, proper ETC function is not only directly relevant for ATP production but also for other cellular processes such as synaptic activity and apoptosis.

Similar electron transport mechanisms are found in all forms of life, either fueled by sunlight in the case of photosynthesis in plants, or by the oxidation of sugars in cellular respiration in animals. Since mitochondria contain their own DNA (mtDNA) that is highly similar to the bacterial genome, and since they are engulfed by a double membrane, these organelles are thought to have arisen through endosymbiosis. As a result of a complex coevolution, part of the protein subunits that constitute the large ETC complexes are encoded by the mitochondrial genome, while others are encoded by the nuclear genome. The proper spatio–temporal assembly of each of the complexes depends on the expression of subunits encoded by the nucleus, their import into the mitochondria, and the step-wise buildup of subcomplexes with mtDNA-encoded proteins. Mutations in both nuclear and mitochondrial ETC-encoding genes have been linked to serious multisystem disorders, including several neurodegenerative diseases. As complex I is the largest of all 4 ETC complexes, comprising 45 subunits,[12] its genetic deficiency is most prevalent among primary neurological mitochondrial conditions.

In textbooks, the ETC complexes are often depicted as separate entities, connected by ubiquinone and cytochrome c mobile electron carriers, but accumulating evidence suggests that they cluster together into supercomplexes or respirasomes.[13,14] While the precise structure and function of these supercomplexes is highly debated, it is hypothesized that they affect electron transport. Further regulation of the ETC occurs through phosphorylation, acetylation, and proteolytic maturation of the complex subunits.[15] Thus, ETC function can be modulated and fine-tuned via posttranslational modification at various levels.

2.2 Metabolic Shifting and the Warburg Effect

Since glycolysis generates only two molecules of ATP for each molecule of glucose, its relative contribution to the overall cellular energy supply is limited. However, under certain conditions, the glycolytic pathway is upregulated. In the absence of oxygen, for example, ATP can no longer be produced via oxidative phosphorylation, and glycolysis is the only pathway that can produce energy anaerobically.

Cancer cells exhibit highly increased levels of glycolysis, displaying glycolytic rates up to 200-fold higher than normal cells.[16] These metabolic changes were first described by Otto Warburg, and are therefore known as the Warburg effect.[17] Although recent developments in the measurement of biochemical parameters have demonstrated that oxidative phosphorylation is not necessarily shut down in these cells, it is clear that the metabolic balance between glycolysis and oxidative phosphorylation is very complex and differs between cancer cell and tissue types.[18] Altered conditions and energy requirements determine the metabolic balance in a given tissue, and the relative contribution of each process has biological and technical consequences. The metabolic state should be taken into account when studying the activity of ETC complexes in different cellular/neuronal populations or in immortal cancer-derived cell lines. Differences in metabolism could also explain specific vulnerabilities of certain cells or tissues under various environmental conditions.

3 IMPACT OF ELECTRON TRANSPORT CHAIN DYSFUNCTION ON NEURONAL FUNCTION

In this chapter we focus on mitochondrial energy production via electron transport and oxidative phosphorylation, and how dysfunction in these processes may lead to neurodegenerative diseases, including Parkinson's disease (PD). As highlighted previously, ATP production is not the only function of mitochondria. However, all other mitochondrial roles, including that of calcium handling and apoptosis regulation, are highly dependent on a functional mitochondrial membrane potential, and thus all are directly or indirectly related to ETC function. It is therefore not surprising that dysfunction of the ETC plays a central role in disease.

3.1 Neuronal Energy Metabolism

The central nervous system has a very high metabolic demand, accounting for almost 20% of the total oxygen consumption and up to 25% of total glucose consumption in an organism.[19] Moreover, neurons are not able to switch to glycolysis and are therefore entirely dependent on oxidative phosphorylation for ATP production. It is therefore not surprising that they are highly susceptible to small perturbations in energy production at the level of the ETC.[20] Since most neurons have a highly extended morphology, not only the efficacy of mitochondrial energy production, but also the distribution and transport of these organelles, is crucial. These aspects are highlighted in later chapters.

One of the brain regions that is majorly—but not exclusively—targeted in the brains of PD patients is the *substantia nigra*. The dopaminergic neurons in this region project to the striatum, and as they gradually degenerate, dopamine signaling to this area is lost. Although other brain regions are also affected, the specific vulnerability of this neuronal population has puzzled researchers for decades. There are some indications that an explanation for this specific vulnerability may be found in the energy demands, as different cellular (and neuronal) populations exhibit different thresholds for efficient functioning of the ETC and oxidative phosphorylation. Dopaminergic neurons have large, thin axons, with little myelination.[21] Moreover, they form a large number of synapses, several orders of magnitude greater than most other neurons.[22] These factors all increase the metabolic demand of dopaminergic neurons. A second atypical feature of dopaminergic neurons that involves mitochondrial function is their handling of calcium fluxes. Contrary to most neurons in the brain, dopaminergic neurons exhibit autonomic pacemaking activity. Since they also engage calcium channels in this process, they have an overall lower capacity for calcium buffering compared to other neurons.[23] Since the electrochemical gradient created by the ETC drives mitochondrial Ca^{2+} uptake, this could again be a reason for higher vulnerability. Thus, in addition to their high energy requirements, minor deficiencies in the mitochondrial membrane potential could also increase the likelihood of calcium handling problems in dopaminergic neurons.

3.2 Oxidative Stress

Reactive oxygen species (ROS) are produced in different cellular locations, for example, in the cytosol, the endoplasmatic reticulum, and perixomes. However, under physiological conditions, the mitochondrial ETC is the major production site of ROS,[24] including superoxide anions (O_2^-), hydrogen peroxide (H_2O_2), and hydroxyl radicals ($OH^.$). In healthy conditions, ROS may play an important role in redox signaling, and antioxidant mechanisms are in place to keep mitochondrial ROS levels in balance. In pathological conditions, however, this balance is lost, resulting in oxidative stress that causes damage to DNA, proteins, and lipids in the cell. Because of their close proximity, mitochondrial ROS are most likely to damage lipids in the inner mitochondrial membrane, the ETC subunits, and mtDNA.

Complex I and III are the major sites of superoxide formation, as electrons leaking from these complexes can react with oxygen. An estimated 1–5% of all oxygen consumed by the ETC is converted to superoxide under physiological conditions.[25] Superoxide is rapidly converted by superoxide dismutase (SOD) to H_2O_2 in the matrix, and further to H_2O by glutathione peroxidase. Interestingly, the production of ROS by ETC enzymes may vary across species and tissues and with age, and it also depends on the energetic state of the cell.[26] Complex I generates large quantities of superoxide when the $NADH/NAD^+$ ratio is high and the electron transfer to ubiquinone is inhibited due to a reduced ubiquinone pool, or when there is reverse electron transport from complex II to I, which blocks electron transfer to ubiquinone as well. This can occur through pharmacological inhibition of ubiquinone binding and/or reduction, or when succinate concentrations are elevated, for example during ischemia.[27,28] Both situations result in electron leakage and the formation of superoxide. In contrast to the ROS produced by complex III, the ROS formation from complex I occurs exclusively on the matrix side since its ubiquinone-binding pocket is oriented toward the matrix.[27]

Neurons in general are very sensitive to the buildup of oxidative damage, since they are postmitotic cells with polyunsaturated fatty acids enriched in their membranes. MtDNA is not protected by histones and has limited repair mechanisms in place, so it is very vulnerable to oxidative damage as well. Several studies have demonstrated the accumulation of mtDNA damage with aging, especially in terminally differentiated tissues with a high metabolic demand, such as the central nervous system.[29] Increased oxidative stress is correlated with aging, accumulation of mtDNA damage, and the functional decline of mitochondria.[30]

Increased levels of oxidative stress have also been identified in PD patient brains. Again, there are hypotheses as to why dopaminergic neurons would be especially vulnerable to oxidative stress. As their name implies, dopaminergic neurons produce dopamine, and oxidation of dopamine and its metabolites may lead to the generation of superoxide radicals and oxidative stress.[31] Moreover, dopaminergic neurons express low levels of the antioxidant glutathione. Defects in dopamine uptake or metabolism, or simply a higher basal exposure to oxidative stress, might render dopaminergic neurons exceptionally vulnerable to additional insults and tilt the balance unfavorably, resulting in massive cell loss.

3.3 Relationship Between the Electron Transport Chain and Parkinson's Disease

Because of the central role of the ETC in energy production, and the potentially harmful effects of some of its ROS side products, small perturbations at any level could have widespread effects. This is especially true for postmitotic neurons with a high energy demand, and mitochondrial dysfunction has indeed been linked to a wide range of neurodegenerative diseases, including PD.

PD is the most common neurodegenerative movement disorder, and it is increasingly prevalent with age.[32] One of its pathological hallmarks is the loss of dopaminergic neurons in the *substantia nigra*. As highlighted earlier, there are a number of hypotheses on why these neurons would be particularly vulnerable to ETC insults. Loss of dopamine signaling leads to movement problems, including bradykinesia, stiffness, postural instability, and tremor. As the disease progresses, patients also experience cognitive decline and loss of autonomic functions. The second pathological hallmark of PD is the formation of protein aggregates called Lewy bodies. One of the major components of these aggregates is α-synuclein.[33] The molecular mechanisms underlying PD have not been fully characterized, but accumulating evidence shows a central role for ETC dysfunction in PD pathogenesis.

3.4 Complex I Dysfunction in Parkinson's Disease Patient Brains

Mitochondrial dysfunction was first associated with PD more than 25 years ago, when a mild complex I deficiency in the *substantia nigra* was described in postmortem tissue from PD patients.[34] Similar observations followed in the frontal cortex and in peripheral tissues, such as skeletal muscle, platelets, and lymphocytes of PD patients.[35–39] In that same decade, N-methyl-4-phenyl 1,2,3,6-tetrahydropyridine (MPTP), one of the byproducts in the synthesis of the heroine-like compound meperidine, was found to induce chronic parkinsonism in drug addicts.[40,41] MPTP can cross the blood–brain barrier and is metabolized to 1-methyl-4-phenylpyridinium (MPP^+) by monoamine oxidase B in astrocytes.[42] MPP^+ is taken up by the dopamine receptor, causing specific cell loss of dopaminergic neurons by inhibition of complex I.[40,43] Several individuals who had intravenously self-administered synthetic meperidine contaminated with MPTP quickly developed severe movement impairments, reminiscent of PD. Moreover, these patients responded to levodopa, suggesting that the underlying mechanisms are similar to those in PD patients.

In addition to MPTP, several other toxins have since been associated with PD, including rotenone.[44] Rotenone is a commonly used pesticide, and its use is believed to explain the increased incidence of PD in rural populations.[45] Rotenone is lipophilic and can readily cross the blood–brain barrier and reach intracellular organelles via passive diffusion. It binds the ubiquinone binding site in Complex I, causing a blockage in the electron transfer, resulting in a strongly reduced ATP production and increase in ROS levels.[46,47]

More recently, transcriptional profiling studies showed that several genes implicated in ETC functioning, including nuclear-encoded complex I subunits, had a decreased expression in dopaminergic neurons in PD patients.[48]

3.5 Oxidative Stress

The herbicide paraquat, on the other hand, is lipophobic, and requires transporters to reach the mitochondria in the brain. Unlike rotenone, paraquat is not a specific complex I inhibitor,[49] and although its structure is similar to that of MPTP, its toxic mechanism is debated. It is hypothesized that one of the ETC complexes, presumably complex I,[50] reduces paraquat to the paraquat radical, which in turn causes oxidative stress by the formation of superoxide. Although several studies have linked paraquat to PD,[51] its precise role in the etiology of the disease remains unclear.[52]

Other evidence for increased oxidative stress in PD patient brains was found in the form of increased lipid and DNA oxidation, as well as decreased levels of gluthatione.[53–56]

3.6 Other Electron Transport Chain Components

While the majority of the studies point to the specific involvement of complex I, several reports show deficits in other components of the ETC.[38,57] Haas et al. and Shults et al. found decreased complex II and complex III activity in mitochondria of platelets, lymphocytes, and *substantia nigra* neurons in PD patients, as well as reduced levels of ubiquinone.[39,58] The transcriptome profiling by Zheng et al. mentioned earlier showed that not only complex I subunits but also subunits of other complexes had a decreased expression in PD patient brains.[48]

Interestingly, the fungicide Maneb preferentially inhibits complex III and has been linked to PD as well.[59,60] Maneb contains the mitotoxin manganese ethylene-bisdithiocarbamate, which blocks complex III, affects apoptosis, and causes oxidative stress.[61] Especially the combination of paraquat and maneb exposure, common in rural areas and during agricultural activities, is linked to an increased risk of PD.[62]

Dysfunction in one complex of the ETC may directly trigger problems at the level of another. This is the case for many mitochondrial diseases caused by a mutation in one of the ETC complex subunits. Although the mutation is specific to one complex, activity deficits are present at different complexes. For example, assembly defects of complex III cause activity defects at the level of complex I.[63] Thus, the ETC deficits observed in PD patient samples may also be interrelated and possibly all converge on complex I.

3.7 Mitochondrial DNA Mutations

Mitochondria in the *substantia nigra* of PD patients harbor an elevated amount of mtDNA mutations.[64,65] Since human cells typically contain an extended network of anywhere between 50 and several hundreds of mitochondria, each containing multiple copies of mtDNA,[66] healthy and mutated genomes can be found in the same cell, a phenomenon known as heteroplasmy. Even in postmitotic cells such as neurons, mtDNA is continuously replicated and destroyed, independent of nuclear division. While the mitochondrial population of dividing cells is randomly passed on to two different daughter cells, in postmitotic cells this is not the case. Interestingly, selection for or against mtDNA mutations does not only occur at the cellular level, but also at the level of the individual organelle. For example, mitochondrial genomes with a deletion have a reduced size and undergo more rapid replication.[67] MtDNA mutations are tolerated to a certain extent, but accumulated damage can eventually result in mitochondrial dysfunction, especially in postmitotic cells and during aging.[68] Although it is difficult to establish a direct link between specific mtDNA mutations and PD because of heteroplasmic variety across a cell population, low-frequency mutations in a specific region of the mtDNA encoding the complex I subunit ND5 have been identified in PD patient brain tissue.[69] In general, complex I mtDNA mutations are found at high frequencies in brain samples from PD patients.[70] Several other studies have reported a correlation between the number of mtDNA mutations and complex I deficiency, but have not identified a direct association with PD.[71]

In total, over 30 different disease-causing mtDNA mutations have been identified in complex I, in addition to mutations in nuclear-encoded subunits. The majority of these diseases are neurodegenerative, most likely because of the increased susceptibility of neurons to even minor defects in energy production. The three most common mtDNA mutations in complex I account for more than 90% of all cases of Leber hereditary optic neuropathy,[72,73] an inherited form of progressive vision loss caused by degeneration of the optic nerve. Other mtDNA complex I mutations cause a variety of clinical manifestations, including Leigh syndrome (a juvenile subacute necrotizing encephalomyelopathy), dystonia, and MELAS (a mitochondrial encephalopathy with stroke-like episodes).[74]

Mutations in the nuclear encoded mtDNA polymerase and proofreading enzyme POLG also result in an increased risk of PD. However, mutator mice, which harbor a mutation in POLG that disables its proofreading ability, accumulate a large amount of mtDNA mutations, without exhibiting respiratory deficiency.[75] It is possible that additional compensatory mechanisms are at play in mutator mice. More research is required to elucidate the mechanisms underlying the links between mtDNA mutations, aging, and ETC dysfunction.

4 WHAT HAVE THE GENETICS TAUGHT US ABOUT ELECTRON TRANSPORT CHAIN DYSFUNCTION AND THE SPORADIC FORMS OF PARKINSON'S DISEASE?

As genetic analysis methods improved, mutations in several genes were linked to familial forms of PD (see also Chapter 1). Although these familial forms account for only a small percentage of all PD cases, the study of the involved genes can shed light on the pathogenic mechanisms of sporadic PD. Several of the identified mutations, especially those causing early-onset recessive PD, are in genes that are directly or indirectly linked to mitochondrial ETC function, and their identification has caused significant advances in our understanding of PD pathogenesis over the past 15 years. As the evidence of ETC dysfunction is well characterized for only a few of the identified genes, we will limit the discussion here to three recessively inherited genes, *PINK1, Parkin*, and *DJ-1*, and one of the autosomal dominant genes, *SNCA*, which encodes the major Lewy body component, α-synuclein.

4.1 PINK1

The *PTEN induced putative kinase 1* (*PINK1*) gene encodes a unique Ser/Thr kinase, with an N-terminal mitochondrial targeting sequence. More than 50 homozygous and compound heterozygous mutations have been identified throughout its sequence that cause recessive familial PD.[76,77] Several putative pathogenic heterozygous mutations have also been detected, both in familial and sporadic PD cases as well as in healthy controls, but their significance remains debated.[78–80] Approximately 6.5% of mutation carriers, depending on ethnicity, carry a mutation in PINK1.[79]

Loss of PINK1 function causes several defects in mitochondrial function, including ETC dysfunction, reduced ATP levels, impaired mitochondrial calcium handling, and increased free radical generation, which in turn results in a fall in mitochondrial membrane potential and an increased susceptibility to apoptosis in neuronal cells and patient-derived fibroblasts.[76,81] PINK1 mutations also lead to perturbations in mitochondrial dynamics, since mitochondrial motility (see also Chapter 4), fission and fusion (see also Chapter 3), and turnover through mitochondrial autophagy (or mitophagy, see also Chapter 5) are all impaired in PINK1 knockout models.[82] The implications of these changes in dynamics are discussed in later chapters.

Specific ETC deficits have been identified not only in *Drosophila* and mouse models,[83–85] but also in human fibroblasts derived from PD patients harboring PINK1 mutations.[85] While Gautier et al. report general ETC deficits in a PINK1-null mouse model, Morais et al. were able to pinpoint these defects to an initial deficit at the level of complex I.[84,85] Vincow et al. documented more recently that PINK1 regulates the protein turnover of multiple subunits of all four ETC complexes and F_1F_0-ATP synthase.[86]

Complex I dysfunction is especially of interest, since it forms a direct link between the observations in sporadic forms of PD with some forms of familial PD. Since complex I deficits result in decreased ATP production, increased ROS generation, and a loss of mitochondrial membrane potential, they could possibly explain other observations of mitochondrial dysfunction in both sporadic and PINK1 familial PD cases as secondary effects, including oxidative stress and impaired mitochondrial transport and quality control.[85] Neurons differentiated from induced pluripotent stem cells derived from PD patients with clinical mutations in PINK1 exhibited a decrease in mitochondrial membrane potential (assessed by TMRE) and decreased ATP content, confirming the clinical presentation of ETC defects in patient cells.[87]

Several putative PINK1 phosphorylation substrates have been identified, located in the cytosol as well as in different mitochondrial compartments.[81] Via these targets, PINK1 regulates a variety of mitochondrial processes, and recently molecular evidence on the link between PINK1 and complex I has emerged. NDUFA10, one of the subunits of complex I, is phosphorylated on Ser 250 in a PINK1-dependent manner.[87] NDUFA10 is located close to the ubiquinone-binding pocket of the complex.[88] In contrast to the ubiquinone-dependent activity of complex I, NADH oxidation is not affected in the absence of PINK1, which indicates that phosphorylation by PINK1 is required for complex I function at the level of ubiquinone binding and/or reduction.[87] Based on the location of this subunit within the structure of complex I,[89,90] it is possible that it aids in the formation of a cavity where the complex can bind the hydrophobic ubiquinone substrate. Introduction of a phosphomimetic version of NDUFA10 rescues the complex I dysfunction observed in PINK1 knockout cells, as well as the membrane potential deficits and ATP content in *Drosophila* pink1 mutants.[87] In an independent screen in *Drosophila*, overexpression of the fly homologue of NDUFA10, ND42, could restore the complex I deficit in *pink1* mutant flies. This effect was independent of ND42 phosphorylation and led to a partial rescue of the locomotor phenotypes in the fly model.[91]

Interestingly, other approaches that boost ETC function also alleviated symptoms of *pink1* mutant flies. Vos et al. screened *Drosophila* mutants and found that mutations in *heixuedian*, which encodes the enzyme synthesizing vitamin K_2, aggravated the mitochondrial and locomotor phenotypes of *pink1* mutant flies, including mitochondrial membrane potential loss and reductions in ATP production.[92] Vitamin K_2 functions as an electron carrier, downstream of complex II, and could alleviate *pink1* mutant symptoms by improving electron transport function through the compensation of upstream respiratory defects.[92] The introduction of a yeast homologue of complex I, Ndi1, also positively affected the phenotype of *pink1* mutant flies, while the use of the sea squirt alternative oxidase (AOX) to bypass complex III and IV did not affect the mitochondrial phenotypes in *pink1* knockouts.[93] These findings all point to a central role for complex I defects in the mitochondrial dysfunction caused by PINK1 mutations.

4.2 Parkin

Mutations in the *Parkin* gene are the most frequent cause of recessive familial PD and account for almost half of these cases.[94-96] Parkin is an

E3-ubiquitin ligase that catalyzes the ubiquitination of a wide variety of cellular proteins,[97] flagging them for removal by the ubiquitin proteasomal system. The majority of Parkin mutations lead to reduced E3 ligase activity. Both cellular and animal Parkin models exhibit mitochondrial dysfunction,[98–101] as Parkin is required for protein ubiquitination on impaired mitochondria, which initiates their turnover by mitophagy.[102] Parkin acts downstream of PINK1, as two landmark studies in *Drosophila* showed that *pink1* knockout phenotypes can be partially rescued by Parkin overexpression, but not vice versa.[83,103] Similar to *pink1* mutants, *parkin* mutant flies display reduced ATP levels;[92] however, they do not exhibit complex I deficiency.[93] In contrast, *parkin* knockdown in zebrafish results in an ETC deficit at the level of complex I, and Parkin regulates the turnover of many ETC subunits by ubiquitination in a human cell line.[86,97]

Intriguingly, while complementation with NDUFA10 can rescue the complex I deficiency observed in *pink1* knockout flies, it has no effect on *parkin* phenotypes or mitophagy.[91] Similarly, the yeast complex I Ndi1 can rescue several phenotypes observed in *pink1* mutant *Drosophila*, while it does not alter *parkin* mutant phenotypes.[93] Similar observations have been made for two putative PINK1 substrates, TRAP1 and HtrA2, but also for other Parkinson-related proteins like DJ-1 and FOXO, or the GDNF-receptor Ret and the ATPase VCP.[104–107] These findings indicate that PINK1's role in ETC regulation and in mitochondrial quality control via Parkin are part of two separate parallel pathways, whose effects partially overlap.

4.3 DJ-1

Mutations in *DJ-1* are very rare but causal of familial PD as well,[108] and they impair cellular protection against oxidative stress in animal models.[109–111] Indeed, DJ-1-deficient mice are more susceptible to MPTP toxicity.[112] DJ-1 is a small protein that forms dimers and is a member of the ThiJ/PfpI family of molecular chaperones, but the actual function of DJ-1 is poorly understood. One possibility is that DJ-1 is an antioxidant scavenger, as DJ-1 mutations lead to an increased susceptibility to oxidative stress. Alternatively it might act as a chaperone or protease.[113] In relation to the ETC, embryonic fibroblast and primary dopaminergic neurons derived from DJ-1 knockout mice display reduced mitochondrial respiration levels and a decreased mitochondrial membrane potential.[114,115] Defects in complex I assembly as well as downregulation of mitochondrial uncoupling proteins leading to increased oxidative stress have also been observed.[114,116] *Drosophila* have two paralogs of *DJ-1*, *DJ-1α*, and *DJ-1β*. When both are deleted, flies display an age-dependent increase in mtDNA mutations and a deficiency in respiration, resulting in decreased ATP levels.[104]

4.4 SNCA

Mutations, duplications, and triplications in *SNCA*, the gene encoding α-synuclein, cause autosomal dominant familial PD.[117] Some mutations have also been identified in sporadic cases, which suggests that de novo mutations in this gene occur at a significant frequency.[118] *SNCA* was the first gene to be linked to familial PD, and as the protein it encodes in the major constituent of Lewy bodies,[33] it must play an important role in the molecular pathogenesis. Despite intensive research, the precise function of α-synuclein remains unclear. Several lines of research indicate that it plays an important role in synaptic transmission,[119] but it has been linked to mitochondrial dysfunction as well.[120] Overexpression of both wild type and mutant α-synuclein lead to elevated ROS levels in mammalian cells.[121,122] Under conditions of oxidative stress or metabolic stress, the lipid composition of the mitochondrial membranes can alter, which in turn could affect the mitochondrial translocation of α-synuclein.[123] It possess a mitochondrial targeting sequence, which traffics the protein to the inner mitochondrial membrane, where it inhibits complex I and accumulates in the *substantia nigra* and striatal mitochondria of PD patients.[124–126] Haelterman et al. propose a model in which misfolded or aggregated α-synuclein stimulates a vicious cycle in which it inhibits complex I, increasing ROS levels and promoting its own translocation to further disrupt ETC function.[82]

4.5 Similarities and Differences Between Familial and Sporadic Parkinson's Disease

With the exception of the early disease onset and mild clinical progression observed in patients with defects in some of the PD-linked genes, the familial cases of PD present the same symptoms as sporadic PD patients. Nevertheless, there has been some debate on whether or not the pathological phenotypes are similar.[127] The presence of Lewy bodies in PINK1 and Parkin patients is unclear (see also next section). Since (1) *SNCA* mutations are a much more common cause of familial PD, (2) accumulation of misfolded α-synuclein is a major pathological hallmark of the disease, and (3) protein aggregation is a general feature of many neurodegenerative diseases, toxic protein aggregation is a primary candidate as a causal factor for PD. It is currently not clear how mitochondrial dysfunction, or ETC dysfunction specifically, is related to α-synuclein accumulation. Intriguingly, several mitochondrial toxins that have been linked to PD, including Maneb, induce α-synuclein misfolding in animal models, while others, such as MPTP do not.[128] While there are several studies suggesting positive enforcement of one another, further research is needed to establish how α-synuclein accumulation and ETC dysfunction are related and which of the two—if any—is the primary culprit in PD onset. Furthermore, the impact of environmental factors will require further elucidation. Large meta-analyses suggest an approximately two-fold increase in the risk of PD due to pesticide exposure,[129] and based on the different age of onset and incomplete penetrance of some of the PD-linked mutations, it is clear that environmental variations play an important role in genetic cases of PD as well. These efforts will help extrapolate some of the findings for specific familial PD cases with relation to ETC dysfunction to the pathogenesis of sporadic PD.

5 CORRELATION BETWEEN FINDINGS OBTAINED USING DIFFERENT ANIMAL MODELS

A large variety of PD animal models are available, which can largely be classified into two main categories: toxin-induced or genetic models (or sometimes a combination of the two). New advances enabling induced and/or cell-specific expression of mutated genes and the advent of induced pluripotent stem cells that can be differentiated into neurons have greatly enhanced the cellular and animal model portfolio for PD.[130] Besides primates and murine models, invertebrate organisms including *Drosophila*, *Caenorhabditis elegans*, and fish are gaining more and more importance, due to the ease and speed of genetic manipulation, reproduction, and screening in these species.[131,132] In addition to providing insight into the disease mechanisms, such animal models can be used to test novel therapeutic approaches.

A good model should exhibit the full pathological and clinical PD phenotype, including (1) age-dependent and progressive loss of dopaminergic neurons, (2) the presence of Lewy bodies and Lewy neuritis, and (3) motor dysfunction including tremor, bradykinesia, postural instability, and rigidity, which is (4) responsive to levodopa replacement treatment.[130] Unfortunately, none of the existing models fulfill all of the previous criteria, and so the ideal animal or cellular model for PD does not exist (Table 2.1). Nevertheless, ETC dysfunction is a recurring theme in many types of models and species, although subtle differences exist at this level as well.

5.1 Toxin-Induced Models and Electron Transport Chain Dysfunction

MPTP induces the specific loss of dopaminergic neurons in a wide range of species, including primates and mice, but not in rats, via the specific inhibition of complex I.[133,134] This cell loss occurs acutely, and not gradually as in PD patients. Although intraneuronal inclusions reminiscent of Lewy bodies have been described in monkeys,[135] other MPTP-induced animal

TABLE 2.1   Electron Transport Chain Defects in Parkinson's Disease Models

	Species	ETC defects	DA cell loss	Lewy bodies	Motor phenotype
TOXIN MODELS					
MPTP	Murine	CI inhibition in DA cells, ↑ ROS, ↓ ATP levels	+	−	+
	Primate	CI inhibition in DA cells	+	±	+
Rotenone	Murine	CI inhibition, ↑ ROS	+	+	+
	Primate	ND	ND	ND	ND
6-OHDA	Murine	↑ ROS, ↓ CI activity	+	−	+
	Primate	↑ ROS	+	±	+
GENETIC LOSS-OF-FUNCTION MODELS					
PINK1	Fly	↑ CI activity, ↓ ATP levels, ↑ mtDNA level, ↓ Δm_ψ	+	−	+
	Murine	↓ CI activity, ↓ respiration, ↓ preprotein import, ↓ ATP levels	−	−	−
	Human	↓ CI activity, ↓ Δm_ψ, ↓ ATP levels, ↓ respiration, ↓ mtDNA level/synthesis	+	?	+
Parkin	Fly	−	+	−	+
	Murine	↓ CI and CIV subunit levels, ↓ ETC activity	−	−	−
	Human	↓ CI activity (in cell models)	+	?	+
DJ-1	Fly	↓ mtDNA level, ↓ respiration, ↓ ATP levels	+	−	+
	Murine cells	↓ CI activity and assembly, ↓ ATP levels, ↓ respiration, ↓ Δm_ψ, ↓ UCP expression	−	−	−
	Human	↓ Δm_ψ (in cell models)	+	−	+
GENETIC MUTATION OR OVEREXPRESSION MODELS					
SNCA	Fly	↑ ATP synthase subunits	+	+	+
	Murine	↑ CIV activity	±	+	±
	Human	↑ CI activity, ↓ Δm_ψ (in cell models)	+	+	+

The most commonly used toxins in murine and nonhuman primate models of PD, MPTP, rotenone, and 6-OHDA all induce dopaminergic cell loss and Parkinsonian phenotypes. While Lewy body pathology is not generally present, all studied models exhibit overt ETC dysfunction. Genetic models for four of the PD-linked genes show a variety of mitochondrial defects, pathological features, and phenotypes across different species. CI, complex I; CIV, complex IV; DA, dopaminergic; Δm_ψ, mitochondrial membrane potential; UCP, uncoupling protein; ND, not determined.

models do not display any Lewy body pathology, which can perhaps be explained by the acute nature of this model. In humans and primates, MPTP induces striking motor symptoms resembling PD. However, only mild motor alterations can be detected in MPTP-treated mice, despite of the profound loss of dopaminergic neurons.[136]

In contrast, the chronic infusion of rotenone in rats or chronic oral administration in mice leads to both selective loss of dopaminergic neurons and the presence of neuronal α-synuclein-positive inclusions.[45,137] Rotenone-treated animals also exhibit behavioral and motor symptoms similar to those of PD patients. Initial studies with high doses of rotenone resulted in widespread brain lesions,[138,139] and animal models with rotenone only became useful after a chronic low-dose regimen was developed.[45,140] This is of significant interest because it demonstrates that selective neurodegeneration can occur even though the complex I inhibition is widespread, demonstrating the inherent vulnerability of dopaminergic neurons to defects in the ETC. It also suggests that chronic effects may be required for the development of Lewy body pathology, and underscores the possibility that chronic exposure to environmental toxins may cause PD.

Another commonly used neurotoxin, 6-OHDA, causes dopaminergic cell death upon injection in rodent brains. Since 6-OHDA cannot cross the blood–brain barrier, its systemic administration has no effect. Intrastriatal injection induces damage of striatal terminals, followed by a delayed, progressive cell loss of dopaminergic neurons, which are secondarily affected through a dying-back mechanism.[141] Injections with 6-OHDA have been applied to mice and rats since the 1950s, leading to the first animal models for PD. Cell toxicity is relatively selective, as 6-OHDA is preferentially taken up by dopamine and noradrenergic transporters.[142] As in the case of MPTP, murine models show specific loss of dopaminergic neurons but no Lewy body pathology. Administration of 6-OHDA causes massive oxidative stress and inhibition of complex I,[143,144] and results in both motor and nonmotor Parkinsonian phenotypes in rodents.

The fungicide Maneb induces oxidative stress and cytoplasmic α-synuclein aggregation in dopaminergic neuron cell lines.[145] Maneb and Paraquat are often used in combination, as the combined effect is greater than that of each toxin alone. Less commonly used toxin models include reserpine, α-methyl-para-tyrosine, lipopolysaccharide, and amphetamines,[134] however, their mode of action is not well understood, and the elicited neurotoxicity is less specific than those of the aforementioned toxins, which is why they have fallen in disuse.

There are additional mitochondrial toxins that cause ETC dysfunction, which are not necessarily linked to PD, and several pharmacological agents have been identified that can specifically block one of the complexes in the respiratory chain. Among these agents, only the complex-I-specific inhibitor rotenone, has been directly linked to PD. Nevertheless, this set of toxins is used to study the molecular mechanisms underlying PD and the potential contribution of each of the components of the ETC.

5.2 Genetic Models and Electron Transport Chain Dysfunction

Not unexpected, the acute nature of many of the toxin models affects their value to model an age-dependent degenerative disease. With the identification of genetic cases of PD, researchers anticipated the development of better disease models, but so far, this approach has been rather disappointing. None of the PD-related genes causes overt parkinsonian phenotypes upon knockout in murine models, including *PINK1*, *Parkin*, and *DJ-1*, genes that are directly involved in mitochondrial dysfunction. In PINK1 knockout mice, only a mild age-dependent loss of dopaminergic levels in the striatum can be observed.[146] Some striatal degeneration has been reported in Parkin knockout mice, but dopaminergic neurons are generally unaffected and behavioral phenotypes are very mild and inconsistent.[147–149] Furthermore, the combined inactivation of *PINK1*, *Parkin*, and *DJ-1* does not lead to significant dopaminergic cell loss,[150] and the occurrence of Lewy bodies in patients harboring PD-related mutations in PINK1 or Parkin is debated.[127,151] However, several genetic models in *Drosophila* do exhibit motor phenotypes as well as dopaminergic cell loss, and because the reproduction, genetic modification, and screening in these animals is much more straightforward than for murine models, flies have gained increasing attention to model PD.[132]

Similar observations have been made in genetic models with regard to mitochondrial deficits: The mitochondrial morphology phenotypes in $PINK1^{-/-}$ mice are subtle and controversial,[84,85] while in *Drosophila*, the effects are more pronounced (Table 2.1). *Pink1* mutant flies display defective neurotransmitter release, ATP depletion, and loss of mitochondrial membrane potential, resulting in thorax muscle degeneration and flight deficits.[83,103,152] Several cellular and animal models have shown that PINK1 dysfunction leads to increased susceptibility to additional insults or mitochondrial toxins.[77,84,153,154] The function of these genes is therefore often evaluated under conditions of stress, or by combining genetic mutations with treatment with any of the mitochondrial toxins listed earlier. An additional mitochondrial toxin that is frequently used to study the role of PINK1 and Parkin in mitochondrial dynamics is the uncoupler carbonyl cyanide m-chlorophenyl hydrazine (CCCP). CCCP is a hydrophobic proton carrier that can freely carry protons across the mitochondrial inner membrane, resulting in loss of the membrane potential. Depolarization leads to the accumulation of PINK1 on the outer mitochondrial membrane, where it is activated by phosphorylation.[155,156] PINK1 subsequently recruits Parkin to the mitochondria[102,157] and phosphorylates both Parkin and ubiquitin,[158–162] leading to the ubiquitination of several outer membrane proteins and the induction of mitochondrial removal through mitophagy (see Chapter 5).[163]

Interestingly, cell-based models as well as several animal models, including flies and mice, devoid of PINK1 show a reduction in complex I activity.[84,85] In contrast to the effects on toxin-induced mitophagy, these complex I defects are observed under resting conditions in mitochondria with normal morphology.

5.3 Convergence on Complex I

The extent and contribution of complex I to pathology differs across different PD models, both in the toxin- and mutation-induced forms. Moreover, while in some cases complex I deficiency is accompanied by dopaminergic cell loss and motor dysfunction, this is not always the case (Table 2.1). The exceptional size and composition of complex I could be one explanation for this discrepancy. Its structure has only recently been revealed and to date, complex I is the least understood complex mechanistically.[164] Electron transfer and proton pumping are spatially separated within this large complex and the mechanisms by which this occurs are only beginning to be unraveled. It is possible that several toxins with different effects on complex I could also lead to different degrees of pathology. Indeed, the same is true for mutations in ETC complex subunits: different mutations in distinct ETC complexes can cause the same phenotypic presentation, while some mutations in complex I cause a different syndrome, for example, MELAS, Leighs disease, or LOHN. Despite specific deficits in complex I, none of the various complex I mutations has been linked to PD. Instead, the effects on complex I appear to be more subtle, depending on posttranslational regulation (eg, phosphorylation of NDUFA10) and/or accumulated secondary insults via environmental exposure to toxins and oxidative stress. We believe that a better mechanistic understanding of the action of several toxins and PD-related proteins on complex I could further improve our understanding of its role in PD development.

The discrepancies between findings in different animal models and between specific mitochondrial defects and overall motor phenotypes also suggest that compensatory mechanisms are in place. This is in line with clinical observations in both sporadic and familial cases, as even in case of early-onset genetic PD (for example in patients harboring mutations in *PINK1* or *Parkin*) mitochondrial dysfunction is present in all cells and from birth. However, only late in life and only in specific neuronal populations does this lead to cell death, demonstrating that compensatory mechanisms allow the patient to cope with these defects.

Although both genetic and toxin-induced models indicate that ETC dysfunction, and specifically complex I dysfunction, plays an important role in the pathogenesis of PD, the lack of a model displaying all

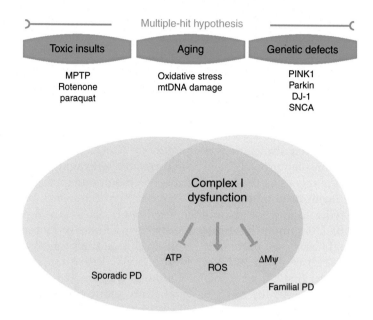

FIGURE 2.3   **Complex I defects and Parkinson's disease.** Both sporadic and familial PD cases have been linked to complex I dysfunction. Impaired complex I activity has been identified in brain and peripheral tissue of PD patients. Several complex I inhibitors, including MPTP and rotenone, cause Parkinsonian phenotypes in humans and animals. Some familial cases of PD (*PINK1*, *Parkin*, *DJ-1*, and *SNCA*) display mitochondrial dysfunction as well. Mutations in the *PINK1* gene result in specific complex I dysfunction, through the loss of phosphorylation of one of its subunits NDUFA10. Complex I dysfunction leads to a reduction in ATP production, increased oxidative stress through the production of ROS, and a decrease in mitochondrial membrane potential. All of these effects have been observed in both sporadic and genetic models of PD as well, putting complex I of the ETC at center stage in PD pathogenesis.

pathological and clinical feature of PD demonstrates the complex etiology of the disease and underlines the convergence of several insults that cause full-blown development of the disease. Such a "multiple-hit" hypothesis is in line with the relatively late onset of the disease, and it is clear that at one point or another complex I and ETC deregulation plays a role in disease development (Fig. 2.3).

6   THERAPEUTIC PATHWAYS TARGETING THE ELECTRON TRANSPORT CHAIN

Almost 200 years after its first description by James Parkinson,[165] there is still no cure for PD. Patients receive symptomatic treatment in the form of levodopa, in combination with other dopamine metabolism regulators, to replenish the lost dopamine signaling in their brain. This approach only provides a temporary solution for the motor symptoms of patients and does not halt further progression of the disease. Although a detailed understanding of the molecular pathways that lead to dopaminergic cell death and the development of PD remains elusive, it is clear that ETC dysfunction has a central role to play, and therefore, a number of promising clinical trials aimed at boosting the activity of the ETC have been set up over the past years (Table 2.2).

One of the first strategies was ubiquinone (CoQ_{10}) supplementation. Decreased ubiquinone levels have been correlated with aging and PD, and the *substantia nigra* has the lowest levels of ubiquinone in the brain.[58,180] Increased levels of ubiquinone can facilitate electron transfer, and moreover, the reduced form of ubiquinone (ubiquinol) also acts as a powerful antioxidant. Ubiquinone has protective effects in cellular models of MPTP toxicity, and has shown promising results in clinical trials for other neurodegenerative diseases, including in Huntingtons disease and Friedreichs ataxia.[181] In the first clinical trials, ubiquinone was found to be safe and tolerable, and the results in these small study groups suggested a possible beneficial disease-modifying effect. However, other studies could not confirm this finding and a larger phase III trial found no evidence for a beneficial effect.[168,170,171] Potentially further stratification of the patient population, focusing only on those with mitochondrial defects, may yield more promising results.

Another promising compound is creatine. This organic acid is synthesized in the liver and kidney and transported to tissues with a high energy demand. The phosphorylated form of creatine, phosphocreatine, acts in concert with creatine kinase to replenish ATP levels under high metabolic demands. Creatine kinase transfers the phosphate of phosphocreatine to ADP, quickly generating a local supply of ATP. This type of energy replenishing is especially important in muscles, which is why athletes and body builders take creatine supplements to improve muscle performance. While creatine was found to be safe and tolerable, a recent large double-blind and placebo-controlled trial was terminated early as interim analysis after 5 years of follow-up revealed no improvements in the clinical outcome of PD patients.[174]

Intranasal administration of the antioxidant glutathione, which reduces oxidative stress by converting H_2O_2, is currently in phase II clinical trials. Previously, intravenous injection of glutathione was not found to be beneficial for patients, even though reduced levels of glutathione have been correlated to the disease.[176] A major question was whether this glutathione could cross the blood–brain barrier, hence the new trial testing a different route of administration.[177]

Urate or uric acid is the end product of pyridine metabolism. Although high urate levels may lead to gout and are associated with an increased

TABLE 2.2   Clinical Trials Directed at Electron Transport Chain Stimulation

	Phase I			Phase II			Phase III		
	Safety and tolerability			Expanded safety and target efficacy			Treatment efficacy		
	Outcome	N	References	Outcome	N	References	Outcome	N	References
CoQ$_{10}$	+	15	[166]	+	80	[167]	−	128	[168]
				+	28	[169]	−	600	[170]
				−	130	[171]			
Creatine	N/A			+	200	[172,173]	−	1741	[174]
Glutathione	+	70	[175]	±	20	[176]			
	+	30	[177]	Ongoing					
Uric acid/Inosine	+	8	[178]	+	27	[179]	Ongoing		

Based on the widespread implication of ETC dysfunction in both sporadic and familial PD, several therapeutic approaches aimed at boosting ETC activity have been tested in clinical trials. Ubiquinone (CoQ$_{10}$) and creatine have shown promising outcomes in preliminary studies, but recent phase III trials have been prematurely terminated. Phase II and phase III trials are currently ongoing to test the efficacy of glutathione and inosine respectively in PD patients. N, number of participants; N/A, not applicable.

risk of diabetes and the formation of kidney stones, they are also correlated with a decreased risk and slower progression of PD.[182,183] In PD rodent models, urate confers neuroprotection because of its antioxidant properties.[184,185] Administration of the urate precursor inosine leads to rapid elevation of urate levels.[179] Phase I and II clinical trials demonstrated that inosine was safe and well-tolerated in PD patients and could induce elevated levels of urate. A phase III trial is currently ongoing to determine the clinical benefit of inosine in a large cohort of PD patients.

Recently, Vos et al. proposed two novel approaches to stimulate ETC activity in PD animal models. In a mechanical approach, irradiation with near-infrared light (808 nm) to activate complex IV improved the mitochondrial function in both *pink1* mutant flies and knockout mice, and also improved ATP levels and locomotor phenotypes in the flies.[186] Although the results in the fly model are promising, the direct stimulation in patients remains a major practical hurdle. A second approach proposed by these authors to improve ETC function is via vitamin K_2 administration. Vitamin K_2 treatment not only improves ETC function, but also the mitochondrial morphology and locomotor activity of *pink1* mutant flies. The effects were equal to those obtained with ubiquinone treatment.[92]

Despite the promising findings in animal models, and the clear mechanistic links between ETC dysfunction and PD, the clinical trials with ETC-stimulating compounds have so far been disappointing. Although methodological difficulties in the measurement of disease progression could hinder the detection of small beneficial effects, one other possibility is that mechanistic heterogeneity of PD pathogenesis leads to conflicting therapeutic outcomes. As highlighted before, the animal models that are currently used in PD research all focus on a limited amount of disease aspects, and none of them reflects the true disease development and pathology as observed in patients. It could be worthwhile to stratify patients into groups that are representative for the pathway that is targeted, especially for those strategies that have shown great potential in certain animal models, as is the case for ETC-related compounds such as ubiquinone and vitamin K_2.[187] To be able to do so, reliable clinical biomarkers for ETC dysfunction are required. In 2015, a first proof-of-concept clinical trial (Mito-PD) will be conducted with ubiquinone and vitamin K_2 to target the specific biochemical deficiency in genetically defined subgroups of PD patients with mitochondrial dysfunction.[187]

7 CONCLUSIONS

The ETC is of vital importance for cellular energy production and its dysfunction leads to neurodegenerative diseases including PD. Accumulating evidence on both sporadic and familial cases indicates that complex I deficiency and concomitant oxidative stress play a central role in the disease pathogenesis. Its strong implication in both toxin-induced and genetic animal models also reinforces the key role of complex I.

Despite its long-established role in disease pathogenesis, therapeutic intervention aimed at stimulation of ETC function had very limited success. Other disease mechanisms including mitochondrial and protein quality control are involved as well, and their interaction with ETC dysfunction remains to be elucidated. The variety in phenotypes exhibited by the different available genetic and toxin-induced PD animal models also indicates this pathogenic diversity. We need to take into account this mechanistic heterogeneity and find better ways to group patients to develop effective treatments at the level of ETC function stimulation. Narrowing down the different pathways and interactions with the ETC will result in a mechanistic understanding of PD pathogenesis and in personalized and effective ways to halt or even reverse the disease.

LIST OF ABBREVIATIONS

ADP/ATP	Adenosine di/tri-phosphate
CCCP	Carbonyl cyanide m-chlorophenyl hydrazine
CoQ$_{10}$	Ubiquinone
ETC	Electron transport chain
IMS	Intermembrane space
MPP$^+$	1-Methyl-4-phenylpyridinium
MPTP	N-Methyl-4-phenyl 1,2,3,6-tetrahydropyridine
mtDNA	Mitochondrial DNA
PD	Parkinson's disease
ROS	Reactive oxygen species
SOD	Superoxide dismutase

References

1. Berg JM, Tymoczko JL, Stryer L. Glycolysis and gluconeogenesis. *Biochemistry*. New York: W.H. Freeman; 2002.
2. Berg JM, Tymoczko JL, Stryer L. The citric acid cycle. *Biochemistry*. 5th ed. New York: W.H. Freeman; 2002.
3. Berg JM, Tymoczko JL, Stryer L. Oxidative phosphorylation. *Biochemistry*. 5th ed. New York: W H Freeman; 2002.
4. Bernardi P, Di Lisa F, Fogolari F, Lippe G. From ATP to PTP and back: a dual function for the mitochondrial ATP synthase. *Circ Res*. 2015;116(11):1850–1862.
5. Takeda Y, Pérez-Pinzón MA, Ginsberg MD, Sick TJ. Mitochondria consume energy and compromise cellular membrane potential by reversing ATP synthetase activity during focal ischemia in rats. *J Cereb Blood Flow Metab*. 2004;24(9):986–992.
6. Das A. Regulation of the mitochondrial ATP-synthase in health and disease. *Mol Genet Metab*. 2003;79(2):71–82.
7. Busiello RA, Savarese S, Lombardi A. Mitochondrial uncoupling proteins and energy metabolism. *Front Physiol*. 2015;6:36.
8. Vos M, Lauwers E, Verstreken P. Synaptic mitochondria in synaptic transmission and organization of vesicle pools in health and disease. *Front Synaptic Neurosci*. 2010;2:139.

9. Newmeyer DD, Farschon DM, Reed JC. Cell-free apoptosis in Xenopus egg extracts: inhibition by Bcl-2 and requirement for an organelle fraction enriched in mitochondria. *Cell.* 1994;79(2):353–364.

10. Wang C, Youle RJ. The role of mitochondria in apoptosis. *Annu Rev Genet.* 2009;43:95–118.

11. Liu X, Kim CN, Yang J, Jemmerson R, Wang X. Induction of apoptotic program in cell-free extracts: requirement for dATP and cytochrome c. *Cell.* 1996;86(1):147–157.

12. Carroll J, Fearnley IM, Skehel JM, Shannon RJ, Hirst J, Walker JE. Bovine complex I is a complex of 45 different subunits. *J Biol Chem.* 2006;281(43):32724–32727.

13. Lapuente-Brun E, Moreno-Loshuertos R, Acín-Pérez R, et al. Supercomplex assembly determines electron flux in the mitochondrial electron transport chain. *Science.* 2013;340(6140):1567–1570.

14. Lenaz G, Genova ML. *Mitochondrial Oxidative Phosphorylation,* vol. 748. In: Kadenbach B, ed. New York: Springer New York; 2012.

15. Koopman WJH, Nijtmans LGJ, Dieteren CEJ, et al. Mammalian mitochondrial complex I: biogenesis, regulation, and reactive oxygen species generation. *Antioxid Redox Signal.* 2010;12(12):1431–1470.

16. Ngo DC, Ververis K, Tortorella SM, Karagiannis TC. Introduction to the molecular basis of cancer metabolism and the Warburg effect. *Mol Biol Rep.* 2015;42(4):819–823.

17. Warberg O. On the origin of cancer cells. *Science.* 1956;123(3191):309–314.

18. Moreno-Sánchez R, Marín-Hernández A, Saavedra E, Pardo JP, Ralph SJ, Rodríguez-Enríquez S. Who controls the ATP supply in cancer cells? Biochemistry lessons to understand cancer energy metabolism. *Int J Biochem Cell Biol.* 2014;50:10–23.

19. Sokoloff L. The metabolism of the central nervous system in vivo. In: Field J, Magoun HW, Hall VE, eds. *Handbook of Physiology.* Washington DC: American Physiological Society; 1960:1843–1864.

20. Pinto M, Pickrell AM, Moraes CT. Regional susceptibilities to mitochondrial dysfunctions in the CNS. *Biol Chem.* 2012;393(4):275–281.

21. Braak H, Del Tredici K, Rüb U, de Vos RA, Jansen SENH, Braak E. Staging of brain pathology related to sporadic Parkinson's disease. *Neurobiol Aging.* 2003;24(2):197–211.

22. Arbuthnott GW, Wickens J. Space, time and dopamine. *Trends Neurosci.* 2007;30(2):62–69.

23. Surmeier JD, Guzman JN, Sanchez-Padilla J. Calcium, cellular aging, and selective neuronal vulnerability in Parkinson's disease. *Cell Calcium.* 2010;47(2):175–182.

24. Boveris A, Oshino N, Chance B. The cellular production of hydrogen peroxide. *Biochem J.* 1972;128(3):617–630.

25. Giorgio M, Trinei M, Migliaccio E, Pelicci PG. Hydrogen peroxide: a metabolic by-product or a common mediator of ageing signals? *Nat Rev Mol Cell Biol.* 2007;8(9):722–728.

26. Barja G. Mitochondrial oxygen radical generation and leak: sites of production in states 4 and 3, organ specificity, and relation to aging and longevity. *J Bioenerg Biomembr.* 1999;31(4):347–366.

27. Jastroch M, Divakaruni AS, Mookerjee S, Treberg JR, Brand MD. Mitochondrial proton and electron leaks. *Essays Biochem.* 2010;47:53–67.

28. Muller FL, Liu Y, Abdul-Ghani MA, et al. High rates of superoxide production in skeletal-muscle mitochondria respiring on both complex I- and complex II-linked substrates. *Biochem J.* 2008;409(2):491–499.

29. Lee HC, Pang CY, Hsu HS, Wei YH. Differential accumulations of 4,977 bp deletion in mitochondrial DNA of various tissues in human ageing. *Biochim Biophys Acta.* 1994;1226(1):37–43.

30. Yan MH, Wang X, Zhu X. Mitochondrial defects and oxidative stress in Alzheimer disease and Parkinson disease. *Free Radic Biol Med.* 2013;62:90–101.

31. Hastings TG. Enzymatic oxidation of dopamine: the role of prostaglandin H synthase. *J Neurochem.* 1995;64(2):919–924.

32. Pringsheim T, Jette N, Frolkis A, Steeves TDL. The prevalence of Parkinson's disease: a systematic review and meta-analysis. *Mov Disord.* 2014;29(13):1583–1590.

33. Spillantini MG, Schmidt ML, Lee VM-Y, Trojanowski JQ, Jakes R, Goedert M. [alpha]-Synuclein in Lewy bodies. *Nature.* 1997;388(6645):839–840.
34. Schapira AHV, Cooper J, Dexter D, Jenner P, Clark J, Marsden C. Mitochondrial complex I deficency in Parkinson's disase. *Lancet.* 1989;1(8649):1269.
35. Mizuno Y, Ohta S, Tanaka M, et al. Deficiencies in complex I subunits of the respiratory chain in Parkinson's disease. *Biochem Biophys Res Commun.* 1989;163(3):1450–1455.
36. Parker WD, Boyson SJ, Parks JK. Abnormalities of the electron transport chain in idiopathic Parkinson's disease. *Ann Neurol.* 1989;26(6):719–723.
37. Schapira AH, Mann VM, Cooper JM, et al. Anatomic and disease specificity of NADH CoQ1 reductase (complex I) deficiency in Parkinson's disease. *J Neurochem.* 1990;55(6):2142–2145.
38. Bindoff LA, Birch-Machin MA, Cartlidge NE, Parker WD, Turnbull DM. Respiratory chain abnormalities in skeletal muscle from patients with Parkinson's disease. *J Neurol Sci.* 1991;104(2):203–208.
39. Haas RH, Nasirian F, Nakano K, et al. Low platelet mitochondrial complex I and complex II/III activity in early untreated Parkinson's disease. *Ann Neurol.* 1995;37(6):714–722.
40. Davis GC, Williams AC, Markey SP, et al. Chronic Parkinsonism secondary to intravenous injection of meperidine analogues. *Psychiatry Res.* 1979;1(3):249–254.
41. Langston J, Ballard P, Tetrud J, Irwin I. Chronic Parkinsonism in humans due to a product of meperidine-analog synthesis. *Science.* 1983;25(2194587):979–980.
42. Nicklas WJ, Vyas I, Heikkila RE. Inhibition of NADH-linked oxidation in brain mitochondria by 1-methyl-4-phenyl-pyridine, a metabolite of the neurotoxin, 1-methyl-4-phenyl-1,2,5,6-tetrahydropyridine. *Life Sci.* 1985;36(26):2503–2508.
43. Langston JW, Forno LS, Tetrud J, Reeves AG, Kaplan JA, Karluk D. Evidence of active nerve cell degeneration in the substantia nigra of humans years after 1-methyl-4-phenyl-1,2,3,6-tetrahydropyridine exposure. *Ann Neurol.* 1999;46(4):598–605.
44. Tanner CM, Kamel F, Ross GW, et al. Rotenone, paraquat, and Parkinson's disease. *Environ Health Perspect.* 2011;119(6):866–872.
45. Betarbet R, Sherer TB, MacKenzie G, Garcia-Osuna M, Panov AV, Greenamyre JT. Chronic systemic pesticide exposure reproduces features of Parkinson's disease. *Nat Neurosci.* 2000;3(12):1301–1306.
46. Sherer TB, Betarbet R, Testa CM, et al. Mechanism of toxicity in rotenone models of Parkinson's disease. *J Neurosci.* 2003;23(34):10756–10764.
47. Sanders LH, Greenamyre JT. Oxidative damage to macromolecules in human Parkinson disease and the rotenone model. *Free Radic Biol Med.* 2013;62:111–120.
48. Zheng B, Liao Z, Locascio JJ, et al. PGC-1α, a potential therapeutic target for early intervention in Parkinson's disease. *Sci Transl Med.* 2010;2(52):52ra73.
49. Richardson JR, Quan Y, Sherer TB, Greenamyre JT, Miller GW. Paraquat neurotoxicity is distinct from that of MPTP and rotenone. *Toxicol Sci.* 2005;88(1):193–201.
50. Cochemé HM, Murphy MP. Complex I is the major site of mitochondrial superoxide production by paraquat. *J Biol Chem.* 2008;283(4):1786–1798.
51. Snyder SH, D'Amato RJ. Predicting Parkinson's disease. *Nature.* 1985;317(6034):198–199.
52. Berry C, La Vecchia C, Nicotera P. Paraquat and Parkinson's disease. *Cell Death Differ.* 2010;17(7):1115–1125.
53. Dexter D, Carter C, Agid F, et al. Lipid peroxidation as cause of nigral cell death in Parkinson's disease. *Lancet.* 1986;2(8507):639–640.
54. Dexter DT, Carter CJ, Wells FR, et al. Basal lipid peroxidation in substantia nigra is increased in Parkinson's disease. *J Neurochem.* 1989;52(2):381–389.
55. Jenner P, Dexter DT, Sian J, Schapira AH, Marsden CD. Oxidative stress as a cause of nigral cell death in Parkinson's disease and incidental Lewy body disease. The Royal Kings and Queens Parkinson's Disease Research Group. *Ann Neurol.* 1992;32 suppl:S82–S87.

56. Jenner P. Altered mitochondrial function, iron metabolism and glutathione levels in Parkinson's disease. *Acta Neurol Scand Suppl.* 1993;146:6–13.
57. Keane PC, Kurzawa M, Blain PG, Morris CM. Mitochondrial dysfunction in Parkinson's disease. *Parkinsons Dis.* 2011;2011:716871.
58. Shults CW, Haas RH, Passov D, Beal MF. Coenzyme Q10 levels correlate with the activities of complexes I and II/III in mitochondria from parkinsonian and nonparkinsonian subjects. *Ann Neurol.* 1997;42(2):261–264.
59. Meco G, Bonifati V, Vanacore N, Fabrizio E. Parkinsonism after chronic exposure to the fungicide maneb (manganese ethylene-bis-dithiocarbamate). *Scand J Work Environ Health.* 1994;20(4):301–305.
60. Zhang J, Fitsanakis VA, Gu G, et al. Manganese ethylene-bis-dithiocarbamate and selective dopaminergic neurodegeneration in rat: a link through mitochondrial dysfunction. *J Neurochem.* 2003;84(2):336–346.
61. Roede JR, Hansen JM, Go Y-M, Jones DP. Maneb and paraquat-mediated neurotoxicity: involvement of peroxiredoxin/thioredoxin system. *Toxicol Sci.* 2011;121(2):368–375.
62. Freire C, Koifman S. Pesticide exposure and Parkinson's disease: epidemiological evidence of association. *Neurotoxicology.* 2012;33(5):947–971.
63. Acín-Pérez R, Bayona-Bafaluy MP, Fernández-Silva P, et al. Respiratory complex III is required to maintain complex I in mammalian mitochondria. *Mol Cell.* 2004;13(6):805–815.
64. Bender A, Krishnan KJ, Morris CM, et al. High levels of mitochondrial DNA deletions in substantia nigra neurons in aging and Parkinson disease. *Nat Genet.* 2006;38(5):515–517.
65. Kraytsberg Y, Kudryavtseva E, McKee AC, Geula C, Kowall NW, Khrapko K. Mitochondrial DNA deletions are abundant and cause functional impairment in aged human substantia nigra neurons. *Nat Genet.* 2006;38(5):518–520.
66. Legros F, Malka F, Frachon P, Lombès A, Rojo M. Organization and dynamics of human mitochondrial DNA. *J Cell Sci.* 2004;117(13):2653–2662.
67. Diaz F, Bayona-Bafaluy MP, Rana M, Mora M, Hao H, Moraes CT. Human mitochondrial DNA with large deletions repopulates organelles faster than full-length genomes under relaxed copy number control. *Nucleic Acids Res.* 2002;30(21):4626–4633.
68. Stewart JB, Chinnery PF. The dynamics of mitochondrial DNA heteroplasmy: implications for human health and disease. *Nat Rev Genet.* 2015;16(9):530–542.
69. Parker WD, Parks JK. Mitochondrial ND5 mutations in idiopathic Parkinson's disease. *Biochem Biophys Res Commun.* 2005;326(3):667–669.
70. Smigrodzki R, Parks J, Parker WD. High frequency of mitochondrial complex I mutations in Parkinson's disease and aging. *Neurobiol Aging.* 2004;25(10):1273–1281.
71. Palin EJH, Paetau A, Suomalainen A. Mesencephalic complex I deficiency does not correlate with parkinsonism in mitochondrial DNA maintenance disorders. *Brain.* 2013;136(Pt 8):2379–2392.
72. Majander A, Huoponen K, Savontaus ML, Nikoskelainen E, Wikström M. Electron transfer properties of NADH:ubiquinone reductase in the ND1/3460 and the ND4/11778 mutations of the Leber hereditary optic neuroretinopathy (LHON). *FEBS Lett.* 1991;292(1–2):289–292.
73. Brown MD, Voljavec AS, Lott MT, Torroni A, Yang CC, Wallace DC. Mitochondrial DNA complex I and III mutations associated with Leber's hereditary optic neuropathy. *Genetics.* 1992;130(1):163–173.
74. Tuppen HAL, Blakely EL, Turnbull DM, Taylor RW. Mitochondrial DNA mutations and human disease. *Biochim Biophys Acta.* 2010;1797(2):113–128.
75. Perier C, Bender A, García-Arumí E, et al. Accumulation of mitochondrial DNA deletions within dopaminergic neurons triggers neuroprotective mechanisms. *Brain.* 2013;136(Pt 8):2369–2378.
76. Deas E, Plun-Favreau H, Wood NW. PINK1 function in health and disease. *EMBO Mol Med.* 2009;1(3):152–165.

77. Valente EM, Abou-Sleiman PM, Caputo V, et al. Hereditary early-onset Parkinson's disease caused by mutations in PINK1. *Science*. 2004;304(5674):1158–1160.
78. Klein C, Lohmann-Hedrich K, Rogaeva E, Schlossmacher MG, Lang AE. Deciphering the role of heterozygous mutations in genes associated with parkinsonism. *Lancet Neurol*. 2007;6(7):652–662.
79. Nuytemans K, Theuns J, Cruts M, Van Broeckhoven C. Genetic etiology of Parkinson disease associated with mutations in the SNCA, PARK2, PINK1, PARK7, and LRRK2 genes: a mutation update. *Hum Mutat*. 2010;31(7):763–780.
80. Madeo G, Schirinzi T, Martella G, et al. PINK1 heterozygous mutations induce subtle alterations in dopamine-dependent synaptic plasticity. *Mov Disord*. 2014;29(1).
81. Aerts L, De Strooper B, Morais VA. PINK1 activation—turning on a promiscuous kinase. *Biochem Soc Trans*. 2015;43(2):280–286.
82. Haelterman NA, Yoon WH, Sandoval H, Jaiswal M, Shulman JM, Bellen HJ. A mitocentric view of Parkinson's disease. *Annu Rev Neurosci*. 2014;37:137–159.
83. Clark IE, Dodson MW, Jiang C, et al. Drosophila pink1 is required for mitochondrial function and interacts genetically with parkin. *Nature*. 2006;441(7097):1162–1166.
84. Gautier CA, Kitada T, Shen J. Loss of PINK1 causes mitochondrial functional defects and increased sensitivity to oxidative stress. *Proc Natl Acad Sci USA*. 2008;105(32):11364–11369.
85. Morais VA, Verstreken P, Roethig A, et al. Parkinson's disease mutations in PINK1 result in decreased complex I activity and deficient synaptic function. *EMBO Mol Med*. 2009;1(2):99–111.
86. Vincow ES, Merrihew G, Thomas RE, et al. The PINK1-Parkin pathway promotes both mitophagy and selective respiratory chain turnover in vivo. *Proc Natl Acad Sci USA*. 2013;110(16):6400–6405.
87. Morais VA, Haddad D, Craessaerts K, et al. PINK1 loss of function mutations affect mitochondrial complex I activity via NdufA10 ubiquinone uncoupling. *Science*. 2014;344(6180):203–207.
88. Vinothkumar KR, Zhu J, Hirst J. Architecture of mammalian respiratory complex I. *Nature*. 2014;515(7525):80–84.
89. Efremov RG, Baradaran R, Sazanov LA. The architecture of respiratory complex I. *Nature*. 2010;465(7297):441–445.
90. Janssen RJRJ, Nijtmans LG, van den Heuvel LP, Smeitink JAM. Mitochondrial complex I: structure, function and pathology. *J Inherit Metab Dis*. 2006;29(4):499–515.
91. Pogson JH, Ivatt RM, Sanchez-Martinez A, et al. The complex I subunit NDUFA10 selectively rescues drosophila pink1 mutants through a mechanism independent of mitophagy. *PLoS Genet*. 2014;10(11):e1004815.
92. Vos M, Esposito G, Edirisinghe JN, et al. Vitamin K2 is a mitochondrial electron carrier that rescues pink1 deficiency. *Science*. 2012;336(6086):1306–1310.
93. Vilain S, Esposito G, Haddad D, et al. The yeast complex I equivalent NADH dehydrogenase rescues pink1 mutants. *PLoS Genet*. 2012;8(1):e1002456.
94. Kitada T, Asakawa S, Hattori N, et al. Mutations in the parkin gene cause autosomal recessive juvenile parkinsonism. *Nature*. 1998;392(6676):605–608.
95. Abbas N, Lücking CB, Ricard S, et al. A wide variety of mutations in the parkin gene are responsible for autosomal recessive parkinsonism in Europe. *Hum Mol Genet*. 1999;8(4):567–574.
96. Lucking CB, Durr A, Bonifati V, et al. Association between early-onset Parkinson's Disease and mutations in the parkin gene. *N Engl J Med*. 2000;342(21):1560–1567.
97. Sarraf Sa, Raman M, Guarani-Pereira V, et al. Landscape of the PARKIN-dependent ubiquitylome in response to mitochondrial depolarization. *Nature*. 2013;496(7445):1–7.
98. Greene JC, Whitworth AJ, Kuo I, Andrews LA, Feany MB, Pallanck LJ. Mitochondrial pathology and apoptotic muscle degeneration in Drosophila parkin mutants. *Proc Natl Acad Sci USA*. 2003;100(7):4078–4083.

99. Palacino JJ, Sagi D, Goldberg MS, et al. Mitochondrial dysfunction and oxidative damage in parkin-deficient mice. *J Biol Chem.* 2004;279(18):18614–18622.
100. Mortiboys H, Thomas KJ, Koopman WJH, et al. Mitochondrial function and morphology are impaired in parkin mutant fibroblasts. *Ann Neurol.* 2008;64(5):555–565.
101. Grünewald A, Voges L, Rakovic A, et al. Mutant Parkin impairs mitochondrial function and morphology in human fibroblasts. *PLoS One.* 2010;5(9):e12962.
102. Narendra DP, Tanaka A, Suen D-F, Youle RJ. Parkin is recruited selectively to impaired mitochondria and promotes their autophagy. *J Cell Biol.* 2008;183(5):795–803.
103. Park J, Lee SB, Lee S, et al. Mitochondrial dysfunction in drosophila PINK1 mutants is complemented by parkin. *Nature.* 2006;441(7097):1157–1161.
104. Hao L-Y, Giasson BI, Bonini NM. DJ-1 is critical for mitochondrial function and rescues PINK1 loss of function. *Proc Natl Acad Sci USA.* 2010;107(21):9747–9752.
105. Kim NC, Tresse E, Kolaitis R-M, et al. VCP is essential for mitochondrial quality control by PINK1/Parkin and this function Is impaired by VCP mutations. *Neuron.* 2013;78(1):1–16.
106. Zhang L, Karsten P, Hamm S, et al. TRAP1 rescues PINK1 loss-of-function phenotypes. *Hum Mol Genet.* 2013;22(14):2829–2841.
107. Klein P, Muller-Rischart AK, Motori E, et al. Ret rescues mitochondrial morphology and muscle degeneration of Drosophila Pink1 mutants. *EMBO.* 2014;30(4):1–15.
108. Bonifati V, Rizzu P, Squitieri F, et al. DJ-1 (PARK7), a novel gene for autosomal recessive, early onset parkinsonism. *Neurol Sci.* 2003;24(3):159–160.
109. Menzies FM, Yenisetti SC, Min K-T. Roles of drosophila DJ-1 in survival of dopaminergic neurons and oxidative stress. *Curr Biol.* 2005;15(17):1578–1582.
110. Park J, Kim SY, Cha G-H, Lee SB, Kim S, Chung J. Drosophila DJ-1 mutants show oxidative stress-sensitive locomotive dysfunction. *Gene.* 2005;361:133–139.
111. Yang Y, Gehrke S, Haque ME, et al. Inactivation of Drosophila DJ-1 leads to impairments of oxidative stress response and phosphatidylinositol 3-kinase/Akt signaling. *Proc Natl Acad Sci USA.* 2005;102(38):13670–13675.
112. Kim RH, Smith PD, Aleyasin H, et al. Hypersensitivity of DJ-1-deficient mice to 1-methyl-4-phenyl-1,2,3,6-tetrahydropyridine (MPTP) and oxidative stress. *Proc Natl Acad Sci USA.* 2005;102(14):5215–5220.
113. Cookson MR. Parkinsonism due to mutations in PINK1, Parkin, and DJ-1 and oxidative stress and mitochondrial pathways. *Cold Spring Harb Perspect Med.* 2012;2(9).
114. Heo JY, Park JH, Kim SJ, et al. DJ-1 null dopaminergic neuronal cells exhibit defects in mitochondrial function and structure: involvement of mitochondrial complex I assembly. *PLoS One.* 2012;7(3):e32629.
115. Krebiehl G, Ruckerbauer S, Burbulla LF, et al. Reduced basal autophagy and impaired mitochondrial dynamics due to loss of Parkinson's disease-associated protein DJ-1. *PLoS One.* 2010;5(2):e9367.
116. Guzman JN, Sanchez-Padilla J, Wokosin D, et al. Oxidant stress evoked by pacemaking in dopaminergic neurons is attenuated by DJ-1. *Nature.* 2010;468(7324):696–700.
117. Polymeropoulos MH, Lavedan C, Leroy E, et al. Mutation in the α-synuclein gene identified in families with Parkinson's disease. *Science.* 1997;276(5321):2045–2047.
118. Lesage S, Brice A. Role of mendelian genes in "sporadic" Parkinson's disease. *Parkinsonism Relat Disord.* 2012;18(Suppl 1):S66–S70.
119. Snead D, Eliezer D. Alpha-synuclein function and dysfunction on cellular membranes. *Exp Neurobiol.* 2014;23(4):292–313.
120. Nakamura K. α-Synuclein and mitochondria: partners in crime? *Neurotherapeutics.* 2013;10(3):391–399.
121. Junn E, Mouradian MM. Human alpha-synuclein over-expression increases intracellular reactive oxygen species levels and susceptibility to dopamine. *Neurosci Lett.* 2002;320(3):146–150.

122. Parihar MS, Parihar A, Fujita M, Hashimoto M, Ghafourifar P. Alpha-synuclein over-expression and aggregation exacerbates impairment of mitochondrial functions by augmenting oxidative stress in human neuroblastoma cells. *Int J Biochem Cell Biol.* 2009;41(10):2015–2024.

123. Cole NB, DiEuliis D, Leo P, Mitchell DC, Nussbaum RL. Mitochondrial translocation of α-synuclein is promoted by intracellular acidification. *Exp Cell Res.* 2008;314(10):2076–2089.

124. Butler EK, Voigt A, Lutz a K, et al. The mitochondrial chaperone protein TRAP1 mitigates α-synuclein toxicity. *PLoS Genet.* 2012;8(2):e1002488.

125. Shavali S, Brown-Borg HM, Ebadi M, Porter J. Mitochondrial localization of alpha-synuclein protein in alpha-synuclein overexpressing cells. *Neurosci Lett.* 2008;439(2):125–128.

126. Devi L, Raghavendran V, Prabhu BM, Avadhani NG, Anandatheerthavarada HK. Mitochondrial import and accumulation of alpha-synuclein impair complex I in human dopaminergic neuronal cultures and Parkinson disease brain. *J Biol Chem.* 2008;283(14):9089–9100.

127. Hardy J, Lees AJ. Parkinson's disease: a broken nosology. *Mov Disord.* 2005;20 (Suppl 1):S2–S4.

128. Martinez TN, Greenamyre JT. Toxin models of mitochondrial dysfunction in Parkinson's disease. *Antioxid Redox Signal.* 2012;16(9):920–934.

129. Priyadarshi A, Khuder SA, Schaub EA, Shrivastava S. A meta-analysis of Parkinson's disease and exposure to pesticides. *Neurotoxicology.* 2000;21(4):435–440.

130. Dawson TM, Ko HS, Dawson VL. Genetic animal models of Parkinson's disease. *Neuron.* 2010;66(5):646–661.

131. Blesa J, Przedborski S. Parkinson's disease: animal models and dopaminergic cell vulnerability. *Front Neuroanat.* 2014;8:155.

132. Vanhauwaert R, Verstreken P. Flies with Parkinson's disease. *Exp Neurol.* 2015;274:42–51.

133. Chiueh CC, Markey SP, Burns RS, Johannessen JN, Jacobowitz DM, Kopin IJ. Neurochemical and behavioral effects of 1-methyl-4-phenyl-1,2,3,6- tetrahydropyridine (MPTP) in rat, guinea pig, and monkey. *Psychopharmacol Bull.* 1984;20(3):548–553.

134. Tieu K. A guide to neurotoxic animal models of Parkinson's disease. *Cold Spring Harb Perspect Med.* 2011;1(1):a009316.

135. Forno LS, Langston JW, DeLanney LE, Irwin I, Ricaurte GA. Locus ceruleus lesions and eosinophilic inclusions in MPTP-treated monkeys. *Ann Neurol.* 1986;20(4):449–455.

136. Taylor TN, Greene JG, Miller GW. Behavioral phenotyping of mouse models of Parkinson's disease. *Behav Brain Res.* 2010;211(1):1–10.

137. Inden M, Kitamura Y, Abe M, Tamaki A, Takata K, Taniguchi T. Parkinsonian rotenone mouse model: reevaluation of long-term administration of rotenone in C57BL/6 mice. *Biol Pharm Bull.* 2011;34(1):92–96.

138. Heikkila RE, Nicklas WJ, Vyas I, Duvoisin RC. Dopaminergic toxicity of rotenone and the 1-methyl-4-phenylpyridinium ion after their stereotaxic administration to rats: implication for the mechanism of 1-methyl-4-phenyl-1,2,3,6-tetrahydropyridine toxicity. *Neurosci Lett.* 1985;62(3):389–394.

139. Ferrante RJ, Schulz JB, Kowall NW, Beal MF. Systemic administration of rotenone produces selective damage in the striatum and globus pallidus, but not in the substantia nigra. *Brain Res.* 1997;753(1):157–162.

140. Greenamyre JT, Betarbet R, Sherer TB. The rotenone model of Parkinson's disease: genes, environment and mitochondria. *Parkinsonism Relat Disord.* 2003;9(Suppl 2):S59–S64.

141. Blandini F, Armentero M-T, Martignoni E. The 6-hydroxydopamine model: news from the past. *Parkinsonism Relat Disord.* 2008;14(Suppl 2):S124–S129.

142. Luthman J, Fredriksson A, Sundström E, Jonsson G, Archer T. Selective lesion of central dopamine or noradrenaline neuron systems in the neonatal rat: motor behavior and monoamine alterations at adult stage. *Behav Brain Res.* 1989;33(3):267–277.

143. Sachs C, Jonsson G. Effects of 6-hydroxydopamine on central noradrenaline neurons during ontogeny. *Brain Res*. 1975;99(2):277–291.

144. Cleeter MW, Cooper JM, Schapira AH. Irreversible inhibition of mitochondrial complex I by 1-methyl-4-phenylpyridinium: evidence for free radical involvement. *J Neurochem*. 1992;58(2):786–789.

145. Zhou Y, Shie F-S, Piccardo P, Montine TJ, Zhang J. Proteasomal inhibition induced by manganese ethylene-bis-dithiocarbamate: relevance to Parkinson's disease. *Neuroscience*. 2004;128(2):281–291.

146. Gispert S, Ricciardi F, Kurz A, et al. Parkinson phenotype in aged PINK1-deficient mice is accompanied by progressive mitochondrial dysfunction in absence of neurodegeneration. *PLoS One*. 2009;4(6):e5777.

147. Goldberg MS, Fleming SM, Palacino JJ, et al. Parkin-deficient mice exhibit nigrostriatal deficits but not loss of dopaminergic neurons. *J Biol Chem*. 2003;278(44):43628–43635.

148. Perez FA, Palmiter RD. Parkin-deficient mice are not a robust model of parkinsonism. *Proc Natl Acad Sci USA*. 2005;102(6):2174–2179.

149. Kitada T, Pisani A, Karouani M, et al. Impaired dopamine release and synaptic plasticity in the striatum of parkin-/- mice. *J Neurochem*. 2009;110(2):613–621.

150. Kitada T, Tong Y, Gautier C, Shen J. Absence of nigral degeneration in aged parkin/ DJ-1/PINK1 triple knockout mice. *J Neurochem*. 2009;111(3):696–702.

151. Samaranch L, Lorenzo-Betancor O, Arbelo JM, et al. PINK1-linked parkinsonism is associated with Lewy body pathology. *Brain*. 2010;133(Pt 4):1128–1142.

152. Yang Y, Gehrke S, Imai Y, et al. Mitochondrial pathology and muscle and dopaminergic neuron degeneration caused by inactivation of Drosophila Pink1 is rescued by Parkin. *Proc Natl Acad Sci USA*. 2006;103(28):10793–10798.

153. Deng H, Jankovic J, Guo Y, Xie W, Le W. Small interfering RNA targeting the PINK1 induces apoptosis in dopaminergic cells SH-SY5Y. *Biochem Biophys Res Commun*. 2005;337:1133–1138.

154. Wood-Kaczmar A, Gandhi S, Yao Z, et al. PINK1 is necessary for long term survival and mitochondrial function in human dopaminergic neurons. *PLoS One*. 2008;3(6):e2455.

155. Okatsu K, Oka T, Iguchi M, et al. PINK1 autophosphorylation upon membrane potential dissipation is essential for Parkin recruitment to damaged mitochondria. *Nat Commun*. 2012;3(1016):1–10.

156. Aerts L, Craessaerts K, De Strooper B, Morais VA. PINK1 catalytic activity is regulated by phosphorylation on serines 228 and 402. *J Biol Chem*. 2015;290(5):2798–2811.

157. Narendra DP, Jin S, Tanaka A, et al. PINK1 is selectively stabilized on impaired mitochondria to activate Parkin. *PLoS Biol*. 2010;8(1):e1000298.

158. Kondapalli C, Kazlauskaite A, Zhang N, et al. PINK1 is activated by mitochondrial membrane potential depolarization and stimulates Parkin E3 ligase activity by phosphorylating serine 65. *Open Biol*. 2012;2(5):120080.

159. Shiba-Fukushima K, Imai Y, Yoshida S, Ishihama Y, Kanao T. PINK1-mediated phosphorylation of the Parkin ubiquitin-like domain primes mitochondrial translocation of Parkin and regulates mitophagy. *Sci Rep*. 2012;2(1002):1–8.

160. Kane LA, Lazarou M, Fogel AI, et al. PINK1 phosphorylates ubiquitin to activate Parkin E3 ubiquitin ligase activity. *J Cell Biol*. 2014;205(2):143–153.

161. Kazlauskaite A, Kondapalli C, Gourlay R, et al. Parkin is activated by PINK1-dependent phosphorylation of ubiquitin at serine65. *Biochem J*. 2014;460(1):127–139.

162. Koyano F, Okatsu K, Kosako H, et al. Ubiquitin is phosphorylated by PINK1 to activate parkin. *Nature*. 2014;510(7503):162–166.

163. Youle RJ, Narendra DP. Mechanisms of mitophagy. *Nat Rev Mol Cell Biol*. 2011;12(1):9–14.

164. Sazanov LA. A giant molecular proton pump: structure and mechanism of respiratory complex I. *Nat Rev Mol Cell Biol*. 2015;16(6):375–388.

165. Parkinson J. *An Essay on the Shaking Palsy*. London: Whittingham and Rowland; 1817.

166. Shults CW, Beal MF, Fontaine D, Nakano K, Haas RH. Absorption, tolerability, and effects on mitochondrial activity of oral coenzyme Q10 in parkinsonian patients. *Neurology*. 1998;50(3):793–795.

167. Shults CW, Oakes D, Kieburtz K, et al. Effects of coenzyme Q10 in early Parkinson disease: evidence of slowing of the functional decline. *Arch Neurol*. 2002;59(10):1541–1550.

168. Snow BJ, Rolfe FL, Lockhart MM, et al. A double-blind, placebo-controlled study to assess the mitochondria-targeted antioxidant MitoQ as a disease-modifying therapy in Parkinson's disease. *Mov Disord*. 2010;25(11):1670–1674.

169. Müller T, Büttner T, Gholipour AF, Kuhn W. Coenzyme Q10 supplementation provides mild symptomatic benefit in patients with Parkinson's disease. *Neurosci Lett*. 2003;341(3):201–204.

170. Beal MF, Oakes D, Shoulson I, et al. A randomized clinical trial of high-dosage coenzyme Q10 in early Parkinson disease: no evidence of benefit. *JAMA Neurol*. 2014;71(5):543–552.

171. Storch A, Jost WH, Vieregge P, et al. Randomized, double-blind, placebo-controlled trial on symptomatic effects of coenzyme Q(10) in Parkinson disease. *Arch Neurol*. 2007;64(7):938–944.

172. NINDSA randomized, double-blind, futility clinical trial of creatine and minocycline in early Parkinson disease. *Neurology*. 2006;66(5):664–671.

173. NINDSA pilot clinical trial of creatine and minocycline in early Parkinson disease: 18-month results. *Clin Neuropharmacol*. 2008;31(3):141–150.

174. Kieburtz K, Tilley BC, Elm JJ, et al. Effect of creatine monohydrate on clinical progression in patients with Parkinson disease: a randomized clinical trial. *JAMA*. 2015;313(6):584–593.

175. Mischley LK, Vespignani MF, Finnell JS. Safety survey of intranasal glutathione. *J Altern Complement Med*. 2013;19(5):459–463.

176. Hauser RA, Lyons KE, McClain T, Carter S, Perlmutter D. Randomized, double-blind, pilot evaluation of intravenous glutathione in Parkinson's disease. *Mov Disord*. 2009;24(7):979–983.

177. Mischley LK, Leverenz JB, Lau RC, et al. A randomized, double-blind phase I/IIa study of intranasal glutathione in Parkinson's disease. *Mov Disord*. 2015;30(12):1696–1701.

178. Waring WS, Webb DJ, Maxwell SR. Systemic uric acid administration increases serum antioxidant capacity in healthy volunteers. *J Cardiovasc Pharmacol*. 2001;38(3):365–371.

179. Schwarzschild MA, Ascherio A, Beal MF, et al. Inosine to increase serum and cerebrospinal fluid urate in Parkinson disease: a randomized clinical trial. *JAMA Neurol*. 2014;71(2):141–150.

180. Hargreaves IP, Lane A, Sleiman PMA. The coenzyme Q10 status of the brain regions of Parkinson's disease patients. *Neurosci Lett*. 2008;447(1):17–19.

181. Spindler M, Beal MF, Henchcliffe C. Coenzyme Q10 effects in neurodegenerative disease. *Neuropsychiatr Dis Treat*. 2009;5:597–610.

182. Ascherio A, LeWitt PA, Xu K, et al. Urate as a predictor of the rate of clinical decline in Parkinson disease. *Arch Neurol*. 2009;66(12):1460–1468.

183. Schwarzschild MA, Schwid SR, Marek K, et al. Serum urate as a predictor of clinical and radiographic progression in Parkinson disease. *Arch Neurol*. 2008;65(6):716–723.

184. Zhu T-G, Wang X-X, Luo W-F, et al. Protective effects of urate against 6-OHDA-induced cell injury in PC12 cells through antioxidant action. *Neurosci Lett*. 2012;506(2):175–179.

185. Chen X, Wu G, Schwarzschild MA. Urate in Parkinson's disease: more than a biomarker? *Curr Neurol Neurosci Rep*. 2012;12(4):367–375.

186. Vos M, Lovisa B, Geens A, et al. Near-Infrared 808 nm light boosts complex IV-dependent respiration and rescues a Parkinson-related pink1 model. *PLoS One*. 2013;8(11):e78562.

187. Vos M, Verstreken P, Klein C. Stimulation of electron transport as potential novel therapy in Parkinson's disease with mitochondrial dysfunction. *Biochem Soc Trans*. 2015;43(2):275–279.

Mitochondrial Fission and Fusion

V.L. Hewitt, A.J. Whitworth

Medical Research Council Mitochondrial Biology Unit, Cambridge Biomedical Campus, Cambridge, United Kingdom

Parkinson's Disease. http://dx.doi.org/10.1016/B978-0-12-803783-6.00003-1

1 INTRODUCTION

While many cellular and molecular processes have been implicated in Parkinsons disease (PD) pathology, mitochondrial dysfunction has long been considered a culprit in the early initiating events contributing to pathogenesis.[1] The evidence implicating a mitochondrial role comes from many different sources. A high proportion of sporadic PD patients exhibit systemic mitochondrial defects, and in particular complex I deficits;[2] several different mitochondrial complex I inhibitors elicit parkinsonian-like syndromes in humans and animal models;[3–5] mitochondrial DNA mutations and deletions are present at high abundance in dopamine neurons from aged individuals and are even more prevalent in PD individuals;[6,7] and functional studies of the genes causative for simple Mendelian forms of PD suggest that many have mitochondrial roles.[8,9]

A central role of mitochondria in the long-term survival or the pathological death of neurons is not surprising given the integral functions of mitochondria in energy generation, calcium regulation, and many metabolic processes, as well as sequestering proapoptotic machinery. On top of this, mitochondria are also the major site where damaging reactive oxygen species (ROS) are generated. So inevitably there are myriad quality control processes that maintain the normal working integrity of the cellular mitochondria.

To provide a ready supply of cellular energy, in the form of adenosine triphosphate (ATP), to all areas of the cell, mitochondria need to be mobile organelles. With improvements in live-cell imaging techniques and the advent of fluorescent markers, we have now come to appreciate that mitochondria are extremely dynamic organelles.[10] Not only are they capable of movement between the extremities of the cell but they frequently undergo fusion and fission events. Hence, any momentary snapshot of a population of mitochondria within a cell reveals a heterogeneous mix of morphologies. In fact, this morphological heterogeneity was a notable early characteristic, and ultimately gave rise to their name, being derived from the Greek "mitos" (thread) and "chondros" (granule or grain). An imbalance in the ongoing cycles of fission and fusion can disturb the overall morphology causing mitochondria to appear as small and fragmented or elongated tubules. In addition to fission/fusion events, mitochondria are transported via both microtubules (long-range) and actin (short-range) cytoskeleton. Collectively these processes are often referred to as mitochondrial dynamics. In light of this inherent dynamism, the mitochondrial population of a cell is now more commonly considered as an inter-related and -mixing network.

Fission and fusion are both vital to maintaining a properly functioning mitochondrial network.[11] Elegant genetic studies in model organisms, yeast in particular, have identified many of the core machinery that mediates mitochondrial fission and fusion.[12] Many of these genes are highly conserved across phyla, highlighting the fundamentally important nature of these processes. As our understanding of the components and regulation of fission and fusion has grown, it has revealed the ongoing cycle of fission and fusion creates a quality control system that is sensitive to the local and specific metabolic requirements of the cell.[13]

A variety of cellular stresses can induce fusion (stress-induced mitochondrial hyperfusion). Fusion allows the mixing of mitochondrial contents, and the resulting hyperfused mitochondria appear to produce ATP more efficiently.[14] Mixing of mitochondrial contents can potentially dilute damaged mitochondrial components, facilitating functional complementation, and helps maintain overall network functioning.[15] During starvation, hyperfusion also occurs in order to prevent the degradation of mitochondria by autophagy—a process where cellular components are digested in bulk to recycle their constituent parts.[16] Together these processes create a stress response mechanism that can help prevent premature activation of cell death pathways.[17] Other potential roles of hyperfused mitochondria have been hypothesized and are all variations on these ideas;[18] a more extensive mitochondrial network may help buffer deleterious effects of damaged components (DNA or protein) or help boost mitochondria functions (eg, calcium exchange or energy transfer), but it is clear that cellular health declines rapidly when mitochondria cannot fuse.[19]

Many forms of cellular stress, especially mitochondrial stress, rapidly induce fission producing small, fragmented mitochondria.[20] The exact purpose of this is not entirely clear, though there is a long-standing link between mitochondrial fission and programmed cell death,[21] and in some instances preventing fission during cell stress also prevents cell death. However, it is also possible that this represents a cell-protective response. The reticular nature of the mitochondrial network would allow the rapid spreading of a toxic stimulus or cellular damage (eg, ROS) via interconnected tubules, which could be counter-acted by fragmentation. Moreover, if irreversible damage does occur, smaller mitochondria are more easily mobilised and degraded, preventing mitochondrial rupture and activation of cell death pathways.[22] Thus, a simple view of this process is that fusion helps mitigate the effects of cellular damage via complementation while fission prepares mitochondria that are beyond repair for efficient degradation.[23]

As well as their dynamic interorganelle connections, mitochondria also form multiple contacts with the ER.[24] These contact sites have been shown to contribute to regulation of diverse processes including mitochondrial trafficking, calcium homeostasis, lipid synthesis, ER stress, inflammasome formation, and apoptosis, as well as regulating fission.[25] These functions clearly have the potential to contribute to neurodegenerative disease mechanisms and ER–mitochondrial interactions and have been shown to be disrupted in Alzheimers disease and motor neuron disease as well as PD.[26]

Thus, regulated mitochondrial dynamics is therefore vital to support the growth, maintenance, and function of healthy neurons.[27] In this chapter, we focus on the roles of fission and fusion in neuronal function and how disrupting these processes results in neurodegeneration and, in particular, the evidence that has linked these processes to PD.

2 FISSION AND FUSION MACHINERY

The fact that mitochondria are dynamic and undergo fission and fusion events had been observed for many years, but over the past couple of decades the machinery that mediates fission and fusion has been exquisitely elucidated primarily through classic genetic techniques. Although many of these components are highly conserved across species, many additional genes have been identified in yeast that appear to be specific to lower organisms. For the purpose of this review we shall focus on the mechanisms known to occur in higher organisms.

2.1 Fission Machinery

The key component that provides the mechanical force that drives mitochondrial fission is dynamin-related protein 1 (Drp1), a large GTPase of the dynamin family.[28–32] Drp1 is a cytosolic protein that is recruited to mitochondria at sites of future scission.[33] Fission occurs by Drp1 oligomerizing and forming spirals around the mitochondrion, which induces scission by constricting and severing the inner and outer membranes (Fig. 3.1).[34,35] Cells lacking Drp1 have mitochondria that are highly elongated and more interconnected,[36] and knockout animal models are lethal, indicating the essential nature of mitochondrial fission.[37]

The mechanisms of Drp1 recruitment that determine the site of scission are still being investigated,[38] but a number of adapter proteins have been identified,[39] including Fis1,[40–42] mitochondrial fission factor (Mff),[43] and the mitochondrial dynamics proteins MiD49 and MiD51/MIEF1.[44,45] There is also growing evidence that dynamin-like proteins may recognize and directly bind specific elements of the lipid membrane.[46] For instance, the recruitment and/or function of Drp1 is modulated by the lipid composition of the membrane.[47] In particular, cardiolipin has recently been shown to stimulate Drp1 GTPase activity and function.[48,49]

Another recent discovery is that tubular extensions of the endoplasmic reticulum (ER) appear to direct sites of fission and help constrict mitochondrial tubules prior to fission in a process called ER-associated mitochondrial division (ERMD).[50] Although the molecular mechanisms of this process are still being investigated, actin polymerization is thought to be a key feature.[51] Modulators of this process have already been identified,

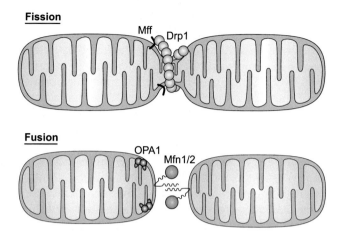

FIGURE 3.1 **The core machinery of mitochondrial fission and fusion.** The large GTPase Drp1 is recruited to mitochondria by Mff, and others such as MiD49 and MiD51 (not shown). Drp1 oligomerizes around the mitochondrion and the GTPase activity drives constriction and scission of the membranes. Fusion is mediated by the action of other large GTPases Mfn1 and 2 on the outer membrane and Opa1 on the inner membrane. Homo- or heterotypic binding of Mfn1/2 via their coil-coiled C-terminal domains anchors the sites of fusion, and GTPase activity drives membrane fusion by an unclear mechanism.

including mitochondrial Spire1C,[52] the ER-bound formin (INF2),[51] and actin-binding proteins such as myosin II,[53,54] cortactin, and cofilin.[55] The discovery of ERMD suggests there could be overlap in pathogenic mechanisms between diseases resulting from disruption of mitochondrial fission and those associated with defects in ER-shaping proteins such as hereditary spastic paraplegia.[56,57]

Various types of posttranslational modification of Drp1, such as phosphorylation, ubiquitylation, sumoylation, and nitrosylation, also regulate fission. For example, Drp1 is phosphorylated at multiple sites by multiple kinases, which can be either activating or inactivating. In a notable link to calcium regulation, relevant to both mitochondrial and neuronal homeostasis, Drp1 is dephosphorylated by the cytosolic calcium-activated phosphatase calcineurin (PP2A).[58,59] Drp1 has also been reported to be both ubiquitylated and sumoylated by the mitochondria-localized E3 ligases MARCH5 (also known as MITOL)[60,61] and MUL1 (also known as MAPL),[62,63] respectively, although these functions are still unclear.[64] While the mechanisms that regulate mitochondrial fission are varied and complex, they all seem to operate via the action of Drp1.[65]

2.2 Fusion Machinery

Mitochondrial fusion is complicated by the requirement to fuse both the inner and outer membranes. While this is often a concerted process, fusion can be limited to the outer membrane in a kiss-and-run interaction.[66–68] Again the key components are a set of large GTPases related to dynamin. Fusion of the outer membrane is mediated by hetero- or homotypic binding of the mitofusins (eg, Mfn1 and Mfn2 in mammals[10]), while Optic atrophy 1 (Opa1) mediates fusion of the inner membrane (Fig. 3.1).[69] Loss of any of these proteins results in small, fragmented mitochondria. Mfn1 and Mfn2 have some functional overlap but removing both Mitofusins or Opa1 results in slower cell growth, reduced respiration and loss of mitochondrial membrane potential.[19]

The Mitofusins are large membrane anchored proteins with a GTPase domain[70] that is essential for their function.[11] Dimerization of a coil–coil domain is part of the process that brings the membranes together[71] but the fusion mechanisms are still being investigated. Mitofusins can be phosphorylated[72] and reversibly ubiquitinated,[73] which can increase or decrease fusion depending on the location of the modification.[74] Such modifications have implications for understanding and treating diseases involving mitochondrial dynamics as they allow the cell to carefully modulate the mitochondrial network to respond to stress conditions.[75]

Like Drp1, the function of Mitofusins also appears to be lipid dependent. The generation of phosphatidic acid from cardiolipin, by mitochondrial phospholipase D, is required for fusion.[76] Since ER–mitochondria contact sites are also implicated in lipid metabolism, this provides an additional connection between ER–mitochondria contacts and mitochondrial dynamics[77]). Moreover, it has been suggested that Mitofusins can form a protein tether at ER–mitochondria contact sites with Mfn2 located on ER membranes binding Mfn1 or 2 on mitochondria,[78] although these results are currently disputed.[79]

Opa1 also exists in a variety of splice-variants that are proteolytically cleaved by a variety of mitochondrial proteases.[80] The exact functions of the different isoforms are not clearly understood but are thought to regulate both the fusibility of the inner membrane and also the architecture of cristae. Acetylation may also be import for regulation of Opa1 function; an acetylated form was recently shown to be less active.[81] Like the other dynamin-like proteins involved in fission and fusion, the lipid composition may be important for the function of the Opa1. In yeast, dimerization of the Opa1 homolog Mgm1 requires cardiolipin, found in high concentrations in the mitochondrial inner membrane.[82]

3 FUNCTIONS OF FISSION AND FUSION IN NEURONS

Why neurons are particularly vulnerable to defects in mitochondrial dynamics is key to our understanding of their role in neurodegenerative disease. The basic function of neurons is extremely energy-demanding, with the continual pumping of ions across cell membranes, and with energy (ATP) production comes ROS. Furthermore, in general adult neurons are postmitotic and nonreplicating. Thus, these cells understandably have a particular requirement for any mechanism that maintains cellular health and fitness.

Due to their highly specialized nature and architecture, neurons represent a particular challenge for many aspects of cell biology and metabolism: (1) elaborate arborization with very long processes, (2) spatially distinct requirements for high ATP and tight regulation of calcium concentration, and (3) constant remodeling in order to ensure the necessary synaptic plasticity required for proper neurotransmission.[83] The precise reasons that disturbed fission/fusion causes neuronal dysfunction and eventual neuronal cell loss are not well understood, but the role of mitochondria in energy generation, calcium homeostasis, and ROS production are likely to impact many of the previous specializations.

When mitochondrial dynamics are impaired the highly branched dendrites and long axons are not appropriately supplied with healthy mitochondria.[27] Moreover, those mitochondria present are not able to undergo the normal quality control processes and will be a heterogeneous population of damaged and dysfunctional mitochondria. In Purkinje cells, Mfn2 appears to be more important than Mfn1 in maintaining healthy mitochondria,[83] with loss of mitochondrial DNA and respiratory complex activity seen in the Mfn2 knockout model and mouse embryonic fibroblasts lacking Mfn2. This difference may be due to different expression levels of the two mitofusins,[83] as suggested by the ability of Mfn1 to rescue the viability defect and mitochondrial distribution in Purkinje cells seen in the Mfn2 knockout mice. The electron transport chain activity is also reduced overall, likely as a result of the distorted inner membrane cristae in these mitochondria. Only 2% of cultured Purkinje cells from these mice survive until P14. The mitochondria in the surviving cells cluster in the cell body and are not properly distributed in the axons.[83] Thus, disruption of fusion contributes to both mitochondrial dysfunction and impaired distribution in neurons. The differences in the functions and regulation of Mfn1 and 2 are still being explored, with potentially important differences in their functions in and localization to mitochondrial–ER contact sites.

Disruption of fission or fusion can lead to defects in transport of mitochondria. There appears to be a size limit on mitochondria that can be transported in neurons,[84] but a range of different length mitochondria can be transported.[85] Mice lacking Drp1 die at an early stage and expressing dominant-negative mutant Drp1 results in depletion of mitochondria in dendrites.[86] A longer, neuron-specific splice variant of Drp1 may also contribute to regulation of fission in neurons.[87] These data show fission is vital for proper distribution of mitochondria in neurons. Increasing fission by overexpression of Drp1 in cultured neurons increased the number of mitochondria in dendrites, the density of dendritic spines formed, and even their activity.[86] Since synaptic activity also results in increased recruitment of mitochondria to more active synapses, synaptic activity is likely to be highly sensitive to changes in fission and fusion. The biological significance of these changes requires further study in model organisms with more accessible neurons.

Neurons can respond to altered levels of fission and fusion machinery to help restore the balance of fission and fusion. In cultured cortical neurons increased Mfn2 lead to increased fission as well as fusion, limiting the changes in mitochondrial length.[88] However increasing fusion can lead to aggregated mitochondria that are not properly transported within neurons.[89] Mitochondria aggregation and impaired mitochondrial transport was observed in primary cultures of dorsal root ganglia from rats with mutant forms of Mfn2 associated with CMT2A.[89] The overall ATP production in these cells was unaffected but local reduction in ATP levels resulting from lack of mitochondria in specific parts of the neuron could contribute to loss of neuronal function. Comparable results in flies reported fewer mitochondria in axons and neuromuscular junctions in Drp1 mutants with reduced fission.[90] This report also suggests the lack of mitochondria impairs synaptic vesicle cycling and hence impairs neuronal function.

A more direct connection between the mitofusins and mitochondrial transport in neurons was discovered when Mfn2 was found to bind to Miro/Milton transport machinery complex.[91] Mfn2 mutants and knockdown both reduced axonal transport of mitochondria[91] and result in axonal degeneration.[92] DISC1, a protein with mutant forms associated with a number of forms of mental illness, was also recently found in this complex and is thought to help regulate the transport and fusion dynamics.[93] Miro also appears to be able to regulate fusion. Overexpression of Miro increases mitochondrial length in axons,[85] and expression of a constitutively active form produces elongated mitochondria in a number of cell types.[94–96] The yeast homolog of Miro (Gem1) is also known to affect mitochondrial morphology,[97] possibly due to a regulatory action on the fungal specific ER–mitochondrial contact site complex ERMES.[98] The role of transport defects in neurodegenerative diseases is discussed in more detail in Chapter 4 of this volume.

4 FISSION AND FUSION IS LINKED TO NEURODEGENERATION

Given the importance of mitochondria to neuronal physiology, it is perhaps not surprising that disturbances in fission and fusion that will negatively impact mitochondrial function will be most prominently felt in neurons. This is most clearly demonstrated with discoveries that mutations in key fission and fusion components cause a variety of neurological diseases. For instance, OPA1 is so named for its causative role in autosomal dominant optic atrophy (ADOA).[99] Loss of visual acuity associated with this disease may also be accompanied by extraocular neurologic features such as deafness and ataxia.[100] Interestingly, OPA1 mutations have also been seen in dominant chronic progressive external ophthalmoplegia (CPEO) with features of parkinsonism and dementia.[101] Mutations in Mfn2 cause Charcot-Marie-Tooth type 2A (CMT2A),[102,103] a peripheral neuropathy characterized by muscle weakness and axonal degeneration of sensory and motor neurons, and hereditary motor and sensory neuropathy type VI (HMSN VI), which is similar to CMT2A but also presents visual impairments including optic atrophy.

Reports of mutations in profission proteins are extremely rare, possibly reflecting the catastrophic impact of disturbed fission. A single case of a de novo heterozygous mutation in Drp1 (also known as DNM1L) was found to cause lethal encephalopathy due to defective mitochondrial and peroxisomal fission (EMPF).[104] The mutation likely caused a dominant negative effect resulting in elongated mitochondria and abnormal peroxisome (peroxisome division also occurs via Drp1). In a syndrome of similar clinical presentation, homozygous truncating mutations have been found in *MFF*, which also cause elongation of mitochondria and peroxisomes.[105]

Another gene linked to mitochondrial fission, *GDAP1*, which encodes ganglioside-induced differentiation associated protein 1, is mutated in CMT type 4A.[106] While GDAP1 is not considered a core mediator of fission, evidence indicates that it acts to promote mitochondrial fission, although the mechanisms are not clear. Overexpression of GDAP1 stimulated Drp1-mediated mitochondrial fission, while its loss does not clearly cause elongation. Interestingly, GDAP1 has been found at mitochondrial–ER contact sites, which may underlie the excess fission caused by its overexpression, and its loss appears to affect mitochondrial mobility, distribution, and co-localization with the ER as well as store-operated Ca^{2+} entry.[107,108]

5 PD GENES AND MITOCHONDRIAL MORPHOLOGY

The etiology of PD is complex and heterogeneous, with contributions from genetic and environmental factors. Over the past two decades tremendous advances have been made in identifying genes responsible for relatively rare dominant and recessive forms of PD (reviewed in Chapter 1). In addition, genome-wide association studies (GWAs) have allowed the identification of a relatively large number of susceptibility loci for sporadic PD. Many of the Mendelian forms are clinically indistinguishable from the more common sporadic PD, supporting the idea that common pathogenic mechanisms are shared between inherited and sporadic cases. Moreover, there is some striking overlap between the Mendelian genes and risk loci identified by GWAs, in particular, *SNCA* and *leucine-rich repeat kinase 2 (LRRK2)*, blurring the traditional boundaries between familial and sporadic PD. In many cases the underlying genetic component(s) of the GWAs risk loci still need to be determined, hence the majority of efforts in modeling PD still focuses on genes linked to the Mendelian forms of PD or related parkinsonism. Functional analyses of these genes has implicated numerous processes in disease pathogenesis, including aberrant protein degradation, mitochondrial dysfunction, calcium imbalance, and inflammation.[109] However, increasing evidence indicates that inactivation of these gene products may also occur in sporadic PD in pathways that converge at mitochondria.[110]

The clearest evidence linking mitochondrial dysfunction to PD comes from genes causing autosomal recessive parkinsonism, such as *Parkin*, *PINK1*, and *DJ-1*. Although a link to mitochondria was not immediately obvious for Parkin and DJ-1 since under basal conditions they are predominantly cytoplasmically localized proteins, the link was more intuitive for PINK1 as it has a clear N-terminal mitochondrial targeting sequence. Nevertheless, a wealth of evidence indicates that while Parkin serves numerous functions in the cell, a predominant role is in mitochondrial homeostasis. Similarly, the exact function of DJ-1 is still unknown, but it appears to have chaperone and antioxidant activities and translocates to mitochondria upon oxidative stress. One mechanism that has received a substantial amount of attention in recent years has been the genetic linkage of Parkin and PINK1 and their molecular interactions in mitochondrial quality control. Given the known impact of mitochondrial dynamics to homeostasis and quality control processes, it was almost inevitable that a link between factors that affect mitochondrial homeostasis and the fission–fusion processes would be made; however, this gave some important clues to the overall picture of major functions for PINK1 and Parkin. Moreover, observations that other PD-genes affect mitochondrial dynamics has given weight that this aspect of cell biology has important implications for the causes, and possibly even therapeutic targeting, of PD.

5.1 Parkin and PINK1

Mutations in *parkin* were the first to be linked to young onset, autosomal recessive PD,[111] implicating loss-of-function as the pathogenic trigger and indicating that the normal function of Parkin somehow promoted neuronal survival. Parkin was rapidly identified as an E3 ubiquitin ligase,[112–114] which initially raised hypotheses that loss of Parkin led to protein aggregation, with α-synuclein as the most obvious suspected target. However, a blow to this hypothesis came when reports began to emerge that postmortem examination of cases of *parkin*-related PD strikingly lacked Lewy bodies.[115,116] This led some to propose that *parkin*-related PD may occur via a different mechanism than idiopathic PD, and even that it should be classified as differently from PD. Nevertheless, the progressive loss of DA neurons was still the apparent cause of motor disturbance, thus understanding Parkin biology should provide valuable insight into the mechanisms that maintain survival.

Early investigations into the function of Parkin-implicated ER stress [112,117] or regulating synaptic proteins.[114,118] However, evidence soon emerged linking it to mitochondria and mechanisms of cell death.[119] This link was cemented with the description of the first gene knockout animal model using *Drosophila*. Flies lacking *parkin* have reduced lifespan, degeneration of DA neurons, and motor deficits.[120,121] More surprisingly, there was widespread apoptosis in the flight muscles of these flies and the mitochondria were severely disrupted, often appearing enlarged with disrupted cristae.[120] Also unexpected was that male *parkin* mutant flies are completely sterile while female fecundity appeared unperturbed. Given these striking effects it was surprising that the mouse knockout revealed only very mild phenotypes with no clear evidence of neurodegeneration or behavioral deficits.

Concurrent to these findings, mutations in *PINK1* were found to also cause autosomal recessive PD. PINK1 was immediately recognized as a mitochondrially targeted serine/threonine kinase, quickly establishing the most plausible pathogenic mechanism to be via mitochondria dysfunction. Again rapid mutational analysis of the *Drosophila* ortholog provided early unequivocal evidence that loss of *Pink1* grossly disrupts mitochondrial homeostasis. Strikingly, *Drosophila Pink1* mutants present phenotypes remarkably reminiscent of *parkin* mutants—shortened lifespan, locomotor deficits, degeneration of flight muscles, and defective spermatid formation with abundant mitochondrial defects.[122–124] The similarity in phenotypes between *Pink1* and *parkin* mutants prompted tests to determine a genetic interaction. First, *Pink1/parkin* double mutants were phenotypically indistinguishable from the respective single mutants,[122,123] consistent with the two mutations affecting a common pathway. Second, genetic epistasis experiments showed that overexpression of *parkin* is able to rescue *Pink1* phenotypes, but not vice versa.[122–124] Together these findings provided compelling evidence that Parkin acts downstream from Pink1 in a common pathway that regulates mitochondrial homeostasis, although at this stage the putative mechanisms were unknown. These findings were subsequently validated in mammalian cells where overexpression of human *parkin* could suppress mitochondrial phenotypes caused by loss of *PINK1*.[125,126]

An important breakthrough came in a series of studies that revealed a genetic link between *Pink1* and *parkin* and the mitochondrial fission/

fusion machinery. The first clue came from an unlikely source; the unexpected observation that *Pink1/parkin* fly mutants are male sterile. The process of spermatogenesis in *Drosophila* involves dramatic remodeling of the gamete mitochondria into the specialized mitochondrial derivative known as the nebenkern. This process involves the aggregation and fusion of mitochondria, which wrap around each other forming the morphologically elaborate "onion-stage" nebenkern. The onion-stage nebenkern contains two topologically distinct compartments that unfurl from each other during flagellar elongation and themselves elongate beside the growing axoneme. Detailed electron-microscopic analysis of *parkin* mutant spermatids revealed that during elongation only one mitochondria derivative unfurls from the nebenkern, thus appearing like excess fusion or aberrant fission.[127] This study was the first to suggest that Parkin may play a role in mitochondrial remodeling and membrane dynamics.

Next came a series of genetic studies, again using *Drosophila*, that clearly established that *Pink1* and *parkin* phenotypes can be modulated by manipulating mitochondrial fission/fusion factors. Genetic combinations between *Pink1* or *parkin* mutants and the fission/fusion genes revealed that removing a single copy of the fission-promoting factor *Drp1* dramatically reduces their viability, while *Drp1* overexpression substantially rescued many of the mutant phenotypes including flight and climbing ability, muscle integrity, and mitochondrial morphology.[128–130] Similarly, loss-of-function mutations in the profusion factors *Opa1* and *Marf* (homologous to Mfn1/2) significantly rescued many of the *Pink1/parkin* mutant defects. Thus, if we consider that mitochondrial dynamics is normally maintained in a fine balance between fission and fusion, these results show that if the balance is tipped toward excess fusion this worsens the fly *Pink1/parkin* phenotypes; however, if the balance is tipped to promote fission this is sufficient to prevent many of the mutant phenotypes. Together, these findings suggested that in flies the PINK1/Parkin pathway normally promotes mitochondrial fission and/or inhibits mitochondrial fusion.

At the same time as these results were reported, other groups were addressing the same questions in mammalian systems; however, a rather different picture was emerging. Using a variety of human cell culture systems a number of studies reported that knockdown of *PINK1* resulted in mitochondrial fragmentation.[125,126,131,132] This situation was also mirrored in fibroblasts from patients with pathogenic mutations in *PINK1*, and also by *parkin* knockdown.[131] Notably, morphological changes induced by loss of PINK1 could be reversed by overexpression of Parkin but not vice versa, consistent with their previously reported genetic hierarchy.[131] In many instances the mitochondrial fragmentation caused by PINK1/Parkin loss of function were rescued by increasing mitochondrial fusion or decreasing fission. In summary, these results are more consistent with the PINK1/Parkin pathway acting to promote mitochondrial fusion—exactly the opposite of the conclusion drawn from studies in flies. To add more complexity, fibroblasts or tissues from *PINK1* knockout mice showed no gross effect on mitochondrial morphology;[133] however, a subtle increased abundance in large mitochondria was noted in another study of *PINK1* knockout mice.[134] However, fibroblasts from *parkin* patients showed an increase in branching and connectivity but not in overall size.[135]

It is hard to immediately reconcile these discordant findings, but the differences in experimental paradigms present a number of considerations. The most obvious is that lower organisms such as *Drosophila* lacks the complexity of higher organisms, and in this context may lack some compensatory mechanisms activated in mammalian cells to help cope with loss of PINK1 or Parkin. Notably, fragmentation was observed in mammalian cells upon transient knockdown of PINK1 or parkin, whereas chronic loss via genetic ablation revealed little morphological disturbance, which could be a consequence of induced compensatory mechanisms. Interestingly, in line with this, *Pink1* or *parkin* silencing in *Drosophila* S2 cells caused an immediate fragmentation of the mitochondrial network, which was rapidly followed by hyperfusion.[131] As discussed earlier, degradation of damaged mitochondria requires a degree of fragmentation,[136] so hyperfused mitochondria will be refractory to this cell-protective mechanism. Hence, it is plausible that the apparent lack of compensatory response in *Drosophila*, and the resulting excessive fusion, may underlie the dramatic phenotypes observed in flies but is not apparent in mammalian models.

Given the propensity for compensatory mechanisms, in particular, those that alter mitochondrial morphology in response to cellular stresses, it is important to determine what may be cause and what is consequence. Certainly the genetic evidence from flies indicates that many of the *Pink1/parkin* phenotypes correlate with excessive fusion but that promoting fission has a beneficial effect, rescuing these phenotypes. However, it isn't immediately clear from cellular studies whether the observed fragmentation is problematic or already a manifestation of the compensatory protection. One observation that may address this from Lutz et al. is that the increase in mitochondrial fragmentation observed in *PINK1*- or *parkin*-deficient cells is not associated with an induction of apoptosis, suggesting that this is more likely to be a protective or beneficial response than a pathologic event. Later studies analyzing up and downregulation of PINK1 and Parkin in rat primary neurons provided further evidence that overall they act to promote mitochondrial fission.[137] While overexpression of PINK1 or Parkin caused fragmentation and *PINK1*-RNAi led to elongated mitochondria, which could be reversed by the activation of Drp1, OPA1, or Mfn1 as expected, increased fusion correlated with greater sensitivity to excitotoxicity, and excess fission was protective.

Thus, the picture developing of the role of PINK1/Parkin in modulating mitochondrial fission and fusion is complex, complicated further by the myriad mechanisms that regulate fission/fusion to different stresses. But overall the most compelling evidence points to a scenario whereby whatever the net effect on mitochondrial morphology by loss of *PINK1* or *parkin*, tipping the balance of fission/fusion toward fission appears to have a beneficial effect on neuronal function and survival. Of course, the final proof lies with determining the molecular mechanisms underlying the observations. In light of the early findings in flies that Pink1/Parkin likely promotes fission and/or inhibits fusion, a putative mechanism that could explain these observations emerged when it was found that Parkin ubiquitylates Marf, the fly homologue of Mfn1/2, in flies,[138,139] and Mfn1 and 2 in mammals.[73,140,141] In flies at least, loss of *Pink1* or *parkin* leads to an increase in Marf steady-state levels, consistent with excess fusion contributing to the organismal phenotypes. Although, as yet untested, this mechanism also presents the possibility that increased Marf/Mfn abundance could have detrimental effects via excess mitochondria–ER contacts.

It is now recognized that Parkin is a promiscuous ubiquitin ligase that has many cellular targets and many of these are mitochondrial proteins, so the regulation of Marf/Mfn levels should not be considered *the* major role of PINK1/Parkin. Indeed, the prevailing view in the field is that Parkin acts with PINK1 to promote the autophagic turnover of mitochondria or mitochondrial components, discussed briefly later and reviewed in-depth in Chapter 5. Although this is one likely endpoint of this mechanism, we contend that an important upstream regulation of this process is in modulating fission and fusion dynamics, not the least to orchestrate appropriately positioned and appropriately sized mitochondria.

5.2 DJ-1

Mutations in DJ-1 are a very rare cause of autosomal recessive PD. DJ-1 encodes a highly conserved, 189 amino acid, ubiquitously expressed protein that belongs to a superfamily of stress-inducible chaperones.[142] DJ-1 has an interesting dynamic relationship with mitochondria; it is normally cytoplasmic but upon oxidative stress the sulphydryl group of the conserved cysteine 106 is oxidized to sulphinic acid, which stimulates DJ-1 translocation to the mitochondrial matrix.[143] The mechanisms are not known but DJ-1 is widely acknowledged to act in some sort of antioxidant capacity. Current hypotheses suggest that DJ-1 acts as a stress sensor, which recognizes and combats oxidative stress. Consistent with this, DJ-1 increases resistance to mitochondrial toxins[143–146] while mitochondria from brain or skeletal muscle of DJ-1 KO mice show increased ROS formation.[147,148] Although DJ-1 can decrease mitochondrial damage-induced fragmentation, and DJ-1-deficient primary neurons and patients' cells show mitochondrial fragmentation and depolarization,[146,148,149] these effects can be prevented by antioxidants, indicating that increased levels of oxidative stress account for the mitochondrial phenotypes.[146,148]

5.3 Alpha-Synuclein

Both coding mutations and extra copies of the *SNCA* gene are both found in genetic cases of autosomal dominant PD.[150,151] *SNCA* encodes the small cytoplasmic protein α-synuclein, which has maintained a prominent position in the pathogenesis of PD with the early observation that this aggregation-prone protein is found in the Lewy body aggregates in sporadic PD patients.[151] The normal function of α-synuclein is still not clear, but it is likely to play a role in synaptic vesicle dynamics.[152] Recent studies have revealed that the primary toxic species of α-synuclein are not likely to be the large aggregates or even fibrils but prefibrillar oligomers. Although the primary toxicity of α-synuclein may stem from oligomer-related cellular stress, a contribution from mitochondrial dysfunction is also well established.

Multiple studies have now reported an association of α-synuclein with mitochondria, including its accumulation within mitochondria.[153–165] α-Synuclein has also been proposed to associate with proteins from complex I and the adenylate translocator,[162] impairing their activity. However, alteration in mitochondrial morphology are a common feature in various models analyzing the pathogenicity of α-synuclein; for instance, overexpression has been found to cause mitochondrial fragmentation.[164] Surprisingly, this fragmentation appears to be independent of the fission/fusion machinery and instead is thought to occur via the direct binding of α-synuclein to bind phospholipid membranes preventing their fusion.[166] Notably, mitochondrial fragmentation induced by overexpression of α-synuclein can be prevented by PINK1, parkin, or DJ-1 but not by their pathogenic mutants.[164]

In vivo evidence further supports the notion that α-synuclein impinges on mitochondrial function. Although loss of α-synuclein appears to have little consequence in vivo, knockout mice are resistant or less sensitive to mitochondrial toxins.[167–169] In contrast, α-synuclein-overexpressing mice are generally found to be more vulnerable to mitochondrial toxins,[161,170,171] although not in all instances.[172,173] Moreover, expression of the mitochondrial chaperone TRAP1 can ameliorate α-synuclein toxicity in cellular models and in *Drosophila*.[174]

Consistent with in vitro data, several studies have found a protective effect of Parkin on α-synuclein-induced toxicity.[175–179] Another mouse model that combined transgenic expression of A30P/A53T α-synuclein and a targeted deletion of *parkin* found that mitochondrial pathology was markedly increased.[180] However, these observations are controversial since another study showed that the absence of *parkin* delayed neurodegeneration caused by A30P α-synuclein overexpression.[181] Supporting this it has been reported that α-synuclein increases autophagy and mitophagy in primary cortical neurons, and depletion of *parkin* improved neuronal survival.[182] The proposed conclusion from this is that α-synuclein is toxic partly by provoking excess mitophagy, and the lack of Parkin prevents this and is protective. Clearly further work needs to be done to resolve the interaction between α-synuclein and Parkin, what mechanism this might be, and whether this has relevance in physiological context.

Nevertheless, reasonable evidence indicates that α-synuclein has the ability to perturb mitochondrial morphology toward fragmentation, likely as a consequence of its membrane-binding capacity rather than an interaction with the fission/fusion machinery. This is likely to contribute to the observed sensitivity to mitochondrial toxins and will undoubtedly render neurons more vulnerable to other stresses.

5.4 LRRK2

Mutations in *LRRK2* are the most common cause of heritable PD, but the *LRRK2* locus is also one of the most robust risk loci identified by GWAS, making *LRRK2* a target of intensive analysis. Pathogenicity has only been proven for relatively few LRRK2 mutations. Of these, G2019S has emerged as the most prevalent LRRK2 mutation in both familial and sporadic cases. *LRRK2* encodes a large, complex, multidomain protein, comprising 2527 amino acids protein (~280 kDa) with both kinase and GTPase activity. Mutation of LRRK2 has the potential to cause a multitude of pathogenic effects and perhaps not surprisingly no consensus has yet emerged over a likely pathogenic mechanism. The most obvious target is its native enzymatic activity. While physiological targets of LRRK2's kinase and GTPase activities are slowly emerging, pathogenic mutations are generally believed to increase kinase activity and/or decrease GTPase activity. The physiological and pathological functions of LRRK2 are still poorly understood, but overexpression of pathogenic LRRK2 mutants in cultured cells, primary neurons, or rodents causes cellular toxicity, which is dependent on the kinase activity.[183–187]

There is mounting evidence that mutations in LRRK2 can affect mitochondrial function, although the relative contribution of this to the pathogenic process is currently unclear. LRRK2 has been observed to localize to mitochondria,[188,189] and mitochondrial pathology has been observed in transgenic or knock-in G2019S LRRK2 mice.[190,191] Moreover, fibroblasts derived from PD patients carrying LRRK2 G2019S exhibited a decrease in both mitochondrial membrane potential (MMP) and total ATP production, and a trend toward increased mitochondrial interconnectivity.[192,193] Notably, although the mechanism causing mitochondrial dysfunction in the patient fibroblasts has not be resolved, the mitochondrial functional deficits could be significantly restored by treatment with several structurally related steroids.[194,195]

However, several recent reports have shown that LRRK2 mutations impact on mitochondrial function by directly affecting mitochondrial dynamics. In cell lines and primary neuronal cultures overexpression of LRRK2 was shown to cause mitochondrial fragmentation, promoting mitochondrial dysfunction, elevated ROS, and neuronal toxicity.[196,197] Moreover, endogenous LRRK2 was found to directly interact with Drp1. This interaction appeared to require GTPase and kinase activity and was exacerbated by pathogenic mutations.[196,197] Further evidence suggests that LRRK2 may even regulate Drp1 activity by direct phosphorylation.[198] Importantly, restoring the fission–fusion balance by inhibition of Drp1 or Mfn1 overexpression reversed the neurotoxicity of LRRK2 mutations in primary cortical neurons.[198,199] Hence, mutations in LRRK2 appear to cause mitochondrial and neuronal dysfunction by aberrantly promoting mitochondrial fission via Drp1, and inhibition of Drp-1-mediated excessive fission represents a therapeutic strategy for this common cause of PD.

An alternative mechanism by which LRRK2 pathology may be linked to mitochondrial dynamics is via the inhibition of axonal transport. LRRK2 has been found to interact with microtubules and influence their acetylation state.[200–202] Mutant LRRK2 shows enhanced microtubule binding and causes excess deacetylation, which is a known effect of axonal transport.[203,204] Consequently, mutant LRRK2 was found to inhibit axonal transport of mitochondria in *Drosophila* and mammalian neurons.[205] Interestingly, mitochondrial transport was inhibited by LRRK2 mutations R1441C and Y1699C that affect the GTPase activity, but not the G2019S mutation that affects the kinase activity. The mechanisms by which this occurs are not yet clear but pharmacological or genetic inhibition of the deacetylase enzymes HDAC6 and SIRT2 restored mitochondrial transport.[205] If this process were proven to contribute to the pathogenic process, such inhibitors would represent possible therapeutics.

6 MITOCHONDRIAL DYNAMICS AND MITOCHONDRIAL TURNOVER

Substantial evidence makes it clear that mitochondrial fission and fusion dynamics impacts on mitochondrial quality control and overall homeostasis. However, examining the end-result does not directly inform the mechanisms that give rise to these outcomes. A seminal study by Twig and colleagues provided some crucial clues to this process. They proposed a model whereby the mitochondrial fission–fusion cycle acts to regulate the sequestration and destruction of damaged organelles.[136,206] Detailed observations of the life-cycle of mitochondria, following individual organelles over time, revealed that mitochondria pass the majority of time in a solitary period. However, when a fusion event occurs this is most often rapidly followed by a fission event. Occasionally, one or other of the daughter mitochondria become or remain depolarized following fission. These organelles were observed to be unable to undergo subsequent fusion events and to be targeted for degradation by autophagy (mitophagy). Thus, this cycle of fusion and fission events was proposed to act as a monitoring process to potentially (by fusion) enrich portions of the network with damaged components and (by fission) segregated these damaged components for degradation (Fig. 3.2).

This study, coupled with the earlier evidence that PINK1 and Parkin affect mitochondrial dynamics, set the perfect backdrop to another seminal study by Narendra and coworkers, which described that Parkin is recruited to depolarized mitochondria and promotes mitophagy.[207] This process has been intensively studied since these initial reports and many of the molecular details have come to light,[208] which are briefly summarized

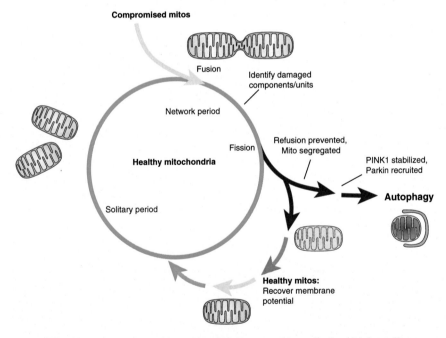

FIGURE 3.2 **Mitochondrial fission and fusion regulate mitochondrial quality control.** Upon fusion, mitochondrial contents can mix to dilute damaged components and provide functional complementation. It is possible that damaged components may be actively segregated for disposal. Fusion is frequently followed by fission, whereupon daughter mitochondria may lose their membrane potential. Healthy mitochondria will regenerate their membrane and be maintained in the network. Terminally damaged mitochondria which cannot regain membrane potential are recognized by the quality control machinery such as PINK1 and Parkin, segregated to prevent refusion, and targeted for degradation by autophagy.

as follows but discussed in greater detail in Chapter 5. In healthy mitochondria (ie, those with a normal membrane potential across their inner membrane), PINK1 abundance is normally kept very low by regulated proteolysis and degradation following import. However, in mitochondria with a severely diminished membrane potential, PINK1's import is disrupted and becomes stabilized on the outer surface where it promotes the recruitment of Parkin and the ubiquitylation of many outer membrane proteins. This occurs by the simultaneous phosphorylation of both Parkin and ubiquitin by PINK1. This both stimulates Parkin's ubiquitin ligase activity and increases its binding capacity to phosphorylated ubiquitin, which acts as a feed-forward mechanism by providing further substrate for PINK1 phosphorylation by local ubiquitylation of membrane proteins. Ultimately, highly ubiquitylated mitochondria are recognized by adapter proteins that bind poly-ubiquitin and trigger engulfment by autophagosomes.[208]

This mechanism is attractive for several reasons: first, it offers an explanation for the variety of mitochondrial defects that have been documented in PINK1 and Parkin-deficient cell models, including decreased membrane potential, deficits in the electron transport chain complexes, reduced ATP synthesis, decreased mitochondrial DNA synthesis, and aberrant mitochondrial calcium efflux,[133,209,210] by suggesting that these pleiotropic phenotypes derive from the accumulation of damaged mitochondria in the absence of a functional mitochondrial quality control system. Second, these findings would explain the protective effects of PINK1 and Parkin overexpression from exposure to mitochondrial toxins[211-213]; and third, the abundant mitochondrial DNA mutational load of substantia nigra DA neurons[6,7] would account for the selective vulnerability of this population of cells to the loss of a mitochondrial quality control system.

While there is considerable debate surrounding the physiological relevance of this mechanism, most notably due to the severe nonphysiological chemical ablation of the membrane potential used to trigger mitophagy, there is some in vivo evidence emerging that supports the model. For instance, in *Drosophila* mutant for *parkin* or *Pink1* the turnover of mitochondrial proteins, monitored by mass spectrometry of stable isotope labeling, was significantly reduced consistent with these proteins acting to promote turnover.[214] In addition, using a mouse model that accumulates mitochondrial DNA mutations, loss of Parkin greatly exaggerated the mutation load consistent with Parkin mediating degradation of defective molecules.[215] Remarkably, this combination led to the selective degeneration of dopaminergic neurons.

While the nature of the physiological stimulus for this process has yet to be elucidated, it remains intuitive that the wholesale mitophagic degradation of entire organelles is likely to be a rare event. However, another mechanism of mitochondrial dynamics has been proposed that offers a more piecemeal process of mitochondria quality control—the formation of cargo-selective mitochondria-derived vesicles (MDVs).[216] The precise function of these MDVs is currently uncertain but evidence implicates a role in the selective transport of oxidized mitochondrial components for degradation in lysosomes.[217–219] This offers a more regulated route by which small amounts of mitochondrial components are degraded, possibly involving an enrichment or selection process.

Importantly, three factors linked to PD have been implicated in the formation and trafficking of MDVs: PINK1, Parkin, and Vps35.[220–222] Recent genetic studies in *Drosophila* have described a striking genetic interaction between *vps35* and *parkin*, providing strong support for a common pathway. Double heterozygotes confer age-related locomotor deficits, loss of DA neurons, and shortened lifespan, whereas Vps35 overexpression was able to rescue several *parkin* mutant phenotypes.[223] Surprisingly, *vps35* did not appear to genetically interact with *Pink1*. Furthermore, *VPS35* mutations have recently been shown to cause mitochondrial dysfunction by affecting the fission–fusion machinery.[222,224]

In addition to modulating mitochondrial fission–fusion dynamics, PINK1 and Parkin have been shown to affect mitochondrial axonal transport (reviewed in Chapter 4). PINK1 was found to bind the Miro/Milton complex,[225] an adapter complex that mediates microtubule-based mitochondrial transport via kinesin motors.[226–228] Recent work has shown that PINK1 phosphorylates Miro, which is then ubiquitylated by Parkin.[84,229,230] This leads to the degradation of Miro and the consequent arrest of mitochondrial motility in neurons.[229] This process is proposed to help in the removal of malfunctioning mitochondria. Similar to the inhibition of fusion, this may act as an early step to quarantine damaged mitochondria and prevent them from fusing with other healthy mitochondria, thereby mixing their damaged components in with the healthy population. Moving mitochondria are much more likely to fuse than stationary mitochondria[68] and therefore, together with the degradation of mitofusin, the loss of Miro will produce a population of stationary, fragmented mitochondria that are ready for engulfment by autophagosomes. Supporting the hypothesis that this mechanism promotes neuronal survival, recent work in *Drosophila* has shown that expression of Miro that lacks the PINK1 phosphorylation sites results in loss of DA neurons.[231]

7 THERAPEUTIC POTENTIAL OF TARGETING FISSION AND FUSION

Due to the long-standing appreciation that mitochondrial fission can contribute to the activation of cell death pathways, considerable effort has been invested into developing ways to inhibit fission, particularly via the direct inhibition of Drp1.[232] A number of inhibitors have been characterized, such as mdivi-1, and reported to offer neuroprotection in various models of neuronal loss or injury, including one study on PINK1 dysfunction.[233]

Since the discovery that Parkin can regulate mitochondrial fission/fusion and quality control, the role of possible alternative ubiquitin ligases that may be able to compensate for loss of Parkin has been explored. Indeed, one group reported that the mitochondrial SUMO/ubiquitin E3 ligase MUL1 can act in a parallel pathway to Parkin. Overexpression of MUL1 promotes mitochondria fission and, in *Drosophila*, was able to suppress many of the *Pink1/parkin* mutant phenotypes.[234] Although the interaction was proposed to occur by MUL1 regulating Marf levels via ubiquitylation, this may also occur via MUL1's known regulation of Drp1.[63]

An alternative approach to this line of inquiry has considered the possible action of opposing deubiquitinase (DUB) enzymes. A number of DUBs that oppose the action of Parkin have been proposed, including USP30,[235–237] USP15,[238] and USP8.[239] As the investigation of these has mainly focused on regulation of mitophagy, it is currently unclear whether they also impinge directly on mitochondrial fission and fusion; however, there is precedent for this mechanism in yeast.[74] Since DUBs are considered highly druggable targets, these mechanisms may yet prove to be an attractive therapeutic avenue.

While this chapter has focused on the interaction of PD genes with the mitochondrial fission/fusion pathways, substantial evidence also links additional pathogenic mechanisms to these genetic perturbations. For example, PINK1 is also considered to affect complex I (CI) of the electron transport chain, with loss of PINK1 leading to CI deficiency in multiple model systems (reviewed in Chapter 2). Compelling evidence that this contributes to the pathogenic process is provided by observations that either the expression of the yeast CI-alternative enzyme, Ndi1, or a specific subunit of CI, NDUFA10, is sufficient to prevent pathogenic outcomes.[240–242] As discussed earlier, promoting fission can rescue *Pink1* mutant phenotypes, at least in flies. Therefore, it was surprising that these same manipulations did not also restore full complex I activity.[240] Likewise, overexpression of Parkin in fly *Pink1* mutants could rescue many of the phenotypes but did not rescue CI activity.[242] The full consequences of this are not yet clear and the lack of full CI capacity may still be detrimental, but one implication of these findings is that manipulating mitochondrial fission/fusion dynamics may be able to circumvent pathogenic events arising from other primary perturbations of mitochondrial functions. These findings further implicate mitochondrial fission–fusion dynamics as an attractive therapeutic target.

8 CONCLUSIONS

Our understanding of the mechanisms of mitochondrial fission and fusion have greatly advanced over the past couple of decades, and it has become clear that mitochondrial dynamics is highly regulated, exquisitely controlled, and kept in a tight balance throughout the life course. It is also clear that disturbances in the balance of mitochondrial fission–fusion have profound effects on mitochondrial quality control and homeostasis, and have been linked to numerous neurodegenerative diseases including PD. While some PD factors have been implicated in directly regulating the fission–fusion machinery, the effect of other mutations are more indirectly linked to mitochondrial dynamics. Nevertheless, the impact of fission–fusion on quality control processes has positioned this as a potential therapeutic avenue. Much more work needs to be done to understand the full implications of manipulating mitochondrial dynamics in vivo or over a long period, but preliminary findings offer some hope for a therapeutic angle.

References

1. Wood-Kaczmar A, Gandhi S, Wood NW. Understanding the molecular causes of Parkinson's disease. *Trends Mol Med*. 2006;12(11):521–528.
2. Schapira AH. Mitochondrial dysfunction in Parkinson's disease. *Cell Death Different*. 2007;14(7):1261–1266.
3. Betarbet R, Sherer TB, Di Monte DA, Greenamyre JT. Mechanistic approaches to Parkinson's disease pathogenesis. *Brain Pathol*. 2002;12(4):499–510.
4. Corti O, Hampe C, Darios F, Ibanez P, Ruberg M, Brice A. Parkinson's disease: from causes to mechanisms. *C R Biol*. 2005;328(2):131–142.
5. Banerjee R, Starkov AA, Beal MF, Thomas B. Mitochondrial dysfunction in the limelight of Parkinson's disease pathogenesis. *Biochimica et Biophysica Acta*. 2009;1792(7):651–663.
6. Bender A, Krishnan KJ, Morris CM, Taylor GA, Reeve AK, Perry RH, Jaros E, Hersheson JS, Betts J, Klopstock T, Taylor RW, Turnbull DM. High levels of mitochondrial DNA deletions in substantia nigra neurons in aging and Parkinson disease. *Nat Genet*. 2006;38(5):515–517.
7. Kraytsberg Y, Kudryavtseva E, McKee AC, Geula C, Kowall NW, Khrapko K. Mitochondrial DNA deletions are abundant and cause functional impairment in aged human substantia nigra neurons. *Nat Genet*. 2006;38(5):518–520.
8. Abou-Sleiman PM, Muqit MM, McDonald NQ, Yang YX, Gandhi S, Healy DG, Harvey K, Harvey RJ, Deas E, Bhatia K, Quinn N, Lees A, Latchman DS, Wood NW. A heterozygous effect for PINK1 mutations in Parkinson's disease? *Ann Neurol*. 2006;60(4):414–419.
9. Marras C, Lohmann K, Lang A, Klein C. Fixing the broken system of genetic locus symbols: Parkinson disease and dystonia as examples. *Neurology*. 2012;78(13):1016–1024.
10. Chen H, Detmer SA, Ewald AJ, Griffin EE, Fraser SE, Chan DC. Mitofusins Mfn1 and Mfn2 coordinately regulate mitochondrial fusion and are essential for embryonic development. *J Cell Biol*. 2003;160(2):189–200.
11. Hoppins S, Lackner L, Nunnari J. The Machines that Divide and Fuse Mitochondria. *Annu Rev Biochem*. 2007;76(1):751–780.
12. Chan DC. Fusion and fission: interlinked processes critical for mitochondrial health. *Ann Rev Genet*. 2012;46:265–287.

13. van der Bliek AM, Shen Q, Kawajiri S. Mechanisms of Mitochondrial Fission and Fusion. *Cold Spring Harb Perspect Biol.* 2013;5(6):a011072–a111072.

14. Tondera D, Grandemange S, Jourdain A, Karbowski M, Mattenberger Y, Herzig S, Da Cruz S, Clerc P, Raschke I, Merkwirth C, Ehses S, Krause F, Chan DC, Alexander C, Bauer C, Youle R, Langer T, Martinou JC. SLP-2 is required for stress-induced mitochondrial hyperfusion. *EMBO J.* 2009;28(11):1589–1600.

15. Detmer SA, Chan DC. Functions and dysfunctions of mitochondrial dynamics. *Nat Rev Mol Cell Biol.* 2007;8(11):870–879.

16. Gomes LC, Di Benedetto G, Scorrano L. During autophagy mitochondria elongate, are spared from degradation and sustain cell viability. *Nat Cell Biol.* 2011;13(5):589–598.

17. Frank S, Gaume B, Bergmann-Leitner ES, Leitner WW, Robert EG, Catez F, Smith CL, Youle RJ. The role of dynamin-related protein 1, a mediator of mitochondrial fission, in apoptosis. *Dev Cell.* 2001;1(4):515–525.

18. Hoitzing H, Johnston IG, Jones NS. What is the function of mitochondrial networks? A theoretical assessment of hypotheses and proposal for future research. *BioEssays.* 2015;37(6):687–700.

19. Chen H, Chomyn A, Chan DC. Disruption of fusion results in mitochondrial heterogeneity and dysfunction. *J Biol Chem.* 2005;280(28):26185–26192.

20. De Vos KJ, Allan VJ, Grierson AJ, Sheetz MP. Mitochondrial function and actin regulate dynamin-related protein 1-dependent mitochondrial fission. *Curr Biol.* 2005;15(7):678–683.

21. Bossy-Wetzel E, Barsoum MJ, Godzik A, Schwarzenbacher R, Lipton SA. Mitochondrial fission in apoptosis, neurodegeneration and aging. *Curr Opin Cell Biol.* 2003;15(6):706–716.

22. Frank M, Duvezin-Caubet S, Koob S, Occhipinti A, Jagasia R, Petcherski A, Ruonala MO, Priault M, Salin B, Reichert AS. Mitophagy is triggered by mild oxidative stress in a mitochondrial fission dependent manner. *Biochimica et Biophysica Acta.* 2012;1823(12):2297–2310.

23. Youle RJ, van der Bliek AM. Mitochondrial fission, fusion, and stress. *Science.* 2012;337(6098):1062–1065.

24. Friedman JR, Nunnari J. Mitochondrial form and function. *Nature.* 2014;505(7483):335–343.

25. Helle SCJ, Kanfer G, Kolar K, Lang A, Michel AH, Kornmann B. Organization and function of membrane contact sites. *Biochim Biophys Acta.* 2013;1833(11):2526–2541.

26. Paillusson S, Stoica R, Gomez-Suaga P, Lau DHW, Mueller S, Miller T, Miller CCJ. There's Something Wrong with my MAM; the ER; Mitochondria Axis and Neurodegenerative Diseases. *Trends Neurosci.* 2016;39(3):146–157.

27. Chen H, Chan DC. Mitochondrial dynamics-fusion, fission, movement, and mitophagy-in neurodegenerative diseases. *Hum Mol Genet.* 2009;18(R2):R169–R176.

28. Shin HW, Shinotsuka C, Torii S, Murakami K, Nakayama K. Identification and subcellular localization of a novel mammalian dynamin-related protein homologous to yeast Vps1p and Dnm1p. *J Biochem.* 1997;122(3):525–530.

29. Yoon Y, Pitts KR, Dahan S, McNiven MA. A novel dynamin-like protein associates with cytoplasmic vesicles and tubules of the endoplasmic reticulum in mammalian cells. *J Cell Biol.* 1998;140(4):779–793.

30. Imoto M, Tachibana I, Urrutia R. Identification and functional characterization of a novel human protein highly related to the yeast dynamin-like GTPase Vps1p. *J Cell Sci.* 1998;111(Pt 10):1341–1349.

31. Kamimoto T, Nagai Y, Onogi H, Muro Y, Wakabayashi T, Hagiwara M. Dymple, a novel dynamin-like high molecular weight GTPase lacking a proline-rich carboxyl-terminal domain in mammalian cells. *J Biol Chem.* 1998;273(2):1044–1051.

32. Pitts KR, Yoon Y, Krueger EW, McNiven MA. The dynamin-like protein DLP1 is essential for normal distribution and morphology of the endoplasmic reticulum and mitochondria in mammalian cells. *Mol Biol Cell.* 1999;10(12):4403–4417.

33. Smirnova E, Shurland DL, Ryazantsev SN, van der Bliek AM. A human dynamin-related protein controls the distribution of mitochondria. *J Cell Biol.* 1998;143(2):351–358.

34. Labrousse AM, Zappaterra MD, Rube DA, van der Bliek AM. *C. elegans* dynamin-related protein DRP-1 controls severing of the mitochondrial outer membrane. *Mol Cell.* 1999;4(5):815–826.
35. Legesse-Miller A, Massol RH, Kirchhausen T. Constriction and Dnm1p recruitment are distinct processes in mitochondrial fission. *Mol Biol Cell.* 2003;14(5):1953–1963.
36. Sesaki H, Jensen RE. Division versus fusion: Dnm1p and Fzo1p antagonistically regulate mitochondrial shape. *J Cell Biol.* 1999;147(4):699–706.
37. Ishihara N, Nomura M, Jofuku A, Kato H, Suzuki SO, Masuda K, Otera H, Nakanishi Y, Nonaka I, Goto Y, Taguchi N, Morinaga H, Maeda M, Takayanagi R, Yokota S, Mihara K. Mitochondrial fission factor Drp1 is essential for embryonic development and synapse formation in mice. *Nat Cell Biol.* 2009;11(8):958–966.
38. Ji WK, Hatch AL, Merrill RA, Strack S, Higgs HN. Actin filaments target the oligomeric maturation of the dynamin GTPase Drp1 to mitochondrial fission sites. *eLife.* 2015;4.
39. Sesaki H, Adachi Y, Kageyama Y, Itoh K, Iijima M. In vivo functions of Drp1: lessons learned from yeast genetics and mouse knockouts. *Biochimica et Biophysica Acta.* 2014;1842(8):1179–1185.
40. Stojanovski D, Koutsopoulos OS, Okamoto K, Ryan MT. Levels of human Fis1 at the mitochondrial outer membrane regulate mitochondrial morphology. *J Cell Sci.* 2004;117(Pt 7):1201–1210.
41. Mozdy AD, McCaffery JM, Shaw JM. Dnm1p GTPase-mediated mitochondrial fission is a multi-step process requiring the novel integral membrane component Fis1p. *J Cell Biol.* 2000;151(2):367–380.
42. Schauss AC, Bewersdorf J, Jakobs S. Fis1p and Caf4p, but not Mdv1p, determine the polar localization of Dnm1p clusters on the mitochondrial surface. *J Cell Sci.* 2006;119(Pt 15):3098–3106.
43. Gandre-Babbe S, van der Bliek AM. The novel tail-anchored membrane protein Mff controls mitochondrial and peroxisomal fission in mammalian cells. *Mol Biol Cell.* 2008;19(6):2402–2412.
44. Palmer CS, Osellame LD, Laine D, Koutsopoulos OS, Frazier AE, Ryan MT. MiD49 and MiD51, new components of the mitochondrial fission machinery. *EMBO Rep.* 2011;12(6):565–573.
45. Zhao J, Liu T, Jin S, Wang X, Qu M, Uhlen P, Tomilin N, Shupliakov O, Lendahl U, Nister M. Human MIEF1 recruits Drp1 to mitochondrial outer membranes and promotes mitochondrial fusion rather than fission. *EMBO J.* 2011;30(14):2762–2778.
46. Frohman MA. Role of mitochondrial lipids in guiding fission and fusion. *J Mol Med.* 2015;93(3):263–269.
47. Ugarte-Uribe B, Muller HM, Otsuki M, Nickel W, Garcia-Saez AJ. Dynamin-related protein 1 (Drp1) promotes structural intermediates of membrane division. *J Biol Chem.* 2014;289(44):30645–30656.
48. Macdonald PJ, Stepanyants N, Mehrotra N, Mears JA, Qi X, Sesaki H, Ramachandran R. A dimeric equilibrium intermediate nucleates Drp1 reassembly on mitochondrial membranes for fission. *Mol Biol Cell.* 2014;25(12):1905–1915.
49. Stepanyants N, Macdonald PJ, Francy CA, Mears JA, Qi X, Ramachandran R. Cardiolipin's propensity for phase transition and its reorganization by dynamin-related protein 1 form a basis for mitochondrial membrane fission. *Mol Biol Cell.* 2015;26(17):3104–3116.
50. Friedman JR, Lackner LL, West M, DiBenedetto JR, Nunnari J, Voeltz GK. ER tubules mark sites of mitochondrial division. *Science.* 2011;334(6054):358–362.
51. Korobova F, Ramabhadran V, Higgs HN. An actin-dependent step in mitochondrial fission mediated by the ER-associated formin INF2. *Science.* 2013;339(6118):464–467.
52. Manor U, Bartholomew S, Golani G, Christenson E, Kozlov M, Higgs H, Spudich J, Lippincott-Schwartz J. A mitochondria-anchored isoform of the actin-nucleating Spire protein regulates mitochondrial division. *eLife.* 2015;4.

53. DuBoff B, Gotz J, Feany MB. Tau promotes neurodegeneration via DRP1 mislocalization in vivo. *Neuron.* 2012;75(4):618–632.
54. Korobova F, Gauvin TJ, Higgs HN. A role for myosin II in mammalian mitochondrial fission. *Curr Biol.* 2014;24(4):409–414.
55. Li S, Xu S, Roelofs BA, Boyman L, Lederer WJ, Sesaki H, Karbowski M. Transient assembly of F-actin on the outer mitochondrial membrane contributes to mitochondrial fission. *J Cell Biol.* 2015;208(1):109–123.
56. Lim Y, Cho IT, Schoel LJ, Cho G, Golden JA. Hereditary Spastic Paraplegia-Linked REEP1 Modulates ER–Mitochondria Contacts. *Ann Neurol.* 2015;78(5):679–696.
57. O'Sullivan NC, Jahn TR, Reid E, O'Kane CJ. Reticulon-like-1, the *Drosophila* orthologue of the hereditary spastic paraplegia gene reticulon 2, is required for organization of endoplasmic reticulum and of distal motor axons. *Human Mol Genet.* 2012;21(15):3356–3365.
58. Cribbs JT, Strack S. Reversible phosphorylation of Drp1 by cyclic AMP-dependent protein kinase and calcineurin regulates mitochondrial fission and cell death. *EMBO Rep.* 2007;8(10):939–944.
59. Cereghetti GM, Stangherlin A, Martins de Brito O, Chang CR, Blackstone C, Bernardi P, Scorrano L. Dephosphorylation by calcineurin regulates translocation of Drp1 to mitochondria. *Proc Natl Acad Sci USA.* 2008;105(41):15803–15808.
60. Nakamura N, Kimura Y, Tokuda M, Honda S, Hirose S. MARCH-V is a novel mitofusin 2- and Drp1-binding protein able to change mitochondrial morphology. *EMBO Rep.* 2006;7(10):1019–1022.
61. Yonashiro R, Ishido S, Kyo S, Fukuda T, Goto E, Matsuki Y, Ohmura-Hoshino M, Sada K, Hotta H, Yamamura H, Inatome R, Yanagi S. A novel mitochondrial ubiquitin ligase plays a critical role in mitochondrial dynamics. *EMBO J.* 2006;25(15):3618–3626.
62. Prudent J, Zunino R, Sugiura A, Mattie S, Shore GC, McBride HM. MAPL SUMOylation of Drp1 stabilizes an ER/mitochondrial platform required for cell death. *Mol Cell.* 2015;59(6):941–955.
63. Braschi E, Zunino R, McBride HM. MAPL is a new mitochondrial SUMO E3 ligase that regulates mitochondrial fission. *EMBO Rep.* 2009;10(7):748–754.
64. Karbowski M, Neutzner A, Youle RJ. The mitochondrial E3 ubiquitin ligase MARCH5 is required for Drp1 dependent mitochondrial division. *J Cell Biol.* 2007;178(1):71–84.
65. Zhang C, Shi Z, Zhang L, Zhou Z, Zheng X, Liu G, Bu G, Fraser PE, Xu H, Zhang YW. Appoptosin interacts with mitochondrial outer-membrane fusion proteins and regulates mitochondrial morphology. *J Cell Science.* 2016;129(5):994–1002.
66. Meeusen S, McCaffery JM, Nunnari J. Mitochondrial fusion intermediates revealed in vitro. *Science.* 2004;305(5691):1747–1752.
67. Malka F, Guillery O, Cifuentes-Diaz C, Guillou E, Belenguer P, Lombes A, Rojo M. Separate fusion of outer and inner mitochondrial membranes. *EMBO Rep.* 2005;6(9):853–859.
68. Liu X, Weaver D, Shirihai O, Hajnoczky G. Mitochondrial 'kiss-and-run': interplay between mitochondrial motility and fusion-fission dynamics. *EMBO J.* 2009;28(20):3074–3089.
69. Olichon A, Baricault L, Gas N, Guillou E, Valette A, Belenguer P, Lenaers G. Loss of OPA1 perturbates the mitochondrial inner membrane structure and integrity, leading to cytochrome c release and apoptosis. *J Biol Chem.* 2003;278(10):7743–7746.
70. Hales KG, Fuller MT. Developmentally regulated mitochondrial fusion mediated by a conserved, novel, predicted GTPase. *Cell.* 1997;90(1):121–129.
71. Koshiba T, Detmer SA, Kaiser JT, Chen H, McCaffery JM, Chan DC. Structural basis of mitochondrial tethering by mitofusin complexes. *Science.* 2004;305(5685):858–862.
72. Leboucher GP, Tsai YC, Yang M, Shaw KC, Zhou M, Veenstra TD, Glickman MH, Weissman AM. Stress-induced phosphorylation and proteasomal degradation of mitofusin 2 facilitates mitochondrial fragmentation and apoptosis. *Mol Cell.* 2012;47(4):547–557.
73. Tanaka A, Cleland MM, Xu S, Narendra DP, Suen DF, Karbowski M, Youle RJ. Proteasome and p97 mediate mitophagy and degradation of mitofusins induced by Parkin. *J Cell Biol.* 2010;191(7):1367–1380.

74. Anton F, Dittmar G, Langer T, Escobar-Henriques M. Two deubiquitylases act on mitofusin and regulate mitochondrial fusion along independent pathways. *Mol Cell.* 2013;49(3):487–498.

75. Wiedemann N, Stiller SB, Pfanner N. Activation and degradation of mitofusins: two pathways regulate mitochondrial fusion by reversible ubiquitylation. *Mol cell.* 2013;49(3):423–425.

76. Choi S-Y, Huang P, Jenkins GM, Chan DC, Schiller J, Frohman MA. A common lipid links Mfn-mediated mitochondrial fusion and SNARE-regulated exocytosis. *Nat Cell Biol.* 2006;8(11):1255–1262.

77. Kornmann B, Currie E, Collins SR, Schuldiner M, Nunnari J, Weissman JS, Walter P. An ER–mitochondria tethering complex revealed by a synthetic biology screen. *Science.* 2009;325(5939):477–481.

78. de Brito OM, Scorrano L. Mitofusin 2 tethers endoplasmic reticulum to mitochondria. *Nature.* 2008;456(7222):605–610.

79. Filadi R, Greotti E, Turacchio G, Luini A, Pozzan T, Pizzo P. Mitofusin 2 ablation increases endoplasmic reticulum-mitochondria coupling. In: *Proc Natl Acad Sci USA.* 2015;112(7):E2174–E2181.

80. Lackner LL. Shaping the dynamic mitochondrial network. *BMC Biol.* 2014;12:35.

81. Samant SA, Zhang HJ, Hong Z, Pillai VB, Sundaresan NR, Wolfgeher D, Archer SL, Chan DC, Gupta MP. SIRT3 deacetylates and activates OPA1 to regulate mitochondrial dynamics during stress. *Mol Cell Biol.* 2014;34(5):807–819.

82. DeVay RM, Dominguez-Ramirez L, Lackner LL, Hoppins S, Stahlberg H, Nunnari J. Coassembly of Mgm1 isoforms requires cardiolipin and mediates mitochondrial inner membrane fusion. *J Cell Biol.* 2009;186(6):793–803.

83. Chen H, McCaffery JM, Chan DC. Mitochondrial fusion protects against neurodegeneration in the cerebellum. *Cell.* 2007;130(3):548–562.

84. Liu S, Sawada T, Lee S, Yu W, Silverio G, Alapatt P, Millan I, Shen A, Saxton W, Kanao T, Takahashi R, Hattori N, Imai Y, Lu B. Parkinson's disease–associated kinase PINK1 regulates miro protein level and axonal transport of mitochondria. *PLoS Genet.* 2012;8(3):e1002537.

85. Russo GJ, Louie K, Wellington A, Macleod GT, Hu F, Panchumarthi S, Zinsmaier KE. Drosophila miro is required for both anterograde and retrograde axonal mitochondrial transport. *J Neurosci.* 2009;29(17):5443–5455.

86. Li Z, Okamoto K-I, Hayashi Y, Sheng M. The importance of dendritic mitochondria in the morphogenesis and plasticity of spines and synapses. *Cell.* 2004;119(6):873–887.

87. Uo T, Dworzak J, Kinoshita C, Inman DM, Kinoshita Y, Horner PJ, Morrison RS. Drp1 levels constitutively regulate mitochondrial dynamics and cell survival in cortical neurons. *Exp Neurol.* 2009;218(2):274–285.

88. Cagalinec M, Safiulina D, Liiv M, Liiv J, Choubey V, Wareski P, Veksler V, Kaasik A. Principles of the mitochondrial fusion and fission cycle in neurons. *J Cell Sci.* 2013;126(10):2187–2197.

89. Baloh RH, Schmidt RE, Pestronk A, Milbrandt J. Altered axonal mitochondrial transport in the pathogenesis of charcot-marie-tooth disease from mitofusin 2 mutations. *J Neurosci.* 2007;27(2):422–430.

90. Verstreken P, Ly CV, Venken KJT, Koh T-W, Zhou Y, Bellen HJ. Synaptic mitochondria are critical for mobilization of reserve pool vesicles at drosophila neuromuscular junctions. *Neuron.* 2005;47(3):365–378.

91. Misko A, Jiang S, Wegorzewska I, Milbrandt J, Baloh RH. Mitofusin 2 Is necessary for transport of axonal mitochondria and interacts with the miro/milton complex. *J Neurosci.* 2010;30(12):4232–4240.

92. Misko AL, Sasaki Y, Tuck E, Milbrandt J, Baloh RH. Mitofusin2 mutations disrupt axonal mitochondrial positioning and promote axon degeneration. *J Neurosci.* 2012;32(12):4145–4155.

93. Norkett R, Modi S, Birsa N, Atkin TA, Ivankovic D, Pathania M, Trossbach SV, Korth C, Hirst WD, Kittler JT. DISC1-dependent regulation of mitochondrial dynamics controls the morphogenesis of complex neuronal dendrites. *J Biol Chem.* 2016;291(2):613–629.

94. Fransson A, Ruusala A, Aspenstrom P. The atypical Rho GTPases Miro-1 and Miro-2 have essential roles in mitochondrial trafficking. *Biochem Biophys Res Commun.* 2006;344(2):500–510.

95. Saotome M, Safiulina D, Szabadkai G, Das S, Fransson A, Aspenstrom P, Rizzuto R, Hajnoczky G. Bidirectional Ca^{2+}-dependent control of mitochondrial dynamics by the Miro GTPase. *Proc Natl Acad Sci USA.* 2008;105(52):20728–20733.

96. MacAskill AF, Brickley K, Stephenson FA, Kittler JT. GTPase dependent recruitment of Grif-1 by Miro1 regulates mitochondrial trafficking in hippocampal neurons. *Mol Cell Neurosci.* 2009:40;301–312.

97. Frederick RL, McCaffery JM, Cunningham KW, Okamoto K, Shaw JM. Yeast Miro GT-Pase, Gem1p, regulates mitochondrial morphology via a novel pathway. *J Cell Biol.* 2004;167(1):87–98.

98. Kornmann B, Osman C, Walter P. The conserved GTPase Gem1 regulates endoplasmic reticulum-mitochondria connections. *Proc Natl Acad Sci USA.* 2011;108(34):14151–14156.

99. Alexander C, Votruba M, Pesch UE, Thiselton DL, Mayer S, Moore A, Rodriguez M, Kellner U, Leo-Kottler B, Auburger G, Bhattacharya SS, Wissinger B. OPA1, encoding a dynamin-related GTPase, is mutated in autosomal dominant optic atrophy linked to chromosome 3q28. *Nat Genet.* 2000;26(2):211–215.

100. Hudson G, Amati-Bonneau P, Blakely EL, Stewart JD, He L, Schaefer AM, Griffiths PG, Ahlqvist K, Suomalainen A, Reynier P, McFarland R, Turnbull DM, Chinnery PF, Taylor RW. Mutation of OPA1 causes dominant optic atrophy with external ophthalmople-gia, ataxia, deafness and multiple mitochondrial DNA deletions: a novel disorder of mtDNA maintenance. *Brain J Neurol.* 2008;131(Pt 2):329–337.

101. Carelli V, Musumeci O, Caporali L, Zanna C, La Morgia C, Del Dotto V, Porcelli AM, Rugolo M, Valentino ML, Iommarini L, Maresca A, Barboni P, Carbonelli M, Trombetta C, Valente EM, Patergnani S, Giorgi C, Pinton P, Rizzo G, Tonon C, Lodi R, Avoni P, Liguori R, Baruzzi A, Toscano A, Zeviani M. Syndromic parkinsonism and dementia associated with OPA1 missense mutations. *Ann Neurol.* 2015;78(1):21–38.

102. Kijima K, Numakura C, Izumino H, Umetsu K, Nezu A, Shiiki T, Ogawa M, Ishizaki Y, Kitamura T, Shozawa Y, Hayasaka K. Mitochondrial GTPase mitofusin 2 mutation in Charcot-Marie-Tooth neuropathy type 2A. *Hum Genet.* 2005;116(1–2):23–27.

103. Zuchner S, Mersiyanova IV, Muglia M, Bissar-Tadmouri N, Rochelle J, Dadali EL, Zap-pia M, Nelis E, Patitucci A, Senderek J, Parman Y, Evgrafov O, Jonghe PD, Takahashi Y, Tsuji S, Pericak-Vance MA, Quattrone A, Battaloglu E, Polyakov AV, Timmerman V, Sch-roder JM, Vance JM. Mutations in the mitochondrial GTPase mitofusin 2 cause Charcot-Marie-Tooth neuropathy type 2A. *Nat Genet.* 2004;36(5):449–451.

104. Waterham HR, Koster J, van Roermund CW, Mooyer PA, Wanders RJ, Leonard JV. A lethal defect of mitochondrial and peroxisomal fission. *N Engl J Med.* 2007;356(17):1736–1741.

105. Koch J, Feichtinger RG, Freisinger P, Pies M, Schrodl F, Iuso A, Sperl W, Mayr JA, Prokisch H, Haack TB. Disturbed mitochondrial and peroxisomal dynamics due to loss of MFF causes Leigh-like encephalopathy, optic atrophy and peripheral neuropathy. *J Med Genet.* 2016;53(4):270–278.

106. Baxter RV, Ben Othmane K, Rochelle JM, Stajich JE, Hulette C, Dew-Knight S, Hentati F, Ben Hamida M, Bel S, Stenger JE, Gilbert JR, Pericak-Vance MA, Vance JM. Ganglioside-induced differentiation-associated protein-1 is mutant in Charcot-Marie-Tooth disease type 4A/8q21. *Nat Genet.* 2002;30(1):21–22.

107. Pla-Martin D, Calpena E, Lupo V, Marquez C, Rivas E, Sivera R, Sevilla T, Palau F, Espinos C. Junctophilin-1 is a modifier gene of GDAP1-related Charcot-Marie-Tooth disease. *Hum Mol Genet.* 2015;24(1):213–229.

108. Barneo-Munoz M, Juarez P, Civera-Tregon A, Yndriago L, Pla-Martin D, Zenker J, Cuevas-Martin C, Estela A, Sanchez-Arago M, Forteza-Vila J, Cuezva JM, Chrast R, Palau F. Lack of GDAP1 induces neuronal calcium and mitochondrial defects in a knockout mouse model of charcot-marie-tooth neuropathy. *PLoS Genet.* 2015;11(4):e1005115.

109. Gupta A, Dawson VL, Dawson TM. What causes cell death in Parkinson's disease? *Ann Neurol.* 2008;64(Suppl 2):S3–S15.

110. Ryan BJ, Hoek S, Fon EA, Wade-Martins R. Mitochondrial dysfunction and mitophagy in Parkinson's: from familial to sporadic disease. *Trends Biochem Sci.* 2015;40(4):200–210.

111. Kitada T, Asakawa S, Hattori N, Matsumine H, Yamamura Y, Minoshima S, Yokochi M, Mizuno Y, Shimizu N. Mutations in the parkin gene cause autosomal recessive juvenile parkinsonism. *Nature.* 1998;392(6676):605–608.

112. Imai Y, Soda M, Takahashi R. Parkin suppresses unfolded protein stress-induced cell death through its E3 ubiquitin-protein ligase activity. *J Biol Chem.* 2000;275(46):35661–35664.

113. Shimura H, Hattori N, Kubo S, Mizuno Y, Asakawa S, Minoshima S, Shimizu N, Iwai K, Chiba T, Tanaka K, Suzuki T. Familial Parkinson disease gene product, parkin, is a ubiquitin-protein ligase. *Nat Genet.* 2000;25(3):302–305.

114. Zhang Y, Gao J, Chung KK, Huang H, Dawson VL, Dawson TM. Parkin functions as an E2-dependent ubiquitin- protein ligase and promotes the degradation of the synaptic vesicle-associated protein, CDCrel-1. *Proc Natl Acad Sci USA.* 2000;97(24):13354–13359.

115. Hayashi S, Wakabayashi K, Ishikawa A, Nagai H, Saito M, Maruyama M, Takahashi T, Ozawa T, Tsuji S, Takahashi H. An autopsy case of autosomal-recessive juvenile parkinsonism with a homozygous exon 4 deletion in the parkin gene. *Mov Disord.* 2000;15(5):884–888.

116. van de Warrenburg BP, Lammens M, Lucking CB, Denefle P, Wesseling P, Booij J, Praamstra P, Quinn N, Brice A, Horstink MW. Clinical and pathologic abnormalities in a family with parkinsonism and parkin gene mutations. *Neurology.* 2001;56(4):555–557.

117. Imai Y, Soda M, Inoue H, Hattori N, Mizuno Y, Takahashi R. An unfolded putative transmembrane polypeptide, which can lead to endoplasmic reticulum stress, is a substrate of Parkin. *Cell.* 2001;105(7):891–902.

118. Shimura H, Schlossmacher MG, Hattori N, Frosch MP, Trockenbacher A, Schneider R, Mizuno Y, Kosik KS, Selkoe DJ. Ubiquitination of a new form of alpha-synuclein by parkin from human brain: implications for Parkinson's disease. *Science.* 2001;293(5528):263–269.

119. Darios F, Corti O, Lucking CB, Hampe C, Muriel MP, Abbas N, Gu WJ, Hirsch EC, Rooney T, Ruberg M, Brice A. Parkin prevents mitochondrial swelling and cytochrome c release in mitochondria-dependent cell death. *Hum Mol Genet.* 2003;12(5):517–526.

120. Greene JC, Whitworth AJ, Kuo I, Andrews LA, Feany MB, Pallanck LJ. Mitochondrial pathology and apoptotic muscle degeneration in *Drosophila* parkin mutants. *Proc Natl Acad Sci USA.* 2003;100(7):4078–4083.

121. Whitworth AJ, Theodore DA, Greene JC, Benes H, Wes PD, Pallanck LJ. Increased glutathione S-transferase activity rescues dopaminergic neuron loss in a *Drosophila* model of Parkinson's disease. *Proc Natl Acad Sci USA.* 2005;102(22):8024–8029.

122. Clark IE, Dodson MW, Jiang C, Cao JH, Huh JR, Seol JH, Yoo SJ, Hay BA, Guo M. Drosophila pink1 is required for mitochondrial function and interacts genetically with parkin. *Nature.* 2006;441(7097):1162–1166.

123. Park J, Lee SB, Lee S, Kim Y, Song S, Kim S, Bae E, Kim J, Shong M, Kim JM, Chung J. Mitochondrial dysfunction in Drosophila PINK1 mutants is complemented by parkin. *Nature.* 2006;441(7097):1157–1161.

124. Yang Y, Gehrke S, Imai Y, Huang Z, Ouyang Y, Wang JW, Yang L, Beal MF, Vogel H, Lu B. Mitochondrial pathology and muscle and dopaminergic neuron degeneration caused by inactivation of Drosophila Pink1 is rescued by Parkin. *Proc Natl Acad Sci USA.* 2006;103(28):10793–10798.

125. Dagda RK, Cherra 3rd SJ, Kulich SM, Tandon A, Park D, Chu CT. Loss of PINK1 function promotes mitophagy through effects on oxidative stress and mitochondrial fission. *J Biol Chem.* 2009;284(20):13843–13855.

126. Exner N, Treske B, Paquet D, Holmstrom K, Schiesling C, Gispert S, Carballo-Carbajal I, Berg D, Hoepken HH, Gasser T, Kruger R, Winklhofer KF, Vogel F, Reichert AS, Auburger G, Kahle PJ, Schmid B, Haass C. Loss-of-function of human PINK1 results in mitochondrial pathology and can be rescued by parkin. *J Neurosci.* 2007;27(45):12413–12418.

127. Riparbelli MG, Callaini G. The Drosophila parkin homologue is required for normal mitochondrial dynamics during spermiogenesis. *Dev Biol.* 2007;303(1):108–120.

128. Poole AC, Thomas RE, Andrews LA, McBride HM, Whitworth AJ, Pallanck LJ. The PINK1/Parkin pathway regulates mitochondrial morphology. *Proc Natl Acad Sci.* 2008;105(5):1638–1643.

129. Yang Y, Ouyang Y, Yang L, Beal MF, McQuibban A, Vogel H, Lu B. Pink1 regulates mitochondrial dynamics through interaction with the fission/fusion machinery. *Proc Natl Acad Sci USA.* 2008;105(19):7070–7075.

130. Deng H, Dodson MW, Huang H, Guo M. The Parkinson's disease genes *pink1* and *parkin* promote mitochondrial fission and/or inhibit fusion in *Drosophila. Proc Natl Acad Sci USA.* 2008;105(38):14503–14508.

131. Lutz AK, Exner N, Fett ME, Schlehe JS, Kloos K, Lammermann K, Brunner B, Kurz-Drexler A, Vogel F, Reichert AS, Bouman L, Vogt-Weisenhorn D, Wurst W, Tatzelt J, Haass C, Winklhofer KF. Loss of parkin or PINK1 function increases Drp1-dependent mitochondrial fragmentation. *J Biol Chem.* 2009;284(34):22938–22951.

132. Sandebring A, Thomas KJ, Beilina A, van der Brug M, Cleland MM, Ahmad R, Miller DW, Zambrano I, Cowburn RF, Behbahani H, Cedazo-Minguez A, Cookson MR. Mitochondrial alterations in PINK1 deficient cells are influenced by calcineurin-dependent dephosphorylation of dynamin-related protein 1. *PloS One.* 2009;4(5):e5701.

133. Morais VA, Verstreken P, Roethig A, Smet J, Snellinx A, Vanbrabant M, Haddad D, Frezza C, Mandemakers W, Vogt-Weisenhorn D, Van Coster R, Wurst W, Scorrano L, De Strooper B. Parkinson's disease mutations in PINK1 result in decreased complex I activity and deficient synaptic function. *EMBO Mol Med.* 2009;1(2):99–111.

134. Gautier CA, Kitada T, Shen J. Loss of PINK1 causes mitochondrial functional defects and increased sensitivity to oxidative stress. *Proc Natl Acad Sci USA.* 2008;105(32):11364–11369.

135. Mortiboys H, Thomas KJ, Koopman WJ, Klaffke S, Abou-Sleiman P, Olpin S, Wood NW, Willems PH, Smeitink JA, Cookson MR, Bandmann O. Mitochondrial function and morphology are impaired in parkin-mutant fibroblasts. *Ann Neurol.* 2008;64(5):555–565.

136. Twig G, Elorza A, Molina AJ, Mohamed H, Wikstrom JD, Walzer G, Stiles L, Haigh SE, Katz S, Las G, Alroy J, Wu M, Py BF, Yuan J, Deeney JT, Corkey BE, Shirihai OS. Fission and selective fusion govern mitochondrial segregation and elimination by autophagy. *EMBO J.* 2008;27(2):433–446.

137. Yu W, Sun Y, Guo S, Lu B. The PINK1/Parkin pathway regulates mitochondrial dynamics and function in mammalian hippocampal and dopaminergic neurons. *Hum Mol Genet.* 2011;20(16):3227–3240.

138. Ziviani E, Tao RN, Whitworth AJ. Drosophila parkin requires PINK1 for mitochondrial translocation and ubiquitinates mitofusin. *Proc Natl Acad Sci USA.* 2010;107(11):5018–5023.

139. Poole AC, Thomas RE, Yu S, Vincow ES, Pallanck L. The mitochondrial fusion-promoting factor mitofusin is a substrate of the PINK1/parkin pathway. *PloS One.* 2010;5(4):e10054.

140. Gegg ME, Cooper JM, Chau KY, Rojo M, Schapira AH, Taanman JW. Mitofusin 1 and mitofusin 2 are ubiquitinated in a PINK1/parkin-dependent manner upon induction of mitophagy. *Hum Mol Genet.* 2010;19(24):4861–4870.

141. Sarraf SA, Raman M, Guarani-Pereira V, Sowa ME, Huttlin EL, Gygi SP, Harper JW. Landscape of the PARKIN-dependent ubiquitylome in response to mitochondrial depolarization. *Nature.* 2013;496(7445):372–376.

142. Lucas JI, Marin I. A new evolutionary paradigm for the Parkinson disease gene DJ-1. *Mol Biol Evol.* 2007;24(2):551–561.
143. Canet-Aviles RM, Wilson MA, Miller DW, Ahmad R, McLendon C, Bandyopadhyay S, Baptista MJ, Ringe D, Petsko GA, Cookson MR. The Parkinson's disease protein DJ-1 is neuroprotective due to cysteine-sulfinic acid-driven mitochondrial localization. *Proc Natl Acad Sci USA.* 2004;101(24):9103–9108.
144. Kim RH, Smith PD, Aleyasin H, Hayley S, Mount MP, Pownall S, Wakeham A, You-Ten AJ, Kalia SK, Horne P, Westaway D, Lozano AM, Anisman H, Park DS, Mak TW. Hypersensitivity of DJ-1-deficient mice to 1-methyl-4-phenyl-1,2,3,6-tetrahydropyridine (MPTP) and oxidative stress. *Proc Natl Acad Sci USA.* 2005;102(14):5215–5220.
145. Zhang L, Shimoji M, Thomas B, Moore DJ, Yu SW, Marupudi NI, Torp R, Torgner IA, Ottersen OP, Dawson TM, Dawson VL. Mitochondrial localization of the Parkinson's disease related protein DJ-1: implications for pathogenesis. *Hum Mol Genet.* 2005;14(14):2063–2073.
146. Thomas KJ, McCoy MK, Blackinton J, Beilina A, van der Brug M, Sandebring A, Miller D, Maric D, Cedazo-Minguez A, Cookson MR. DJ-1 acts in parallel to the PINK1/parkin pathway to control mitochondrial function and autophagy. *Hum Mol Genet.* 2011;20(1):40–50.
147. Andres-Mateos E, Perier C, Zhang L, Blanchard-Fillion B, Greco TM, Thomas B, Ko HS, Sasaki M, Ischiropoulos H, Przedborski S, Dawson TM, Dawson VL. DJ-1 gene deletion reveals that DJ-1 is an atypical peroxiredoxin-like peroxidase. *Proc Natl Acad Sci USA.* 2007;104(37):14807–14812.
148. Irrcher I, Aleyasin H, Seifert EL, Hewitt SJ, Chhabra S, Phillips M, Lutz AK, Rousseaux MW, Bevilacqua L, Jahani-Asl A, Callaghan S, MacLaurin JG, Winklhofer KF, Rizzu P, Rippstein P, Kim RH, Chen CX, Fon EA, Slack RS, Harper ME, McBride HM, Mak TW, Park DS. Loss of the Parkinson's disease-linked gene DJ-1 perturbs mitochondrial dynamics. *Hum Mol Genet.* 2010;19(19):3734–3746.
149. Krebiehl G, Ruckerbauer S, Burbulla LF, Kieper N, Maurer B, Waak J, Wolburg H, Gizatullina Z, Gellerich FN, Woitalla D, Riess O, Kahle PJ, Proikas-Cezanne T, Kruger R. Reduced basal autophagy and impaired mitochondrial dynamics due to loss of Parkinson's disease-associated protein DJ-1. *PloS One.* 2010;5(2):e9367.
150. Eriksen JL, Przedborski S, Petrucelli L. Gene dosage and pathogenesis of Parkinson's disease. *Trends Mol Med.* 2005;11(3):91–96.
151. Cookson MR. The biochemistry of Parkinson's disease. *Ann Rev Biochem.* 2005;74:29–52.
152. Tofaris GK, Spillantini MG. Physiological and pathological properties of alpha-synuclein. *Cell Mol Life Sci.* 2007;64(17):2194–2201.
153. Chinta SJ, Andersen JK. Redox imbalance in Parkinson's disease. *Biochimica et Biophysica Acta.* 2008;1780(11):1362–1367.
154. Cole NB, Dieuliis D, Leo P, Mitchell DC, Nussbaum RL. Mitochondrial translocation of alpha-synuclein is promoted by intracellular acidification. *Exp Cell Res.* 2008;314(10):2076–2089.
155. Devi L, Raghavendran V, Prabhu BM, Avadhani NG, Anandatheerthavarada HK. Mitochondrial import and accumulation of alpha-synuclein impair complex I in human dopaminergic neuronal cultures and Parkinson disease brain. *J Biol Chem.* 2008;283(14):9089–9100.
156. Li WW, Yang R, Guo JC, Ren HM, Zha XL, Cheng JS, Cai DF. Localization of alpha-synuclein to mitochondria within midbrain of mice. *Neuroreport.* 2007;18(15):1543–1546.
157. Liu G, Zhang C, Yin J, Li X, Cheng F, Li Y, Yang H, Ueda K, Chan P, Yu S. Alpha-synuclein is differentially expressed in mitochondria from different rat brain regions and dose-dependently down-regulates complex I activity. *Neurosci Lett.* 2009;454(3):187–192.
158. Martin LJ, Pan Y, Price AC, Sterling W, Copeland NG, Jenkins NA, Price DL, Lee MK. Parkinson's disease alpha-synuclein transgenic mice develop neuronal mitochondrial degeneration and cell death. *J Neurosci.* 2006;26(1):41–50.

159. Nakamura K, Nemani VM, Wallender EK, Kaehlcke K, Ott M, Edwards RH. Optical reporters for the conformation of alpha-synuclein reveal a specific interaction with mitochondria. *J Neurosci.* 2008;28(47):12305–12317.
160. Parihar MS, Parihar A, Fujita M, Hashimoto M, Ghafourifar P. Mitochondrial association of alpha-synuclein causes oxidative stress. *Cell Mol Life Sci.* 2008;65(7–8):1272–1284.
161. Shavali S, Brown-Borg HM, Ebadi M, Porter J. Mitochondrial localization of alpha-synuclein protein in alpha-synuclein overexpressing cells. *Neurosci Lett.* 2008;439(2):125–128.
162. Zhu Y, Duan C, Lu L, Gao H, Zhao C, Yu S, Ueda K, Chan P, Yang H. Alpha-synuclein overexpression impairs mitochondrial function by associating with adenylate translocator. *Int J Biochemi Cell Biol.* 2011;43(5):732–741.
163. Zhang L, Zhang C, Zhu Y, Cai Q, Chan P, Ueda K, Yu S, Yang H. Semi-quantitative analysis of alpha-synuclein in subcellular pools of rat brain neurons: an immunogold electron microscopic study using a C-terminal specific monoclonal antibody. *Brain Res.* 2008;1244:40–52.
164. Kamp F, Exner N, Lutz AK, Wender N, Hegermann J, Brunner B, Nuscher B, Bartels T, Giese A, Beyer K, Eimer S, Winklhofer KF, Haass C. Inhibition of mitochondrial fusion by alpha-synuclein is rescued by PINK1, Parkin and DJ-1. *EMBO J.* 2010;29(20):3571–3589.
165. Nakamura K, Nemani VM, Azarbal F, Skibinski G, Levy JM, Egami K, Munishkina L, Zhang J, Gardner B, Wakabayashi J, Sesaki H, Cheng Y, Finkbeiner S, Nussbaum RL, Masliah E, Edwards RH. Direct membrane association drives mitochondrial fission by the Parkinson Disease-associated protein α-synuclein. *J Biol Chem.* 2011;286(23):20710–20726.
166. Kamp F, Beyer K. Binding of alpha-synuclein affects the lipid packing in bilayers of small vesicles. *J Biol Chem.* 2006;281(14):9251–9259.
167. Dauer W, Kholodilov N, Vila M, Trillat AC, Goodchild R, Larsen KE, Staal R, Tieu K, Schmitz Y, Yuan CA, Rocha M, Jackson-Lewis V, Hersch S, Sulzer D, Przedborski S, Burke R, Hen R. Resistance of alpha-synuclein null mice to the parkinsonian neurotoxin MPTP. *Proc Natl Acad Sci USA.* 2002;99(22):14524–14529.
168. Schluter OM, Fornai F, Alessandri MG, Takamori S, Geppert M, Jahn R, Sudhof TC. Role of alpha-synuclein in 1-methyl-4-phenyl-1,2,3,6-tetrahydropyridine-induced parkinsonism in mice. *Neuroscience.* 2003;118(4):985–1002.
169. Klivenyi P, Siwek D, Gardian G, Yang L, Starkov A, Cleren C, Ferrante RJ, Kowall NW, Abeliovich A, Beal MF. Mice lacking alpha-synuclein are resistant to mitochondrial toxins. *Neurobiol Dis.* 2006;21(3):541–548.
170. Orth M, Tabrizi SJ, Schapira AH, Cooper JM. Alpha-synuclein expression in HEK293 cells enhances the mitochondrial sensitivity to rotenone. *Neurosci Lett.* 2003;351(1):29–32.
171. Song DD, Shults CW, Sisk A, Rockenstein E, Masliah E. Enhanced substantia nigra mitochondrial pathology in human alpha-synuclein transgenic mice after treatment with MPTP. *Exp Neurol.* 2004;186(2):158–172.
172. Rathke-Hartlieb S, Kahle PJ, Neumann M, Ozmen L, Haid S, Okochi M, Haass C, Schulz JB. Sensitivity to MPTP is not increased in Parkinson's disease-associated mutant alpha-synuclein transgenic mice. *J Neurochem.* 2001;77(4):1181–1184.
173. Dong Z, Ferger B, Feldon J, Bueler H. Overexpression of Parkinson's disease-associated alpha-synucleinA53T by recombinant adeno-associated virus in mice does not increase the vulnerability of dopaminergic neurons to MPTP. *J Neurobiol.* 2002;53(1):1–10.
174. Butler EK, Voigt A, Lutz AK, Toegel JP, Gerhardt E, Karsten P, Falkenburger B, Reinartz A, Winklhofer KF, Schulz JB. The mitochondrial chaperone protein TRAP1 mitigates alpha-synuclein toxicity. *PLoS Genet.* 2012;8(2):e1002488.
175. Petrucelli L, O'Farrell C, Lockhart PJ, Baptista M, Kehoe K, Vink L, Choi P, Wolozin B, Farrer M, Hardy J, Cookson MR. Parkin protects against the toxicity associated with mutant alpha-synuclein: proteasome dysfunction selectively affects catecholaminergic neurons. *Neuron.* 2002;36(6):1007–1019.
176. Yang Y, Nishimura I, Imai Y, Takahashi R, Lu B. Parkin suppresses dopaminergic neuron-selective neurotoxicity induced by Pael-R in Drosophila. *Neuron.* 2003;37(6):911–924.

177. Lo Bianco C, Schneider BL, Bauer M, Sajadi A, Brice A, Iwatsubo T, Aebischer P. Lentiviral vector delivery of parkin prevents dopaminergic degeneration in an {alpha}-synuclein rat model of Parkinson's disease. *Proc Natl Acad Sci USA.* 2004;101(50):17510–17515.

178. Yasuda T, Miyachi S, Kitagawa R, Wada K, Nihira T, Ren YR, Hirai Y, Ageyama N, Terao K, Shimada T, Takada M, Mizuno Y, Mochizuki H. Neuronal specificity of alpha-synuclein toxicity and effect of Parkin co-expression in primates. *Neuroscience.* 2007;144(2):743–753.

179. Khandelwal PJ, Dumanis SB, Feng LR, Maguire-Zeiss K, Rebeck G, Lashuel HA, Moussa CE. Parkinson-related parkin reduces alpha-synuclein phosphorylation in a gene transfer model. *Mol Neurodegener.* 2010;5:47.

180. Zhu XR, Maskri L, Herold C, Bader V, Stichel CC, Gunturkun O, Lubbert H. Non-motor behavioural impairments in parkin-deficient mice. *Eur J Neurosci.* 2007;26(7):1902–1911.

181. Fournier M, Vitte J, Garrigue J, Langui D, Dullin JP, Saurini F, Hanoun N, Perez-Diaz F, Cornilleau F, Joubert C, Ardila-Osorio H, Traver S, Duchateau R, Goujet-Zalc C, Paleologou K, Lashuel HA, Haass C, Duyckaerts C, Cohen-Salmon C, Kahle PJ, Hamon M, Brice A, Corti O. Parkin deficiency delays motor decline and disease manifestation in a mouse model of synucleinopathy. *PloS One.* 2009;4(8):e6629.

182. Choubey V, Safiulina D, Vaarmann A, Cagalinec M, Wareski P, Kuum M, Zharkovsky A, Kaasik A. Mutant A53T alpha-synuclein induces neuronal death by increasing mitochondrial autophagy. *J Biol Chem.* 2011;286(12):10814–10824.

183. Greggio E, Jain S, Kingsbury A, Bandopadhyay R, Lewis P, Kaganovich A, van der Brug MP, Beilina A, Blackinton J, Thomas KJ, Ahmad R, Miller DW, Kesavapany S, Singleton A, Lees A, Harvey RJ, Harvey K, Cookson MR. Kinase activity is required for the toxic effects of mutant LRRK2/dardarin. *Neurobiol Dis.* 2006;23(2):329–341.

184. Smith WW, Pei Z, Jiang H, Dawson VL, Dawson TM, Ross CA. Kinase activity of mutant LRRK2 mediates neuronal toxicity. *Nat Neurosci.* 2006;9(10):1231–1233.

185. West AB, Moore DJ, Choi C, Andrabi SA, Li X, Dikeman D, Biskup S, Zhang Z, Lim KL, Dawson VL, Dawson TM. Parkinson's disease-associated mutations in LRRK2 link enhanced GTP-binding and kinase activities to neuronal toxicity. *Hum Mol Genet.* 2007;16(2):223–232.

186. Ho CC, Rideout HJ, Ribe E, Troy CM, Dauer WT. The Parkinson disease protein leucine-rich repeat kinase 2 transduces death signals via Fas-associated protein with death domain and caspase-8 in a cellular model of neurodegeneration. *J Neurosci.* 2009;29(4):1011–1016.

187. Dusonchet J, Kochubey O, Stafa K, Young Jr SM, Zufferey R, Moore DJ, Schneider BL, Aebischer P. A rat model of progressive nigral neurodegeneration induced by the Parkinson's disease-associated G2019S mutation in LRRK2. *J Neurosci.* 2011;31(3):907–912.

188. West AB, Moore DJ, Biskup S, Bugayenko A, Smith WW, Ross CA, Dawson VL, Dawson TM. Parkinson's disease-associated mutations in leucine-rich repeat kinase 2 augment kinase activity. *Proc Natl Acad Sci USA.* 2005;102(46):16842–16847.

189. Biskup S, Moore DJ, Celsi F, Higashi S, West AB, Andrabi SA, Kurkinen K, Yu SW, Savitt JM, Waldvogel HJ, Faull RL, Emson PC, Torp R, Ottersen OP, Dawson TM, Dawson VL. Localization of LRRK2 to membranous and vesicular structures in mammalian brain. *Ann Neurol.* 2006;60(5):557–569.

190. Ramonet D, Daher JP, Lin BM, Stafa K, Kim J, Banerjee R, Westerlund M, Pletnikova O, Glauser L, Yang L, Liu Y, Swing DA, Beal MF, Troncoso JC, McCaffery JM, Jenkins NA, Copeland NG, Galter D, Thomas B, Lee MK, Dawson TM, Dawson VL, Moore DJ. Dopaminergic neuronal loss, reduced neurite complexity and autophagic abnormalities in transgenic mice expressing G2019S mutant LRRK2. *PloS One.* 2011;6(4):e18568.

191. Yue M, Hinkle KM, Davies P, Trushina E, Fiesel FC, Christenson TA, Schroeder AS, Zhang L, Bowles E, Behrouz B, Lincoln SJ, Beevers JE, Milnerwood AJ, Kurti A, McLean PJ, Fryer JD, Springer W, Dickson DW, Farrer MJ, Melrose HL. Progressive dopaminergic alterations and mitochondrial abnormalities in LRRK2 G2019S knock-in mice. *Neurobiol Dis.* 2015;78:172–195.

192. Mortiboys H, Johansen KK, Aasly JO, Bandmann O. Mitochondrial impairment in patients with Parkinson disease with the G2019S mutation in LRRK2. *Neurology*. 2010;75(22):2017–2020.

193. Papkovskaia TD, Chau KY, Inesta-Vaquera F, Papkovsky DB, Healy DG, Nishio K, Staddon J, Duchen MR, Hardy J, Schapira AH, Cooper JM. G2019S leucine-rich repeat kinase 2 causes uncoupling protein-mediated mitochondrial depolarization. *Hum Mol Genet*. 2012;21(19):4201–4213.

194. Mortiboys H, Aasly J, Bandmann O. Ursocholanic acid rescues mitochondrial function in common forms of familial Parkinson's disease. *Brain J Neurol*. 2013;136(Pt 10):3038–3050.

195. Mortiboys H, Furmston R, Bronstad G, Aasly J, Elliott C, Bandmann O. UDCA exerts beneficial effect on mitochondrial dysfunction in LRRK2G2019S carriers and in vivo. *Neurology*. 2015;85(10):846–852.

196. Wang X, Yan MH, Fujioka H, Liu J, Wilson-Delfosse A, Chen SG, Perry G, Casadesus G, Zhu X. LRRK2 regulates mitochondrial dynamics and function through direct interaction with DLP1. *Hum Mol Genet*. 2012;21(9):1931–1944.

197. Niu J, Yu M, Wang C, Xu Z. Leucine-rich repeat kinase 2 disturbs mitochondrial dynamics via dynamin-like protein. *J Neurochem*. 2012;122(3):650–658.

198. Su YC, Qi X. Inhibition of excessive mitochondrial fission reduced aberrant autophagy and neuronal damage caused by LRRK2 G2019S mutation. *Hum Mol Genet*. 2013;22(22):4545–4561.

199. Stafa K, Tsika E, Moser R, Musso A, Glauser L, Jones A, Biskup S, Xiong Y, Bandopadhyay R, Dawson VL, Dawson TM, Moore DJ. Functional interaction of Parkinson's disease-associated LRRK2 with members of the dynamin GTPase superfamily. *Hum Mol Genet*. 2014;23(8):2055–2077.

200. Caesar M, Zach S, Carlson CB, Brockmann K, Gasser T, Gillardon F. Leucine-rich repeat kinase 2 functionally interacts with microtubules and kinase-dependently modulates cell migration. *Neurobiol Dis*. 2013;54:280–288.

201. Kett LR, Boassa D, Ho CC, Rideout HJ, Hu J, Terada M, Ellisman M, Dauer WT. LRRK2 Parkinson disease mutations enhance its microtubule association. *Hum Mol Genet*. 2012;21(4):890–899.

202. Law BM, Spain VA, Leinster VH, Chia R, Beilina A, Cho HJ, Taymans JM, Urban MK, Sancho RM, Ramirez MB, Biskup S, Baekelandt V, Cai H, Cookson MR, Berwick DC, Harvey K. A direct interaction between leucine-rich repeat kinase 2 and specific beta-tubulin isoforms regulates tubulin acetylation. *J Biol Chem*. 2014;289(2):895–908.

203. Reed NA, Cai D, Blasius TL, Jih GT, Meyhofer E, Gaertig J, Verhey KJ. Microtubule acetylation promotes kinesin-1 binding and transport. *Curr Biol*. 2006;16(21):2166–2172.

204. Dompierre JP, Godin JD, Charrin BC, Cordelieres FP, King SJ, Humbert S, Saudou F. Histone deacetylase 6 inhibition compensates for the transport deficit in Huntington's disease by increasing tubulin acetylation. *J Neurosci*. 2007;27:3571–3583.

205. Godena VK, Brookes-Hocking N, Moller A, Shaw G, Oswald M, Sancho RM, Miller CCJ, Whitworth AJ, De Vos KJ. Increasing microtubule acetylation rescues axonal transport and locomotor deficits caused by LRRK2 Roc-COR domain mutations. *Nat Commun*. 2014;5:1–11.

206. Twig G, Hyde B, Shirihai OS. Mitochondrial fusion, fission and autophagy as a quality control axis: the bioenergetic view. *Biochimica et Biophysica Acta*. 2008;1777(9):1092–1097.

207. Narendra D, Tanaka A, Suen DF, Youle RJ. Parkin is recruited selectively to impaired mitochondria and promotes their autophagy. *J Cell Biol*. 2008;183(5):795–803.

208. Pickrell AM, Youle RJ. The roles of PINK1, parkin, and mitochondrial fidelity in Parkinson's disease. *Neuron*. 2015;85(2):257–273.

209. Gandhi S, Wood-Kaczmar A, Yao Z, Plun-Favreau H, Deas E, Klupsch K, Downward J, Latchman DS, Tabrizi SJ, Wood NW, Duchen MR, Abramov AY. PINK1-associated Parkinson's disease is caused by neuronal vulnerability to calcium-induced cell death. *Mol Cell*. 2009;33(5):627–638.

210. Gegg ME, Cooper JM, Schapira AH, Taanman JW. Silencing of PINK1 expression affects mitochondrial DNA and oxidative phosphorylation in dopaminergic cells. *PloS One.* 2009;4(3):e4756.
211. Haque ME, Thomas KJ, D'Souza C, Callaghan S, Kitada T, Slack RS, Fraser P, Cookson MR, Tandon A, Park DS. Cytoplasmic Pink1 activity protects neurons from dopaminergic neurotoxin MPTP. *Proc Natl Acad Sci USA.* 2008;105(5):1716–1721.
212. Paterna JC, Leng A, Weber E, Feldon J, Bueler H. DJ-1 and Parkin modulate dopamine-dependent behavior and inhibit MPTP-induced nigral dopamine neuron loss in mice. *Mol Ther.* 2007;15(4):698–704.
213. Christodoulou S, Lockyer AE, Foster JM, Hoheisel JD, Roberts DB. Nucleotide sequence of a Drosophila melanogaster cDNA encoding a calnexin homologue. *Gene.* 1997;191(2):143–148.
214. Vincow ES, Merrihew G, Thomas RE, Shulman NJ, Beyer RP, MacCoss MJ, Pallanck LJ. The PINK1-Parkin pathway promotes both mitophagy and selective respiratory chain turnover in vivo. *Proc Natl Acad Sci USA.* 2013;110(16):6400–6405.
215. Pickrell AM, Huang CH, Kennedy SR, Ordureau A, Sideris DP, Hoekstra JG, Harper JW, Youle RJ. Endogenous parkin preserves dopaminergic substantia nigral neurons following mitochondrial dna mutagenic stress. *Neuron.* 2015;87(2):371–381.
216. Neuspiel M, Schauss AC, Braschi E, Zunino R, Rippstein P, Rachubinski RA, Andrade-Navarro MA, McBride HM. Cargo-selected transport from the mitochondria to peroxisomes is mediated by vesicular carriers. *Curr Biol.* 2008;18(2):102–108.
217. Soubannier V, McBride HM. Positioning mitochondrial plasticity within cellular signaling cascades. *Biochimica et Biophysica Acta.* 2009;1793(1):154–170.
218. Sugiura A, McLelland GL, Fon EA, McBride HM. A new pathway for mitochondrial quality control: mitochondrial-derived vesicles. *EMBO J.* 2014;33(19):2142–2156.
219. Soubannier V, Rippstein P, Kaufman BA, Shoubridge EA, McBride HM. Reconstitution of mitochondria derived vesicle formation demonstrates selective enrichment of oxidized cargo. *PloS One.* 2012;7(12):e52830.
220. Braschi E, Goyon V, Zunino R, Mohanty A, Xu L, McBride HM. Vps35 mediates vesicle transport between the mitochondria and peroxisomes. *Curr Biol.* 2010;20(14):1310–1315.
221. McLelland GL, Soubannier V, Chen CX, McBride HM, Fon EA. Parkin and PINK1 function in a vesicular trafficking pathway regulating mitochondrial quality control. *EMBO J.* 2014;33(4):282–295.
222. Wang W, Wang X, Fujioka H, Hoppel C, Whone AL, Caldwell MA, Cullen PJ, Liu J, Zhu X. Parkinson's disease-associated mutant VPS35 causes mitochondrial dysfunction by recycling DLP1 complexes. *Nat Med.* 2016;22(1):54–63.
223. Malik BR, Godena VK, Whitworth AJ. VPS35 pathogenic mutations confer no dominant toxicity but partial loss of function in Drosophila and genetically interact with parkin. *Hum Mol Genet.* 2015;24(21):6106–6117.
224. Tang FL, Liu W, Hu JX, Erion JR, Ye J, Mei L, Xiong WC. VPS35 deficiency or mutation causes dopaminergic neuronal loss by impairing mitochondrial fusion and function. *Cell Rep.* 2015;12(10):1631–1643.
225. Weihofen A, Thomas KJ, Ostaszewski BL, Cookson MR, Selkoe DJ. Pink1 forms a multiprotein complex with Miro and Milton, linking Pink1 function to mitochondrial trafficking. *Biochemistry.* 2009;48(9):2045–2052.
226. Guo X, Macleod GT, Wellington A, Hu F, Panchumarthi S, Schoenfield M, Marin L, Charlton MP, Atwood HL, Zinsmaier KE. The GTPase dMiro is required for axonal transport of mitochondria to *Drosophila* synapses. *Neuron.* 2005;47(3):379–393.
227. Glater EE, Megeath LJ, Stowers RS, Schwarz TL. Axonal transport of mitochondria requires milton to recruit kinesin heavy chain and is light chain independent. *J Cell Biol.* 2006;173(4):545–557.
228. Stowers RS, Megeath LJ, Górska-Andrzejak J, Meinertzhagen IA, Schwarz TL. Axonal transport of mitochondria to synapses depends on milton, a novel *Drosophila* protein. *Neuron.* 2002;36(6):1063–1077.

229. Wang X, Winter D, Ashrafi G, Schlehe J, Wong YL, Selkoe D, Rice S, Steen J, LaVoie MJ, Schwarz TL. PINK1 and Parkin target miro for phosphorylation and degradation to arrest mitochondrial Motility. *Cell*. 2011;147(4):893–906.

230. Birsa N, Norkett R, Wauer T, Mevissen TET, Wu HC, Foltynie T, Bhatia K, Hirst WD, Komander D, Plun-Favreau H, Kittler JT. Lysine 27 ubiquitination of the mitochondrial transport protein miro is dependent on serine 65 of the parkin ubiquitin ligase. *J Biol Chem*. 2014;289(21):14569–14582.

231. Tsai PI, Course MM, Lovas JR, Hsieh CH, Babic M, Zinsmaier KE, Wang X. PINK1-mediated phosphorylation of Miro inhibits synaptic growth and protects dopaminergic neurons in Drosophila. *Sci Rep*. 2014;4:6962.

232. Cassidy-Stone A, Chipuk JE, Ingerman E, Song C, Yoo C, Kuwana T, Kurth MJ, Shaw JT, Hinshaw JE, Green DR, Nunnari J. Chemical inhibition of the mitochondrial division dynamin reveals its role in Bax/Bak-dependent mitochondrial outer membrane permeabilization. *Dev Cell*. 2008;14(2):193–204.

233. Cui M, Tang X, Christian WV, Yoon Y, Tieu K. Perturbations in mitochondrial dynamics induced by human mutant PINK1 can be rescued by the mitochondrial division inhibitor mdivi-1. *J Biol Chem*. 2010;285(15):11740–11752.

234. Yun J, Puri R, Yang H, Lizzio MA, Wu C, Sheng Z-H, Guo M. MUL1 acts in parallel to the PINK1/parkin pathway in regulating mitofusin and compensates for loss of PINK1/parkin. *eLife*. 2014;3.

235. Bingol B, Tea JS, Phu L, Reichelt M, Bakalarski CE, Song Q, Foreman O, Kirkpatrick DS, Sheng M. The mitochondrial deubiquitinase USP30 opposes parkin-mediated mitophagy. *Nature*. 2014;510(7505):370–375.

236. Liang JR, Martinez A, Lane JD, Mayor U, Clague MJ, Urbe S. USP30 deubiquitylates mitochondrial Parkin substrates and restricts apoptotic cell death. *EMBO Rep*. 2015;16(5):618–627.

237. Wang Y, Serricchio M, Jauregui M, Shanbhag R, Stoltz T, Di Paolo CT, Kim PK, McQuibban GA. Deubiquitinating enzymes regulate PARK2-mediated mitophagy. *Autophagy*. 2015;11(4):595–606.

238. Cornelissen T, Haddad D, Wauters F, Van Humbeeck C, Mandemakers W, Koentjoro B, Sue C, Gevaert K, De Strooper B, Verstreken P, Vandenberghe W. The deubiquitinase USP15 antagonizes Parkin-mediated mitochondrial ubiquitination and mitophagy. *Hum Mol Genet*. 2014;23(19):5227–5242.

239. Durcan TM, Tang MY, Perusse JR, Dashti EA, Aguileta MA, McLelland GL, Gros P, Shaler TA, Faubert D, Coulombe B, Fon EA. USP8 regulates mitophagy by removing K6-linked ubiquitin conjugates from parkin. *EMBO J*. 2014;33(21):2473–2491.

240. Vilain S, Esposito G, Haddad D, Schaap O, Dobreva MP, Vos M, Van Meensel S, Morais VA, De Strooper B, Verstreken P. The yeast complex I equivalent NADH dehydrogenase rescues pink1 mutants. *PLoS Genet*. 2012;8(1):e1002456.

241. Morais VA, Haddad D, Craessaerts K, De Bock PJ, Swerts J, Vilain S, Aerts L, Overbergh L, Grunewald A, Seibler P, Klein C, Gevaert K, Verstreken P, De Strooper B. PINK1 loss-of-function mutations affect mitochondrial complex I activity via NdufA10 ubiquinone uncoupling. *Science*. 2014;344(6180):203–207.

242. Pogson JH, Ivatt RM, Sanchez-Martinez A, Tufi R, Wilson E, Mortiboys H, Whitworth AJ. The complex I subunit *NDUFA10* selectively rescues *Drosophila pink1* mutants through a mechanism independent of mitophagy. *PLoS Genet*. 2014;10(11):e1004815.

CHAPTER

4

Axonal Mitochondrial Transport

E. Shlevkov*,**, T.L. Schwarz*,**

*F. M. Kirby Neurobiology Center, Boston Children's Hospital, Boston, MA, United States; **Department of Neurobiology, Harvard Medical School, Harvard University, Boston, MA, United States

OUTLINE

1 PRELUDE

The extraordinary architecture of neurons—vast arbors of axons and dendrites—puts an extraordinary demand on the machinery of motor-based transport. Consequently, evidence has accumulated in many neurodegenerative disorders to implicate transport defects and, in some cases,

Parkinson's Disease. http://dx.doi.org/10.1016/B978-0-12-803783-6.00004-3

those defects may be primary causes of the disorder. Indeed, mutations of motor proteins that are found throughout the body can give rise selectively to neurological defects, presumably because neurons are more dependent on efficient transport than other cells. Mitochondrial transport may be particularly essential because of the intense reliance of axons and synapses on mitochondria as a source of ATP and Ca^{2+}-buffering. In the case of Parkinson's disease (PD), many signposts point to disruption of mitochondrial health as a major factor and thus the importance of mitochondrial transport in maintaining mitochondrial health takes on great relevance. Moreover, there is evidence for the direct involvement of PINK1 and Parkin in the regulation of mitochondrial motility and defects in mitochondrial motility may contribute to other forms of the disease. In this Chapter we will review and examine that evidence as well as the overall importance of mitochondrial movement for mitochondrial health and neuronal survival.

In Section 2 we present as background the severity of the problem that neurons face in matching the demand for energy to its supply. We review the importance of mitochondrial respiration and Ca^{2+}-buffering for the survival and functioning of the neuron and the extreme need for axonal transport of mitochondrial transport that this implies. We highlight the particular properties of the Substantia Nigra pars compacta neurons (SNpc) that make their need for mitochondrial transport particularly acute. In Section 3 of this chapter we examine the importance of axonal transport in keeping the mitochondrial population healthy. Mitochondrial transport is not a one-time event to populate the periphery but rather a constant need. While mitochondrial proteins likely last a matter of days or weeks, neurons need to survive for the lifetime of the organism and that requires a constant resupply of material from the soma. In Section 4 we turn to the mechanism of mitochondrial motility: what are the motors on mitochondria, how are they attached, and how are they regulated? This section introduces the crucial motor adaptor proteins Milton and Miro (also called TRAK1/2 and RhoT1/2) which figure prominently in Section 5 where the interactions of PINK1 and Parkin with the adaptor complex are presented. The ability of PINK1 and Parkin to arrest the movement of mitochondria offers the most explicit link of mitochondrial transport to PD. Other PD-associated genes and proteins have been linked to mitochondrial motility as well, although here the evidence is not so abundant and in most if not all cases, indirect. Section 6 will examine these additional PD-associated genes while bearing in mind that virtually any toxic insult to the cell will eventually cause mitochondria to stop. This section therefore weighs the evidence for primary versus indirect involvement of a transport defect in PD. Section 7 summarizes these concepts and proposes a model in which the interrelationship of mitochondrial motility and mitochondrial health is key. In healthy neurons, mitochondrial transport and mitochondrial health are tightly linked in a virtuous cycle that sustains the mitochondrial population and provides energy to the cell. But defects in either transport or mitochondrial functioning can trigger a vicious cycle in which both transport and function decline and compromise the viability of the neuron. Understanding the interdependency of transport and mitochondrial function may prove important for sustaining the viability of neurons in PD.

2 ON THE DISTRIBUTION OF POWER: MITOCHONDRIA TO MATCH SUPPLY AND DEMAND

The nervous system has an impressive appetite for energy: a single resting cortical neuron consumes over 4.7 billion ATP molecules per second.[1] Since neurons rely heavily on oxidative phosphorylation in mitochondria to produce most of their ATP,[2] our brains account for 20% of the oxygen we consume at rest,[3,4] although the brain is only 2% of our body mass. Within a neuron, the need for ATP is not uniform; pre- and postsynaptic mechanisms are particularly energetically expensive. The ions that enter through transmitter receptors and voltage-gated ion channels must be pumped out at a high cost of ATP [Ref. 5 and references therein].

Therein lies a neuron-specific problem: the extremities of neurons, far from the cell body where most proteins and organelles are made, have an intense requirement for mitochondria-derived ATP. The exceptional morphology of a neuron is designed for the rapid transmission of electrical signals down long processes, but this necessitates the more difficult task of transporting mitochondria to generate ATP. Whereas most non-neuronal cells are measured in hundreds of microns, neurons extend their axons and dendrites for millimeters or centimeters or, in the case of motor and sensory neurons, for up to 1 m. Consequently for most neurons, the majority of their cellular volume resides in their axons. The volume of a human sensory axon that is 10 µ in diameter and a meter long, for example, is more than 100-fold greater than that of the soma on which it depends for supplies. The same is true for the neurons of the *Substania Nigra pars compacta* (SNpc) which are critical in PD. Although the SNpc somata are just centimeters apart from the terminals in the striatum, the axonal arbor is still vast because the axons branch so extensively within the striatum. It has been calculated that all the axon branches of a single dopaminergic neuron of the rat SNpc add up to 78 cm (~30 in.) of axon.[6] If the branching patterns in human SNpc DA neurons are similar to those in rats, it has been estimated that a single human neuron should give rise to 4.5 m of axon.[7] Studies in rats have estimated that the axon of a single SNpc DA neuron can harbor from 100,000 to over 250,000 synapses.[7] These values are in stark contrast with other cell types in the basal ganglia such as *globus pallidus* (~2,000 synapses), medium spiny neuron (~300–500 synapses) and fast-spiking GABA in the *striatum* (~5,000 synapses), cell types usually spared in PD. This immense architecture will impose a tremendous demand on mitochondrial transport as well for mitochondrial ATP. The same authors have estimated that the energy cost of action potential propagation in a human SNpc DA neuron would be ~9.3×10^{10} ATP molecules. Moreover, the SNpc neurons have a well-described pacemaker activity[8] and the L-type Ca^{2+} channel Cav1.3.[9–11] Each Ca^{2+} ion that enters require approximately one ATP to be moved back across the plasma membrane and thus the demand for ATP to power ion pumps and preserve Ca^{2+} homeostasis is intense and unrelenting.

These features of the SNpc may partially explain one of the abiding puzzles of PD: why is this population of neurons particularly vulnerable to degenerate in PD; their vast size and energy requirements may combine to make them exceptionally susceptible to any alterations of mitochondrial function, distribution, or turnover. A recent study has provided experimental evidence supporting the idea that the arbor size of PD neurons is a critical determinant of vulnerability.[12] This study found that mouse SNc DA neurons, but not ventral tegmental area DA neurons, which are spared in the disease,[13,14] have very large axonal arbors when grown in culture, and this difference is accompanied by a larger mitochondrial density, respiration rate, and ROS production. SNpc DA neurons had lower respiratory reserve. Respiratory reserve is defined as the difference between the maximum capacity for ATP generation by oxidative phosphorylation and the basal consumption of ATP. The smaller the respiratory reserve, the greater the likelihood that episodic demands on metabolism will cause cellular ATP to fall, creating a failure in many important ATP-dependent processes.[15] Reducing the arbor of cultured SNpc DA neurons increased their survival after exposure to mitochondrial poisons. Thus the extent of the axonal arbor is a main determinant of their selective vulnerability to mitochondrial pathology.

These studies illuminate a fundamental problem faced in all neurons, but particularly those of the SNpc: the distribution of power. How does one maintain an adequate supply of healthy, functioning mitochondria in vast arbors of axon and dendrite? How do these cells or any large neuron match their energy demands with supply? If all ATP were to be synthesized by mitochondria in the cell body, it would take several years for ATP to diffuse to the extremities of the cell. Diffusion is obviously inadequate and, not surprisingly, instead of shipping the final product, neurons evolved to send the ATP factories themselves. As neurons develop, mitochondria are transported into the growing axons and dendrites[16–20] and positioned so as to match the energy supply to energy demand throughout the arbor. This distribution must involve still unknown mechanisms for localizing mitochondria to synapses. However, like all factories, mitochondria have a limited useful lifetime and their protein components must turn over on a regular basis. The half-lives of mitochondrial proteins in the brain fall in the range of 1 week to 2 months,[21] thus mitochondrial replenishing is a constant and ongoing need. At present, there is some evidence for synthesis of nuclear-encoded mitochondrial proteins outside the neuronal soma,[22-24] but most mitochondrial biogenesis is still presumed to be in the soma. Consequently, without efficient, high volume, motor-based movement, the neuron cannot provide newly synthesized components from the protein synthesis apparatus in the soma to the energy-hungry axon, dendrites, and synapses. Mitochondrial densities in axons in vivo range from 65/mm in zebrafish[25] to about 200/mm in the mouse corpus collosum.[26] That would amount to between 300,000 and 1,000,000 mitochondria in the axons of a human SNpc neuron, without taking into account the mitochondria at the millions of synapses formed by each axon. With a conservative estimate of 2,000,000 mitochondria per neuron of the SNpc and a very generous estimate of 60 days for the lifetime of each mitochondrion, that implies a replacement rate of 33,000/day or the equivalent of over 20 newly synthesized mitochondria leaving the soma every minute for the long journey to their destination in the periphery. Efficient mitochondrial transport is indeed essential for the maintenance of these cells.

3 MITOCHONDRIAL MOVEMENT IS ESSENTIAL FOR MAINTAINING A HEALTHY MITOCHONDRIAL POPULATION IN NEURONS

The "back of the napkin" calculation in the preceding section and its estimate of the rate of mitochondrial transport in an axon is of course an oversimplification of the mechanisms involved in maintaining healthy mitochondria in a neuron. In this section we will consider the role of mitochondrial dynamics in mitochondrial health in more detail. Four main processes are likely to act in a coordinated fashion in axons and dendrites: mitochondrial movement, mitochondrial fission and fusion, mitochondrial biogenesis, and mitophagy. The balance of these processes will govern the rate at which neuronal mitochondria are replaced and, as we have seen, when we are discussing neuronal mitochondria it is overwhelmingly a question of axonal mitochondria.

The preceding section already introduced the importance of mitochondrial movement for moving mitochondria into the axon and dendrites, but the phenomenology of mitochondrial movement is quite complex. Although in many cell types and within the neuronal soma mitochondria may form a highly branched and interconnected reticulum, axonal mitochondria exist as discrete organelles roughly 1–3 μm long. From numerous studies, both in vivo and in vitro, it is clear that this axonal population reflects several distinct states or classes. Some mitochondria appear not to move at all, at least for the feasible duration of experimental observation and these are referred to as the stationary pool. Other mitochondria form a motile pool and these mitochondria may move either anterograde (toward the periphery of the neuron) or retrograde (toward the soma) or pause for seconds or minutes. The temporary pauses of the motile pool, however, are clearly distinct from the long-term immobility of the stationary pool. The biochemical correlates of these distinct states are not completely worked out, but it is apparent that all mitochondria, including the stationary pool, are associated with motor proteins[27] but stationary mitochondria also contain an anchoring protein, syntaphilin, that can tether them to microtubules.[28] The percent of mitochondria that are in the motile pool may vary during development and between cell types. Measurements from cultured mammalian hippocampal neurons, or in vivo measurements from mammalian PNS or CNS neurons, or an identified neuron in zebrafish larvae, found that the motile pool in primary axons represents between 10% and 40% of the mitochondrial population.[27,29–32] This number may be higher in developing neurons and young animals and decline with age[18,33] and once synapses are established they often acquire stationary mitochondria and this pool will therefore dominate, both pre and postsynaptically, in synapse rich regions.[19,20]

The motile pool itself may represent two distinct subdivisions: those with an overall anterograde trajectory and those whose movement is overall retrograde. Though mitochondria may upon occasion reverse directions, it has been observed that a mitochondrion moving in a given direction is most likely to continue in the same direction after a pause or after a transient reversal.[30] On the basis of measurements of mitochondrial membrane potential, it has been suggested that mitochondria travelling in the retrograde direction may be older, or damaged, and have compromised membrane potentials[34,35] and therefore are returning to the soma, but others have failed to see any difference in membrane potential for mitochondria moving anterograde versus retrograde.[36] One experiment on cultured neurons employed mito-Timer, a mitochondrially targeted fluorescent protein that undergoes a time-dependent shift from green to red.[37,38] As expected, peripheral mitochondria appeared older (more red), but the investigators never observed an old, (red) mitochondrion moving back into the neuronal soma as would be expected if the retrograde population were the aging organelles from the periphery. Thus, at present, it is unclear if there is a large distinction between the mitochondria with anterograde versus retrograde bias, nor is it understood how that bias is produced.

The stationary pool of mitochondria presents a different challenge. Since, as stated in section 2, mitochondrial proteins have a half-life measured in weeks, how are the contents of stationary mitochondria refreshed? One possibility is that the stationary mitochondria convert to the motile pool and are replaced by motile mitochondria that are then captured into the stationary pool. These events might be sufficiently rare as to be invisible to standard imaging experiments. A more likely explanation is presented by the fact that mitochondria can undergo fusion and fission, processes that allow the exchange of materials between them[39–41] (and Chapter 3). Even brief moments of contact can allow exchange of proteins.[41] Mitochondria appear to go through cycles of fusion and fission and undergoing fusion is a strong predictor of a subsequent fission event. Thus if a motile mitochondrion fuses in passing with a stationary mitochondrion the exchange of material between the two, while fused, can cause the stationary organelle to acquire recently synthesized protein. Thus, the motile mitochondrial population can be compared to the space shuttle that provides essential replacement parts to the international space station, which remains in orbit.

Evidence for this model is abundant. Fusion events between moving and stationary mitochondria are frequently observed during live video imaging of axons and were particularly evident when a photoactivatable protein was used to mark one of the mitochondria.[42] Moreover, when neurons express a mitochondrially targeted fluorescent protein, such as mito-GFP or mito-RFP, all the mitochondria in the axon are labeled; one never observes a subset of unlabeled stationary mitochondria that failed to receive the marker. Finally, when mitoTimer was used to distinguish recently synthesize mitochondria from older ones, those closest to the soma were the youngest and the apparent age of the organelles increased linearly with the length of the axon. The smooth gradient of age, in contrast to intermingled populations of markedly older and younger mitochondria, implied the continuous mixing of components between the pools, as expected from ongoing fission and fusion. That those distal mitochondria were, on average, older likely represents both the aging of the mitoTimer while it was transported and the dilution of "young" mitoTimer in the motile pool due to fusions with mitochondria en route. When mitochondrial transport was interrupted, "old" mitoTimer proteins were seen more proximal to the soma than in the control, supporting a role for mitochondrial transport in mitochondrial renewal.[43] Fusion and fission are not independent, of course, of mitochondrial movement; some movement is necessary if the mitochondria are to encounter one another and it has been observed that, in axons, the probability of a single mitochondrion fusing is dictated primarily by its motility.[42] Moreover, the mechanisms of movement and fusion may be mechanistically coupled.[44]

Genetics has demonstrated the importance of fission and fusion for the axonal pool of mitochondria. Fission is mediated by Drp1 and *Drosophila* neurons lacking this factor cannot transport mitochondria to nerve terminals, presumably because the mitochondria cannot split away from the reticulum found in the soma.[45] Mutations in the human Mitofusin2 gene cause a form of Charcot-Marie-Tooth peripheral neuropathy[46,47] and mutations in mice cause a loss of mitochondrial DNA and membrane potential and consequent defects in dendritic outgrowth, spine formation and neuronal survival.[48] Since fission and fusion cannot occur without mitochondrial movement, their link to neuropathy reinforces the importance of mitochondrial transport to neuronal survival.

Movement and fusions alone, however, cannot replenish peripheral mitochondria if no new mitochondrial components are being synthesized. The third factor to consider, therefore is mitochondrial biogenesis which we have so far presumed to occur in the soma which is indeed the major site of synthesis of mitochondrial proteins.[49] Most mitochondrial proteins are encoded by nuclear genes and evidence for protein synthesis in the axons of mature neurons remains a contentious issue. Nevertheless, it is possible that some local biogenesis of mitochondria can, in part, reduce the requirement for transport from the neuronal soma. The ability of mitochondrial DNA to replicate in the periphery was demonstrated by BrdU incorporation in severed axons.[40] In addition, axonal synthesis of mitochondrial proteins has been observed for several proteins, both in vitro and in vivo.[23,24] Moreover, in analyses of the mRNA pool found in isolated axons, mRNA coding for mitochondrial proteins is the predominant category, and accounts for between 20% and 30% of the total.[24,50–52] Inhibition of axonal protein synthesis can reduce axonal mitochondrial membrane potential.[22] Thus, it is likely that at least some parts of the mitochondrial proteome can be replaced in situ by local translation. The extent to which this replaces mitochondrial transport from the soma is not yet clear, nor it is understood how mitochondrial mRNA are transported to enable local translation.

The final component for understanding how mitochondrial health is maintained in the periphery is the removal of old mitochondria or mitochondrial components. This is a complex subject, reviewed elsewhere (Chapter 5), but also closely linked to the question of mitochondrial movement. In brief, mitochondrial proteins can be degraded individually by mitochondrial proteases, or mitochondria can fragment and be engulfed en masse by mitophagy. An intermediate process, involving the shedding of mitochondrial-derived vesicles for targeting to lysosomes also occurs in cell lines and potentially in neurons.[53,54]

The question arises, however, whether these processes occur locally in the axon or require transport to the soma. The answer may depend on the circumstances under which the mitochondria are being cleared. One study examined autophagy in the growth cones of cultured DRG neurons and observed engulfment of mitochondria in the terminal, which was followed by the rapid retrograde movement of the autophagosome. As the autophagosome moved back toward the soma, it appeared to acidify and acquire lysosomal properties.[55,56] A contrasting study examined acutely damaged or depolarized mitochondria in the axons of cultured hippocampal neurons and found that the damaged mitochondria did not move and underwent mitophagy and degradation locally. The damaged mitochondria were enclosed in a stationary autophagosomal membrane and motile Lamp1-positive lysosomes then fused with the autophagosome and mitochondrial markers were lost.[57] The process was rapid: within 30–60 min local mitophagy had eliminated the mitochondrial marker. Thus, although axonal lysosomes are sparser than in neuronal cell bodies and may contain lower levels of cathepsins,[58] both growth cones and axons are capable of very rapid local degradation that does not require transport prior to mitophagy. Whether the autophagosome is motile or stationary may depend on whether it is formed in the growth cone or midaxon and whether or not it is part of a constitutive process or triggered by acute damage. These findings of local mitophagy are consistent with the experiments with mitoTimer mentioned previously which did not see older mitochondria arriving in the cell soma.[38] Local mitophagy may also explain an imbalance in anterograde and retrograde movement that was observed in a very quantitative study of mitochondrial dynamics and flux in zebrafish.[25] The number of mitochondria traveling down the primary axon was not matched by those returning, suggesting that some may have been destroyed in the periphery. On the other hand, a longer-term exposure of entire neurons to depolarizing agents observed a preponderance mitophagic profiles in the cell body.[59,60] With these longer-term treatments it is possible that the rapid phase of local clearance was already completed but that the remaining mitochondria underwent retrograde transport to the soma for degradation. Thus for mitophagy, the movement of mitochondria may not be as crucial as the movement of autophagosomes and lysosomes.

Thus, mitochondrial trafficking is critical to the maintenance of a healthy pool of mitochondria in axons, dendrites, and synapses (Fig. 4.1).

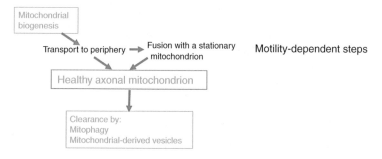

FIGURE 4.1 **Maintenance of healthy axonal mitochondria.** Supporting the mitochondrial population of axons over the life of an organism is a complex process that requires biogenesis of new mitochondrial proteins, lipids, and DNA, and the clearance of older or damaged components. Movement of mitochondria is almost certainly a crucial part of the quality control and resupply mechanism and the steps most dependent on mitochondrial motility (purple) include transport from the major site of biogenesis in the soma and fusion with existing stationary mitochondria in the axon.

Transport of mitochondria will bring newly synthesized proteins from the cell body to the periphery and, by fusion and fission events, can refresh the components of the stationary mitochondrial pool. Although some local biogenesis may complement what is delivered by movement from the soma, transport remains crucial. For clearance of damaged or old mitochondria, the picture is more complicated and what movement occurs may take place largely after the mitochondrion has been enclosed in an autophagosomal membrane. Indeed, as will be reviewed later, there is evidence that mitochondrial damage will stop mitochondrial transport as a preliminary step toward mitophagy and this process may be impaired in forms of PD.[57,61]

4 A MOTOR/ADAPTOR COMPLEX ENABLES AND REGULATES MITOCHONDRIAL TRAFFICKING

Mitochondrial movement requires motor proteins to be attached to the surface of the mitochondrion. This is accomplished by a protein complex that mediates mitochondrial transport in neurons and probably in most animal cells (Fig. 4.2). The complex is both essential for the maintenance of neuronal health and also the target of the PD genes PINK1 and Parkin.

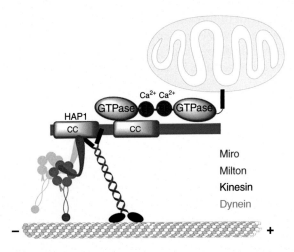

FIGURE 4.2 **The motor/adaptor complex for mitochondrial transport.** A single adaptor complex mediates the attachment of both the anterograde kinesin motor and the retrograde dynein motor. The motors are attached to the surface of the mitochondrion by binding to milton and Miro. Milton has two coiled-coil domains (CC), one of which has homology to huntingtin-associated protein 1 (HAP1). Miro has two GTPase domains separated by Ca^{2+}-binding EF hands. The C-terminal transmembrane domain of Miro tethers this complex to the outer mitochondrial membrane. *Source: Image courtesy of Himanish Basu.*

Its core is composed of two proteins, a mitochondrial GTPase Miro (also called RhoT1 and RhoT2), and the adaptor protein milton (also known as TRAK1 and TRAK2).[62-67] Milton and Miro form the mitochondrial receptors or adaptors that bind the necessary motors: conventional kinesin-1 heavy chain, which moves the mitochondria in the direction of the plus ends of microtubules (ie, anterograde, toward axon terminals) and the dynein/dynactin complex which provides retrograde transport in axons back to the soma.[68-70] The stoichiometry of the complex is not yet known, but the properties of the individual proteins and their relevance to mitochondrial transport has been elucidated through a combination of *Drosophila* genetics, biochemistry, and imaging.

The kinesin heavy chain (KHC) motor has three isoforms in mammals: Kif5A, B, and C and, though it may vary with cell type and species, all have been implicated in mitochondrial transport.[70-72] Conventionally, the Kinesin-1 motor is a tetramer in which two of these heavy chains form a dimer and bind to two kinesin light chains (KLC). The cargo associates with the motor via the KLC chains. In the case of mitochondria, however, milton replaces the light chain as the adaptor protein.[67] Milton also coprecipitates with components of the dynein/dynactin complex[73] and this serves to link both classes of motor to mitochondria. Although milton is necessary for mitochondrial movement into axons and dendrites, it is not otherwise required for mitochondrial function; in *Drosophila* photoreceptors homozygous for the mutation, for example, axons and nerve terminals are devoid of mitochondria, but functional and morphologically normal mitochondria remain in the soma.[62,74] The two mammalian isoforms of milton are often called TRAK1 and TRAK2 and the ability of milton to alter mitochondrial distribution and associate with motors is shared by the fly and mammalian homologs.[62,65,67,75,76] Distinct isoforms in mammals, and alternatively spliced isoforms in both mammals and flies, appear to be important in regulating mitochondrial motility.[67,73] Milton lacks an obvious predicted structure other than an extensive coiled-coil region with a significant region of homology to huntingtin-associated protein (HAP-1), a protein also implicated in organelle traffic.[77] Although Milton cofractionates with mitochondria, it lacks a transmembrane domain. Instead, its association with the mitochondrion is mediated by its binding to Miro.[62,67]

Miro is both the membrane anchor for the motor adaptor complex[67] and a regulator of motility. Consequently, as discussed later, the ability of PINK1 and Parkin to modify Miro and cause it to be degraded is of considerable importance. Miro has two domains of homology with small GTPases separated by a linker with four EF hand motifs and a transmembrane domain at the carboxyl terminus that tethers it to the outer mitochondrial membrane.[63,64,78] There is a single fly Miro[66] but two mammalian isoforms of Miro termed RhoT1 and RhoT2.[79] *Drosophila* without Miro lack axonal and synaptic mitochondria, phenotypes resembling milton mutant phenotypes.[66] Milton and Miro are direct binding partners, and a truncated Miro lacking its transmembrane anchor can serve as a dominant negative construct[64,67] by preventing milton and the motor proteins from associating with the mitochondrial surface. Both *milton* and *miro* fly mutants show bidirectional transport defects.[30,62,66,80]

Despite decades of research, we still know comparatively little about the exact mechanisms that orchestrate mitochondrial motility in axons and dendrites, what determines in which direction they will move or under what circumstances they pause, nor how they cluster at synapses,[68,81–83] in growth cones and branches,[17,84] at Nodes of Ranvier,[85] and at myelination boundaries.[86] Nevertheless, several signaling pathways within neurons can control mitochondrial transport by interacting with milton and Miro. These pathways include increases in cytosolic Ca^{2+},[87–90] the glucose-responsive enzyme O-GlcNAc Transferase,[31] the axon-specific protein syntaphilin which is largely responsible for the stationary pool in vertebrate axons,[91] trophic factors such as NGF,[16,92] the hypoxia-responsive protein HUMMR[93] and a cluster of genes collectively known as the Armcx family.[94] This extensive regulatory control of mitochondrial dynamics underscores its overall importance to neuronal function and why misregulation or disruption of mitochondrial motility can be deleterious for a neuron in PD or other neurodegenerative disorders.

5 LINKING PD AND MITOCHONDRIAL TRANSPORT: MITOCHONDRIAL DYSFUNCTION CAUSES MITOCHONDRIAL ARREST

Having reviewed the importance of mitochondrial motility to neuronal health, this chapter can now turn to the critical question of how the agents and genes that cause PD interact with mitochondrial transport. Mitochondria themselves have been the suspect in PD for long. One of the initial links was the observation that 1-methyl-4-phenyl-1,2,3,4-tetrahydropyridine (MPTP), caused DOPA- responsive Parkinsonism with dopaminergic neurons loss[95] and was a potent inhibitor of complex I.[96] As reviewed in Chapter 2, several other mitochondrial toxins are now known to cause PD-like symptoms including paraquat, 1-methyl-4-phenylpyridinium (MPP+), and rotenone. Recent data have provided a direct link between mitotoxins and mitochondrial transport deficits: MPP+[97] affects axonal transport of mitochondria when the drug is applied specifically in axons using microfluidic devices. Mitochondrial arrest was clearly secondary to the compromised mitochondrial function.

The most direct and mechanistic link between PD, mitochondrial damage, and mitochondrial arrest comes from the study of PINK1 (PTEN-induced putative kinase; encoded by the human *PARK6* gene) and Parkin (encoded by *PARK2*). As presented in detail in Chapter 5 and reviewed elsewhere[98–100] these proteins are part of a pathway that triggers mitophagic clearance of mitochondria with compromised membrane potentials or accumulations of misfolded proteins. PINK1 is a mitochondrial kinase located on chromosome 1p36 and linked to autosomal recessive juvenile parkinsonism (ARJP).[101] More than 70 PINK1 mutations have been identified that can cause parkinsonism. Many PINK1 mutations alter or eliminate the kinase domain and are therefore loss of function mutations, and at least some disease-causing mutations affect the mitochondria-targeting domain as reviewed in Refs.[102–104] Parkin is a cytosolic E3 ligase and is also a causative gene for ARJP,[105] with over 100 pathogenic mutations reported to date.[103] Pioneer genetic studies in *Drosophila* and subsequent analysis in mammalian cells and rodent models have demonstrated that PINK1 and Parkin act in a linear pathway for mitochondrial quality control. This pathway also triggers mitochondrial arrest.[61]

In brief, PINK1 is normally kept at low levels and rapidly inactivated in healthy cells by being imported into mitochondria via the TOM/TIM complexes. This import triggers its cleavage and degradation by the mitochondrial proteases such as Mitochondrial Processing Protease (MPP) and Presenilin-Associated Rhomboid Like protease (PARL), and, after retrotranslocation of fragments back to the cytosol, by the N-rule pathway.[106–111] Since mitochondrial import requires the mitochondrial membrane potential and can be inhibited by misfolded proteins, these conditions will prevent degradation of PINK1 and stabilize it on the surface of the mitochondrion. Thus the import and degradation of PINK1 becomes a sensor detecting dysfunctional mitochondria. Among the major effects of PINK1 stabilization is phosphorylation and activation of Parkin, recruitment of Parkin from the cytosol to the mitochondrial surface, and phosphorylation of ubiquitin on proteins of the outer mitochondrial membrane.[112–114] The consequential increase in phosphoubiquitination activates the pathway for mitophagy.[115,116] Parkin is also important for mitochondrial quality control through the more restricted mechanism of producing mitochondrial-derived vesicles.[54]

PINK1 and Parkin also regulate mitochondrial transport (Fig. 4.3) and this provides an intriguingly direct link between PD and mitochondrial transport. An unbiased approach to identify PINK1 interactors by coimmunoprecipitation indicated that Miro2/RhoT2 and Milton formed a complex with PINK1.[117] It was later found that Parkin can also bind this complex, and the association of both PINK1 and Parkin with the motor/adaptor complex can be greatly increased if mitochondria are depolarized to activate PINK1.[61] PINK1 can phosphorylate Miro and Parkin can ubiquitinate Miro and together these actions cause the proteasomal degradation of Miro.[61,118] The ability of proteasomal inhibitors to prevent the degradation of Miro distinguishes it from the mitophagic destruction of mitochondrial proteins, which depends on lysosomes and not the proteasome.

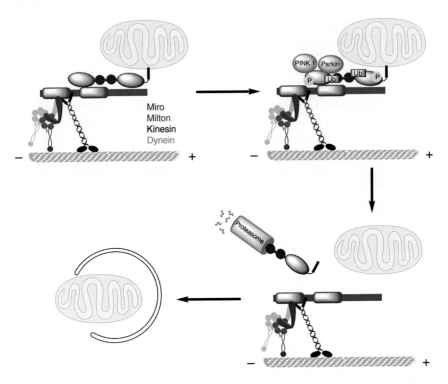

FIGURE 4.3 **PINK1 and Parkin stop damaged mitochondria prior to mitophagy.** Healthy mitochondria express the motor/adaptor complex of Miro, milton, kinesin, and dynein on their outer membrane. Depolarization of the mitochondrion causes PINK1 and Parkin to bind and phosphorylate Miro. The subsequent degradation of Miro by the proteasome releases milton and the motors from the organelle which can subsequently be surrounded by autophagosomal membranes. *Source: Image courtesy of Himanish Basu.*

Moreover, the loss of Miro occurs in conditions and at time points when mitophagy is not observed. Since Miro is required to anchor milton and the motor proteins to the mitochondrial membrane, degradation of Miro releases them to the cytosol and thereby arrests mitochondrial movement irreversibly (Fig. 4.3). Overexpression of either PINK1 or Parkin induces mitochondrial transport arrest in rat hippocampal neurons and in *Drosophila* motor neurons.[61] PINK1 action requires the presence of Parkin, but Parkin overexpression can arrest mitochondrial motility even in a PINK1 null genetic background in either *Drosophila* or rodent models, as expected for a pathway in which PINK1 acts upstream of Parkin.

It is not clear at present whether this pathway is responsible for an ongoing regulation of mitochondrial motility or is only brought into play as a prelude to mitophagy or upon overexpression of the proteins. Cultured hippocampal neurons from PINK1 or Parkin knockout animals had normal levels of mitochondrial movement and normal velocities.[61]

In *Drosophila*, RNAi knockdown of PINK1 and Parkin increased mitochondrial motility in one study[61] but not in another.[119] The significance of *Drosophila* Miro as a PINK1 target, however, was demonstrated by the observation that downregulation of Miro was enough to rescue PINK1 mutant-induced degeneration of DA neurons.[118] This finding suggests that the misregulated movement of mitochondria in the absence of PINK1 can lead to neuronal death. Similarly, in flies where Miro has been rendered phosphorylation-resistant by mutating putative PINK1 phosphorylation sites to alanines, mitochondrial transport was increased and dopaminergic neurons were decreased.[120] Taken together, these observations suggest that the regulation of Miro by the PINK1/Parkin pathway regulates mitochondrial dynamics in *Drosophila* neurons, and that this control is necessary for neuronal survival.

The most important function of Miro degradation by PINK1 and Parkin may be to arrest the movement of damaged mitochondria as a prelude to mitophagy. Ashrafi et al.[57] found that damaging a subset of mitochondria within axons—either through ROS production by mitoKillerRed or by exposure to antimycinA—caused mitochondria to stop moving, become fragmented, and subsequently engulfed by autophagosomes. All of this occurred locally within the axon and all required both PINK1 and Parkin. Arrest of movement by Miro degradation may work in parallel with another downstream action of PINK1 and Parkin: arrest of mitochondrial fusion by degradation of mitofusin. Mitofusin, like Miro, is a substrate for both PINK1 and Parkin and is rapidly degraded by the proteasome.[121–124] Thus it appears likely that, in damaged mitochondria with poor membrane potentials or misfolded proteins, PINK1 and Parkin cause mitochondria to stop moving and fusing with one another by quickly targeting Miro and mitofusin for degradation.

How is PINK1/Parkin control of mitochondrial dynamics thus linked to neuronal survival? The most straightforward hypothesis is that degradation of Miro and mitofusin serve to quarantine damaged mitochondria to prepare them for mitophagy. In the absence of PINK1-driven mitochondrial arrest, damaged mitochondria travel along the axon. The propagation of damaged mitochondria can cause damage by itself, as they are likely to act as sources of oxidative stress. Motile damaged mitochondria also can fuse with healthy organelles, effectively spreading the contamination of mitochondrial damage. Over time, damage can build up to a level that will compromise axonal survival.

The lessons from toxins, PINK1, and Parkin are thus twofold. Toxins may compromise mitochondrial health and hence neuronal survival by arresting mitochondria that should be moving. Mutations in PINK1 and Parkin, in contrast, may damage neurons by allowing damaged mitochondria to continue to move when they should be stopped and quarantined. On a mechanistic level, the two phenomena may be identical; it is not known if the arrest in mitochondrial transport observed when MPP+ or

6-OHDA is applied[97,125] is due to activation of PINK1 and Parkin and the degradation of Miro. Nevertheless, mitochondrial motility and the development of PD are clearly linked.

6 DO OTHER FORMS OF PD INVOLVE DEFECTS IN MITOCHONDRIAL MOVEMENT?

The pathological hallmark of PD is the formation of Lewy bodies, protein aggregates composed primarily of α-synuclein (α-syn), a small protein encoded by the SNCA gene.[126–128] A recent study on cortical neurons dissected from transgenic mice carrying the A53T mutation in α-syn showed pronounced bidirectional mitochondrial arrest.[129] The arrest was specific to mitochondria, as other organelle movement was not altered. However, mitochondria in these neurons also had reduced membrane potential and respiratory rate. Similarly, high levels of ROS in mitochondria were observed in iPSC-derived neurons from a patient with the A53T mutation.[130] Excessive ROS production can depolarize the mitochondria and cause mitochondrial arrest via the PINK1/Parkin pathway discussed earlier.[57,131] Therefore, until more is known about the mechanism of A53T it is not clear whether the mitochondrial arrest induced by the A53T mutation reflects a direct influence on mitochondrial motility, an indirect consequence of protein aggregates that physically hinder mitochondrial movement, or an indirect by-product of mitochondrial depolarization and consequent Miro degradation. However, because poor mitochondrial trafficking by itself will slow the replenishing of peripheral mitochondria, it can contribute to the decline of the mitochondrial population and generate a deleterious positive feed forward loop.

Mutations in the gene encoding leucine-rich repeat kinase 2 (LRRK2) are the most frequent cause of familiar PD.[132] At present there is less to link LRRK2 directly to mitochondria than has been found for other PD associated genes. Nevertheless, LRRK2 R1441C and Y1699C (Roc-COR mutations), but not G2019S mutations, sharply inhibited axonal transport of mitochondria in primary rodent neurons as well as in Drosophila neurons.[133] This effect can be at least partially explained by the reported interactions of LRRK2 with the tubulin cytoskeleton. The GTPase domain of LRRK2 can pull-down α/β tubulin from cell lysates and LRRK2 coprecipitates with β tubulin from brains.[134–136] LRRK2 with pathogenic Roc-COR mutations decorate deacetylated microtubules,[133,137] whereas the wild type or the pathogenic LRRK2-G2019S mutants do not.[133] Roc-COR mutations also cause deficits in locomotor behavior of flies. Impaired axonal transport of mitochondria in the presence of these mutations can at least be partially explained by their propensity to bind deacetylated tubulin: increasing α-tubulin acetylation, either pharmacologically or by using flies harboring mutations in microtubule deacetylases, inhibited the association of the pathogenic proteins with microtubules and rescued axonal transport of mitochondria and locomotor deficits.[133] Tubulin binding by LRRK2 mutations, whether acting as simple obstructions or by altering microtubule stability, might be expected to cause a general transport deficit, not specific to mitochondria. Nevertheless, the lack of mitochondrial trafficking may be one of the most pathological aspects by diminishing mitochondrial quality and energy supply.

Mutations in DJ-1 (*PARK7*) cause recessive familiar parkinsonism.[138] Over 20 pathogenic deletions and point mutations have been identified in the PARK7 gene.[103] DJ-1 is a multifunctional, ~22 kDa protein that works as a dimer, and is proposed to participate in antioxidative stress responses, among many other functions (reviewed in ref. 139). A portion of DJ-1 localizes to mitochondria under normal conditions,[140,141] and oxidative stress stimulates its mitochondrial translocation.[142] DJ-1 in the substantia nigra can limit excessive ROS production.[143] Pathogenic mutations prevent dimerization of DJ-1 and thereby interfere with its localization on mitochondria and antioxidative role. Aged flies mutant for DJ-1 show mitochondrial defects.[144] Interestingly, increased expression of DJ-1 can complement the loss of PINK1, but not of Parkin[144] suggesting that DJ-1 acts in parallel or downstream to PINK1. Collectively, it seems likely that loss of DJ-1 can cause mild mitochondrial depolarization due to excessive ROS production, which would then lead to PINK/Parkin-dependent mitochondrial arrest.

In considering whether reported defects in mitochondrial motility are contributing to neuronal degeneration there is an important caveat. Mitochondrial motility can be affected by many stressors, and of course will be impaired when a cell is dying. Therefore, whenever a mitochondrial transport defect is observed, it is necessary to determine whether the change in transport is a cause of the degeneration or a late defect. In dying cells with low ATP or high cytosolic Ca^{2+} mitochondria will not move. There are several approaches to resolve this question. One method is to simultaneously track mitochondria and another axonal organelle, for example by coexpressing synaptophysin-GFP and mito-RFP to get a readout of the health of the axon; in a dying cell not only mitochondria but multiple cargoes will no longer be transported. Late endosome markers such as Rab7-GFP also give a strong signal to noise ratio and can be used to monitor the viability of the axon. In addition, it can be useful to monitor mitochondrial membrane potential using potentiometric dyes such as tetramethylrhodamine (TMRM). Membrane potential depends on the physical integrity of mitochondria as well as on mitochondrial function, and a significant drop in signal can be an early indicator of cell death. Ideally, both methods should be employed and attention should also be paid to the timing of the onset of transport defects relative to other signs of degeneration such as microtubule fragmentation or axon blebbing.

7 SUMMARY: THE VIRTUOUS AND VICIOUS CYCLES THAT LINK MITOCHONDRIAL TRANSPORT AND MITOCHONDRIAL HEALTH

How then are mitochondrial transport defects involved in the development of PD? A useful model to consider is one that illustrates the tight interrelationship of mitochondrial movement to mitochondrial quality (Fig. 4.4). This interrelationship can function either as a virtuous cycle that supports mitochondrial health and thus neuronal survival or as a vicious cycle in which defects in movement and quality reinforce one another in a downward spiral that ultimately can kill a neuron. In the healthy cell, mitochondria have good membrane potentials and few misfolded proteins. Consequently, the PINK1/Parkin pathway is switched off unless a particularly old mitochondrion needs to be targeted for mitophagy. If the pathway is switched off, Miro will remain on the mitochondrial surface and mitochondria will move both anterograde and retrograde. This movement will support resupply of fresh mitochondria from the soma to the vast regions of axons and dendrites and will permit the fission and fusion reactions that keep even the stationary pool of mitochondria healthy. Healthy movement permits healthy mitochondria to be maintained in axons, which in turn permits continued axonal transport of mitochondria. However, if either mitochondrial quality or motility is compromised,

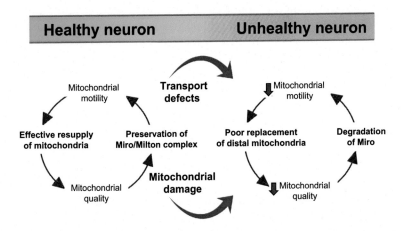

FIGURE 4.4 **Virtuous and Vicious Cycles Couple Mitochondrial Health and Mitochondrial Motility.** The interdependence of mitochondrial health and movement is illustrated by two cycles representing either a mutually reinforcing beneficial effect in healthy neurons or a downward spiral in unhealthy neurons. An initial decrease in either mitochondrial quality or mitochondrial transport can shift a cell to the unhealthy cycle. In PD, this can occur either because of toxic or genetic damage to mitochondria or by a disruption of axonal transport and will involve the degradation of Miro by the PINK1/Parkin pathway.

whether by a toxin, a mutation, ROS, DNA damage, or just the aging of the mitochondrial pool, the cycle will operate in reverse and reinforce the decline of both quality and motility. Some stressors may act first on transport and thereby inhibit the replenishing of axonal mitochondria, whose lower membrane potentials will trigger Miro degradation and the further inhibition of motility. Other stressors may first compromise mitochondrial protein folding or membrane potential and thereby inhibit motility as a secondary effect. But once a threshold is crossed, decreased motility will further diminish mitochondrial quality and the vicious cycle toward neuronal degeneration will continue. The larger the cell, the greater the requirement for axonal transport of mitochondria and hence the risk of slipping into the vicious cycle. It is very tempting to hypothesize that this contributes significantly to the vulnerability of the dopaminergic SNpc neurons in particular and, given the association of PD mutations with changes in mitochondrial transport, to the broader development of PD.

References

1. Zhu X-H, et al. Quantitative imaging of energy expenditure in human brain. *NeuroImage*. 2012;60(4):2107–2117.
2. Erecinska M, Silver IA. Ions and energy in mammalian brain. *Prog Neurobiol*. 1994;43(1):37–71.
3. Mink JW, Blumenschine RJ, Adams DB. Ratio of central nervous system to body metabolism in vertebrates: its constancy and functional basis. *Am J Physiol*. 1981;241(3):R203–R212.
4. Silver I, Erecinska M. Oxygen and ion concentrations in normoxic and hypoxic brain cells. *Adv Exp Med Biol*. 1998;454:7–16.
5. Harris JJ, Jolivet R, Attwell D. Synaptic energy use and supply. *Neuron*. 2012;75(5):762–777.
6. Matsuda W, et al. Single nigrostriatal dopaminergic neurons form widely spread and highly dense axonal arborizations in the neostriatum. *J Neurosci*. 2009;29(2):444–453.
7. Bolam JP, Pissadaki EK. Living on the edge with too many mouths to feed: why dopamine neurons die. *J Mov Diord*. 2012;27(12):1478–1483.
8. Silva NL, Bunney BS. Intracellular studies of dopamine neurons in vitro: pacemakers modulated by dopamine. *Eur J Pharmacol*. 1988;149(3):307–315.
9. Puopolo M, Raviola E, Bean BP. Roles of subthreshold calcium current and sodium current in spontaneous firing of mouse midbrain dopamine neurons. *J Neurosci*. 2007;27(3):645–656.
10. Guzman JN, Sanchez-Padilla J, Chan CS, Surmeier DJ. Robust pacemaking in substantia nigra dopaminergic neurons. *J Neurosci*. 2009;29(35):11011–11019.
11. Surmeier DJ, Schumacker PT. Calcium, bioenergetics, and neuronal vulnerability in Parkinson's disease. *J Biol Chem*. 2013;288(15):10736–10741.
12. Pacelli C, et al. Elevated mitochondrial bioenergetics and axonal arborization size are key contributors to the vulnerability of dopamine neurons. *Curr Biol*. 2015;25(18):2349–2360.
13. German DC, Manaye KF, Sonsalla PK, Brooks BA. Midbrain dopaminergic cell loss in Parkinson's disease and MPTP-induced parkinsonism: sparing of calbindin-D28k-containing cells. *Ann NY Acad Sci*. 1992;648:42–62.

14. Kish SJ, Shannak K, Hornykiewicz O. Uneven pattern of dopamine loss in the striatum of patients with idiopathic Parkinson's disease. Pathophysiologic and clinical implications. *N Engl J Med*. 1988;318(14):876–880.
15. Nicholls DG. Oxidative stress and energy crises in neuronal dysfunction. *Ann NY Acad Sci*. 2008;1147:53–60.
16. Chada SR, Hollenbeck PJ. Mitochondrial movement and positioning in axons: the role of growth factor signaling. *J Exp Biol*. 2003;206(Pt 12):1985–1992.
17. Morris RL, Hollenbeck PJ. The regulation of bidirectional mitochondrial transport is coordinated with axonal outgrowth. *J Cell Sci*. 1993;104(Pt 3):917–927.
18. Chang DT, Honick AS, Reynolds IJ. Mitochondrial trafficking to synapses in cultured primary cortical neurons. *J Neurosci*. 2006;26(26):7035–7045.
19. Courchet J, et al. Terminal axon branching is regulated by the LKB1-NUAK1 kinase pathway via presynaptic mitochondrial capture. *Cell*. 2013;153(7):1510–1525.
20. Faits MC, Zhang C, Soto F, Kerschensteiner D. Dendritic mitochondria reach stable positions during circuit development. *Elife*. 2016;5:e11583.
21. Vincow ES, et al. The PINK1-Parkin pathway promotes both mitophagy and selective respiratory chain turnover in vivo. *Proc Natl Acad Sci*. 2013;110(16):6400–6405.
22. Hillefors M, Gioio AE, Mameza MG, Kaplan BB. Axon viability and mitochondrial function are dependent on local protein synthesis in sympathetic neurons. *Cell Mol Neurobiol*. 2007;27(6):701–716.
23. Yoon BC, et al. Local translation of extranuclear lamin B promotes axon maintenance. *Cell*. 2012;148(4):752–764.
24. Gioio AE, et al. Local synthesis of nuclear-encoded mitochondrial proteins in the presynaptic nerve terminal. *J Neurosci Res*. 2001;64(5):447–453.
25. Plucinska G, et al. In vivo imaging of disease-related mitochondrial dynamics in a vertebrate model system. *J Neurosci*. 2012;32(46):16203–16212.
26. Ohno N, et al. Mitochondrial immobilization mediated by syntaphilin facilitates survival of demyelinated axons. *Proc Natl Acad Sci*. 2014;111(27):9953–9958.
27. Wang X, Schwarz TL. The mechanism of Ca^{2+}-dependent regulation of kinesin-mediated mitochondrial motility. *Cell*. 2009;136(1):163–174.
28. Kang J-S, et al. Docking of axonal mitochondria by syntaphilin controls their mobility and affects short-term facilitation. *Cell*. 2008;132(1):137–148.
29. Misgeld T, Kerschensteiner M, Bareyre FM, Burgess RW, Lichtman JW. Imaging axonal transport of mitochondria in vivo. *Nature Methods*. 2007;4(7):559–561.
30. Russo GJ, et al. *Drosophila* Miro is required for both anterograde and retrograde axonal mitochondrial transport. *J Neurosci*. 2009;29(17):5443–5455.
31. Pekkurnaz G, Trinidad JC, Wang X, Kong D, Schwarz TL. Glucose regulates mitochondrial motility via milton modification by O-GlcNAc transferase. *Cell*. 2014;158(1):54–68.
32. Sorbara CD, et al. Pervasive axonal transport deficits in multiple sclerosis models. *Neuron*. 2014;84(6):1183–1190.
33. Vagnoni A, Hoffmann PC, Bullock SL. Reducing Lissencephaly-1 levels augments mitochondrial transport and has a protective effect in adult *Drosophila* neurons. *J Cell Sci*. 2016;129(1):178–190.
34. Fahim MA, Lasek RJ, Brady ST, Hodge AJ. AVEC-DIC and electron microscopic analyses of axonally transported particles in cold-blocked squid giant axons. *J Neurocytol*. 1985;14(5):689–704.
35. Miller KE, Sheetz MP. Axonal mitochondrial transport and potential are correlated. *J Cell Sci*. 2004;117(Pt 13):2791–2804.
36. Verburg J, Hollenbeck PJ. Mitochondrial membrane potential in axons increases with local nerve growth factor or semaphorin signaling. *J Neurosci*. 2008;28(33):8306–8315.
37. Hernandez G, et al. MitoTimer: a novel tool for monitoring mitochondrial turnover. *Autophagy*. 2013;9(11):1852–1861.

38. Ferree AW, et al. MitoTimer probe reveals the impact of autophagy, fusion, and motility on subcellular distribution of young and old mitochondrial protein and on relative mitochondrial protein age. *Autophagy.* 2013;9(11):1887–1896.

39. Detmer SA, Chan DC. Functions and dysfunctions of mitochondrial dynamics. *Nat Rev Mol Cell Biol.* 2007;8(11):870–879.

40. Amiri M, Hollenbeck PJ. Mitochondrial biogenesis in the axons of vertebrate peripheral neurons. *Dev Neurobiol.* 2008;68(11):1348–1361.

41. Liu X, Weaver D, Shirihai O, Hajnoczky G. Mitochondrial 'kiss-and-run': interplay between mitochondrial motility and fusion-fission dynamics. *EMBO J.* 2009;28(20):3074–3089.

42. Cagalinec M, et al. Principles of the mitochondrial fusion and fission cycle in neurons. *J Cell Sci.* 2013;126(Pt 10):2187–2197.

43. Ferree AW, et al. MitoTimer probe reveals the impact of autophagy, fusion, and motility on subcellular distribution of young and old mitochondrial protein and on relative mitochondrial protein age. *Autophagy.* 2013;9(11):1887–1896.

44. Misko A, Jiang S, Wegorzewska I, Milbrandt J, Baloh RH. Mitofusin 2 is necessary for transport of axonal mitochondria and interacts with the Miro/Milton complex. *J Neurosci.* 2010;30(12):4232–4240.

45. Verstreken P, et al. Synaptic mitochondria are critical for mobilization of reserve pool vesicles at *Drosophila* neuromuscular junctions. *Neuron.* 2005;47(3):365–378.

46. Zuchner S, et al. Mutations in the mitochondrial GTPase mitofusin 2 cause Charcot-Marie-Tooth neuropathy type 2A. *Nat Genet.* 2004;36(5):449–451.

47. Kijima K, et al. Mitochondrial GTPase mitofusin 2 mutation in Charcot-Marie-Tooth neuropathy type 2A. *Hum Genet.* 2005;116(1-2):23–27.

48. Chen H, McCaffery JM, Chan DC. Mitochondrial fusion protects against neurodegeneration in the cerebellum. *Cell.* 2007;130(3):548–562.

49. Davis AF, Clayton DA. In situ localization of mitochondrial DNA replication in intact mammalian cells. *J Cell Biol.* 1996;135(4):883–893.

50. Taylor AM, et al. Axonal mRNA in uninjured and regenerating cortical mammalian axons. *J Neurosci.* 2009;29(15):4697–4707.

51. Andreassi C, et al. An NGF-responsive element targets myo-inositol monophosphatase-1 mRNA to sympathetic neuron axons. *Nat Neurosci.* 2010;13(3):291–301.

52. Minis A, et al. Subcellular transcriptomics-dissection of the mRNA composition in the axonal compartment of sensory neurons. *Dev Neurobiol.* 2014;74(3):365–381.

53. Soubannier V, et al. A vesicular transport pathway shuttles cargo from mitochondria to lysosomes. *Cur Biol.* 2012;22(2):135–141.

54. McLelland GL, Soubannier V, Chen CX, McBride HM, Fon EA. Parkin and PINK1 function in a vesicular trafficking pathway regulating mitochondrial quality control. *EMBO J.* 2014;33(4):282–295.

55. Maday S, Wallace KE, Holzbaur EL. Autophagosomes initiate distally and mature during transport toward the cell soma in primary neurons. *J Cell Biol.* 2012;196(4):407–417.

56. Maday S, Holzbaur EL. Autophagosome biogenesis in primary neurons follows an ordered and spatially regulated pathway. *Dev Cell.* 2014;30(1):71–85.

57. Ashrafi G, Schlehe JS, LaVoie MJ, Schwarz TL. Mitophagy of damaged mitochondria occurs locally in distal neuronal axons and requires PINK1 and Parkin. *J Cell Biol.* 2014;206(5):655–670.

58. Gowrishankar S, et al. Massive accumulation of luminal protease-deficient axonal lysosomes at Alzheimer's disease amyloid plaques. *Proc Natl Acad Sci.* 2015;112(28):E3699–E3708.

59. Cai Q, Zakaria HM, Simone A, Sheng ZH. Spatial parkin translocation and degradation of damaged mitochondria via mitophagy in live cortical neurons. *Cur Biol.* 2012;22(6):545–552.

60. Bingol B, et al. The mitochondrial deubiquitinase USP30 opposes parkin-mediated mitophagy. *Nature*. 2014;510(7505):370–375.

61. Wang X, et al. PINK1 and parkin target Miro for phosphorylation and degradation to arrest mitochondrial motility. *Cell*. 2011;147(4):893–906.

62. Stowers RS, Megeath LJ, Gorska-Andrzejak J, Meinertzhagen IA, Schwarz TL. Axonal transport of mitochondria to synapses depends on milton, a novel *Drosophila* protein. *Neuron*. 2002;36(6):1063–1077.

63. Fransson A, Ruusala A, Aspenstrom P. Atypical Rho GTPases have roles in mitochondrial homeostasis and apoptosis. *J Biol Chem*. 2003;278(8):6495–6502.

64. Fransson S, Ruusala A, Aspenstrom P. The atypical Rho GTPases Miro-1 and Miro-2 have essential roles in mitochondrial trafficking. *Biochem Biophys Res Commun*. 2006;344(2):500–510.

65. Brickley K, Smith MJ, Beck M, Stephenson FA. GRIF-1 and OIP106, members of a novel gene family of coiled-coil domain proteins: association in vivo and in vitro with kinesin. *J Biol Chem*. 2005;280(15):14723–14732.

66. Guo X, et al. The GTPase dMiro is required for axonal transport of mitochondria to *Drosophila* synapses. *Neuron*. 2005;47(3):379–393.

67. Glater EE, Megeath LJ, Stowers RS, Schwarz TL. Axonal transport of mitochondria requires milton to recruit kinesin heavy chain and is light chain independent. *J Cell Biol*. 2006;173(4):545–557.

68. Hollenbeck PJ, Saxton WM. The axonal transport of mitochondria. *J Cell Sci*. 2005;118(Pt 23):5411–5419.

69. Pilling AD, Horiuchi D, Lively CM, Saxton WM. Kinesin-1 and Dynein are the primary motors for fast transport of mitochondria in *Drosophila* motor axons. *Mol Biol Cell*. 2006;17(4):2057–2068.

70. Tanaka Y, et al. Targeted disruption of mouse conventional kinesin heavy chain, kif5B, results in abnormal perinuclear clustering of mitochondria. *Cell*. 1998;93(7):1147–1158.

71. Kanai Y, et al. KIF5C, a novel neuronal kinesin enriched in motor neurons. *J Neurosci*. 2000;20(17):6374–6384.

72. Campbell PD, et al. Unique function of Kinesin Kif5A in localization of mitochondria in axons. *J Neurosci*. 2014;34(44):14717–14732.

73. van Spronsen M, et al. TRAK/Milton motor-adaptor proteins steer mitochondrial trafficking to axons and dendrites. *Neuron*. 2013;77(3):485–502.

74. Gorska-Andrzejak J, et al. Mitochondria are redistributed in *Drosophila* photoreceptors lacking milton, a kinesin-associated protein. *J Comp Neurol*. 2003;463(4):372–388.

75. Brickley K, Stephenson FA. Trafficking kinesin protein (TRAK)-mediated transport of mitochondria in axons of hippocampal neurons. *J Biol Chem*. 2011;286(20):18079–18092.

76. Smith MJ, Pozo K, Brickley K, Stephenson FA. Mapping the GRIF-1 binding domain of the kinesin, KIF5C, substantiates a role for GRIF-1 as an adaptor protein in the anterograde trafficking of cargoes. *J Biol Chem*. 2006;281(37):27216–27228.

77. Gauthier LR, et al. Huntingtin controls neurotrophic support and survival of neurons by enhancing BDNF vesicular transport along microtubules. *Cell*. 2004;118(1):127–138.

78. Klosowiak JL, et al. Structural coupling of the EF hand and C-terminal GTPase domains in the mitochondrial protein Miro. *EMBO Rep*. 2013;14(11):968–974.

79. Aspenstrom P, Ruusala A, Pacholsky D. Taking Rho GTPases to the next level: the cellular functions of atypical Rho GTPases. *Exp Cell Res*. 2007;313(17):3673–3679.

80. Babic M, et al. Miro's N-terminal GTPase domain is required for transport of mitochondria into axons and dendrites. *J Neurosci*. 2015;35(14):5754–5771.

81. Palay SL. Synapses in the central nervous system. *J Biophys Biochem Cytol*. 1956;2(4 Suppl):193–202.

82. Treeck HH, Pirsig W. Differentiation of nerve endings in the cochlear nucleus on morphological and experimental basis. *Acta Otolaryngol*. 1979;87(1-2):47–60.

83. Gotow T, Miyaguchi K, Hashimoto PH. Cytoplasmic architecture of the axon terminal: filamentous strands specifically associated with synaptic vesicles. *Neuroscience.* 1991;40(2):587–598.

84. Povlishock JT. The fine structure of the axons and growth cones of the human fetal cerebral cortex. *Brain Res.* 1976;114(3):379–1379.

85. Fabricius C, Berthold CH, Rydmark M. Axoplasmic organelles at nodes of Ranvier. II. Occurrence and distribution in large myelinated spinal cord axons of the adult cat. *J Neurocytol.* 1993;22(11):941–954.

86. Bristow EA, Griffiths PG, Andrews RM, Johnson MA, Turnbull DM. The distribution of mitochondrial activity in relation to optic nerve structure. *Arch Ophthalmol.* 2002;120(6):791–796.

87. Yi M, Weaver D, Hajnoczky G. Control of mitochondrial motility and distribution by the calcium signal: a homeostatic circuit. *J Cell Biol.* 2004;167(4):661–672.

88. Wang X, Schwarz TL. The mechanism of Ca2+-dependent regulation of kinesin-mediated mitochondrial motility. *Cell.* 2009;136(1):163–174.

89. Saotome M, et al. Bidirectional Ca2+-dependent control of mitochondrial dynamics by the Miro GTPase. *Proc Natl Acad Sci.* 2008;105(52):20728–20733.

90. Macaskill AF, et al. Miro1 is a calcium sensor for glutamate receptor-dependent localization of mitochondria at synapses. *Neuron.* 2009;61(4):541–555.

91. Kang JS, et al. Docking of axonal mitochondria by syntaphilin controls their mobility and affects short-term facilitation. *Cell.* 2008;132(1):137–148.

92. Chada SR, Hollenbeck PJ. Nerve growth factor signaling regulates motility and docking of axonal mitochondria. *Curr Biol.* 2004;14(14):1272–1276.

93. Li Y, et al. HUMMR, a hypoxia- and HIF-1alpha-inducible protein, alters mitochondrial distribution and transport. *J Cell Biol.* 2009;185(6):1065–1081.

94. Lopez-Domenech G, et al. The Eutherian Armcx genes regulate mitochondrial trafficking in neurons and interact with Miro and Trak2. *Nat Commun.* 2012;3:814.

95. Langston JW, Ballard P, Tetrud JW, Irwin I. Chronic Parkinsonism in humans due to a product of meperidine-analog synthesis. *Science.* 1983;219(4587):979–980.

96. Nicklas WJ, Youngster SK, Kindt MV, Heikkila RE. MPTP, MPP+ and mitochondrial function. *Life Sci.* 1987;40(8):721–729.

97. Kim-Han JS, Antenor-Dorsey JA, O'Malley KL. The parkinsonian mimetic, MPP+, specifically impairs mitochondrial transport in dopamine axons. *J Neurosci.* 2011;31(19):7212–7221.

98. Guo M. *Drosophila* as a model to study mitochondrial dysfunction in Parkinson's disease. *Cold Spring Harb Perspect Med.* 2012;2(11):1–17(a009944).

99. Jin SM, Youle RJ. PINK1- and Parkin-mediated mitophagy at a glance. *J Cell Sci.* 2012;125(Pt 4):795–799.

100. Pickrell AM, Youle RJ. The roles of PINK1, parkin, and mitochondrial fidelity in Parkinson's disease. *Neuron.* 2015;85(2):257–273.

101. Valente EM, et al. Hereditary early-onset Parkinson's disease caused by mutations in PINK1. *Science.* 2004;304(5674):1158–1160.

102. Deas E, Plun-Favreau H, Wood NW. PINK1 function in health and disease. *EMBO Mol Med.* 2009;1(3):152–165.

103. Nuytemans K, Theuns J, Cruts M, Van Broeckhoven C. Genetic etiology of Parkinson disease associated with mutations in the SNCA, PARK2, PINK1, PARK7, and LRRK2 genes: a mutation update. *Hum Mutat.* 2010;31(7):763–780.

104. Corti O, Lesage S, Brice A. What genetics tells us about the causes and mechanisms of Parkinson's disease. *Physiol Rev.* 2011;91(4):1161–1218.

105. Kitada T, et al. Mutations in the parkin gene cause autosomal recessive juvenile parkinsonism. *Nature.* 1998;392(6676):605–608.

106. Jin SM, et al. Mitochondrial membrane potential regulates PINK1 import and proteolytic destabilization by PARL. *J Cell Biol.* 2010;191(5):933–942.

107. Deas E, et al. PINK1 cleavage at position A103 by the mitochondrial protease PARL. *Hum Mol Genet.* 2011;20(5):867–879.

108. Meissner C, Lorenz H, Weihofen A, Selkoe DJ, Lemberg MK. The mitochondrial intramembrane protease PARL cleaves human Pink1 to regulate Pink1 trafficking. *J Neurochem.* 2011;117(5):856–867.

109. Shi G, et al. Functional alteration of PARL contributes to mitochondrial dysregulation in Parkinson's disease. *Hum Mol Genet.* 2011;20(10):1966–1974.

110. Greene AW, et al. Mitochondrial processing peptidase regulates PINK1 processing, import and Parkin recruitment. *EMBO Rep.* 2012;13(4):378–385.

111. Yamano K, Youle RJ. PINK1 is degraded through the N-end rule pathway. *Autophagy.* 2013;9(11):1758–1769.

112. Narendra D, Tanaka A, Suen DF, Youle RJ. Parkin is recruited selectively to impaired mitochondria and promotes their autophagy. *J Cell Biol.* 2008;183(5):795–803.

113. Narendra DP, et al. PINK1 is selectively stabilized on impaired mitochondria to activate Parkin. *PLoS Biol.* 2010;8(1):e1000298.

114. Kane LA, et al. PINK1 phosphorylates ubiquitin to activate Parkin E3 ubiquitin ligase activity. *J Cell Biol.* 2014;205(2):143–153.

115. Lazarou M, et al. The ubiquitin kinase PINK1 recruits autophagy receptors to induce mitophagy. *Nature.* 2015;524(7565):309–314.

116. Wong YC, Holzbaur EL. Optineurin is an autophagy receptor for damaged mitochondria in parkin-mediated mitophagy that is disrupted by an ALS-linked mutation. *Proc Natl Acad Sci.* 2014;111(42):E4439–E4448.

117. Weihofen A, Thomas KJ, Ostaszewski BL, Cookson MR, Selkoe DJ. Pink1 forms a multiprotein complex with Miro and Milton, linking Pink1 function to mitochondrial trafficking. *Biochemistry.* 2009;48(9):2045–2052.

118. Liu S, et al. Parkinson's disease-associated kinase PINK1 regulates Miro protein level and axonal transport of mitochondria. *PLoS Genet.* 2012;8(3):e1002537.

119. Devireddy S, Liu A, Lampe T, Hollenbeck PJ. The organization of mitochondrial quality control and life cycle in the nervous system in vivo in the absence of PINK1. *J Neurosci.* 2015;35(25):9391–9401.

120. Tsai P-I, et al. PINK1-mediated phosphorylation of Miro inhibits synaptic growth and protects dopaminergic neurons in *Drosophila. Sci Rep.* 2014;4:6962.

121. Poole AC, et al. The PINK1/Parkin pathway regulates mitochondrial morphology. *Proc Natl Acad Sci.* 2008;105(5):1638–1643.

122. Tanaka A, et al. Proteasome and p97 mediate mitophagy and degradation of mitofusins induced by Parkin. *J Cell Biol.* 2010;191(7):1367–1380.

123. Chan NC, et al. Broad activation of the ubiquitin-proteasome system by Parkin is critical for mitophagy. *Hum Mol Genet.* 2011;20(9):1726–1737.

124. Chen Y, Dorn GW. PINK1-phosphorylated mitofusin 2 is a Parkin receptor for culling damaged mitochondria. *Science.* 2013;340(6131):471–475.

125. Lu X, Kim-Han JS, Harmon S, Sakiyama-Elbert SE, O'Malley KL. The Parkinsonian mimetic, 6-OHDA, impairs axonal transport in dopaminergic axons. *Mol Neurodegener.* 2014;9:17.

126. Spillantini MG, et al. Alpha-synuclein in Lewy bodies. *Nature.* 1997;388(6645):839–840.

127. Braak H, Sandmann-Keil D, Gai W, Braak E. Extensive axonal Lewy neurites in Parkinson's disease: a novel pathological feature revealed by alpha-synuclein immunocytochemistry. *Neurosci Lett.* 1999;265(1):67–69.

128. Braak H, et al. Staging of brain pathology related to sporadic Parkinson's disease. *Neurobiol Aging.* 2003;24(2):197–211.

129. Li L, et al. Human A53T alpha-synuclein causes reversible deficits in mitochondrial function and dynamics in primary mouse cortical neurons. *PLoS One.* 2013;8(12):e85815.

130. Ryan SD, et al. Isogenic human iPSC Parkinson's model shows nitrosative stress-induced dysfunction in MEF2-PGC1alpha transcription. *Cell.* 2013;155(6):1351–1364.

131. Richter C, et al. Oxidants in mitochondria: from physiology to diseases. *Biochim Biophys Acta*. 1995;1271(1):67–74.
132. Paisan-Ruiz C, Lewis PA, Singleton AB. LRRK2: cause, risk, and mechanism. *J Parkinsons Dis*. 2013;3(2):85–103.
133. Godena VK, et al. Increasing microtubule acetylation rescues axonal transport and locomotor deficits caused by LRRK2 Roc-COR domain mutations. *Nat Commun*. 2014;5:5245.
134. Gandhi PN, Chen SG, Wilson-Delfosse AL. Leucine-rich repeat kinase 2 (LRRK2): a key player in the pathogenesis of Parkinson's disease. *J Neurosci Res*. 2009;87(6):1283–1295.
135. Gillardon F. Leucine-rich repeat kinase 2 phosphorylates brain tubulin-beta isoforms and modulates microtubule stability—a point of convergence in parkinsonian neurodegeneration?. *J Neurochem*. 2009;110(5):1514–1522.
136. Meixner A, et al. A QUICK screen for Lrrk2 interaction partners--leucine-rich repeat kinase 2 is involved in actin cytoskeleton dynamics. *Mol Cell Proteomics*. 2011;10(1):M110 001172.
137. Kett LR, et al. LRRK2 Parkinson disease mutations enhance its microtubule association. *Hum Mol Genet*. 2012;21(4):890–899.
138. Bonifati V, et al. Mutations in the DJ-1 gene associated with autosomal recessive early-onset parkinsonism. *Science*. 2003;299(5604):256–259.
139. Cookson MR. Parkinsonism due to mutations in PINK1, parkin, and DJ-1 and oxidative stress and mitochondrial pathways. *Cold Spring Harb Perspect Med*. 2012;2(9):a009415.
140. Zhang L, et al. Mitochondrial localization of the Parkinson's disease related protein DJ-1: implications for pathogenesis. *Hum Mol Genet*. 2005;14(14):2063–2073.
141. Hayashi T, et al. DJ-1 binds to mitochondrial complex I and maintains its activity. *Biochem Biophys Res Commun*. 2009;390(3):667–672.
142. Canet-Aviles RM, et al. The Parkinson's disease protein DJ-1 is neuroprotective due to cysteine-sulfinic acid-driven mitochondrial localization. *Proc Natl Acad Sci*. 2004;101(24):9103–9108.
143. Guzman JN, et al. Oxidant stress evoked by pacemaking in dopaminergic neurons is attenuated by DJ-1. *Nature*. 2010;468(7324):696–700.
144. Hao LY, Giasson BI, Bonini NM. DJ-1 is critical for mitochondrial function and rescues PINK1 loss of function. *Proc Natl Acad Sci*. 2010;107(21):9747–9752.

5

Mitophagy

L. King, H. Plun-Favreau

University College London Institute of Neurology, Queen Square,
London, United Kingdom

OUTLINE

Parkinson's Disease. http://dx.doi.org/10.1016/B978-0-12-803783-6.00005-5

1 INTRODUCTION

The importance of mitochondrial abnormalities in Parkinson's disease (PD) pathogenesis was first observed in drug users accidentally exposed to 1-methyl-4-phenyl-1,2,3,4-tetrahydropyridine (MPTP), an inhibitor of mitochondrial electron transport chain (ETC) complex I, found to cause parkinsonism and degeneration of dopaminergic (DA) neurons.[1] A link between mitochondrial dysfunction and PD was further established in experimental models of PD induced by neurotoxic compounds, such as 6-hydroxydopamine, MPTP, or rotenone,[2–4] and in postmortem tissues from sporadic PD patients.[5–9] The relevance of mitochondrial dysfunction in PD etiology was reinforced by the identification of specific gene mutations (in *PINK1* and *Parkin* in particular) playing a central role in mitochondrial function (Chapter 1). While defects in mitochondrial respiration have long been implicated in the etiology and pathogenesis of PD (Chapter 2), the role of mitochondria in PD extends well beyond defective respiration and also involves perturbations in mitochondrial dynamics, leading to alterations in mitochondrial morphology (Chapter 3), intracellular trafficking (Chapter 4), or quality control, the focus of this chapter.

2 MITOCHONDRIAL QUALITY CONTROL

Mitochondria play key roles in numerous cellular processes, including ATP production, calcium homeostasis, and cell death pathways. In no cell type is their function more vital than in long-lived neurons, where mitochondria must be maintained for an entire lifetime, and where energy requirements are very high and long axons necessitate energy transport over large distances. While functioning mitochondria are key for maintaining cellular health, dysfunctional mitochondria can generate excessive amounts of reactive oxygen species (ROS) that contribute to oxidative damage in a range of neurodegenerative conditions including PD.[10] In order to maintain a healthy population of mitochondria and prevent cellular damage, several mitochondrial quality control (MQC) mechanisms have evolved.[11]

2.1 Mitochondrial Proteolytic and Proteasomal Degradation Pathways

The first line of defense involves mitochondrial chaperones and proteases, which maintain an optimal amount of folded proteins in the mitochondria. In addition, more recent studies have highlighted the importance of the cytosolic ubiquitin proteasome system (UPS) and a mitochondria-specific unfolded protein response (^{mt}UPR) in MQC.

Correct protein folding is crucial to maintain mitochondrial homeostasis. In cases where proteins become misfolded or damaged, they need to be degraded to prevent their toxic accumulation, hence the role of mitochondrial chaperones and proteases. The vast majority of mitochondrial proteins are encoded by the nucleus, synthesized in the cytosol, and subsequently imported into the mitochondria. For efficient translocation through the narrow pore of the translocases of the mitochondrial inner and outer membranes (TIM and TOM), most of these proteins are transported in their unfolded state. Members of the heat shock protein (HSP) family drive their import to the mitochondrial surface and facilitate their folding once inside the mitochondria.[12] One such mitochondrial chaperone is mtHsp70 (also called mortalin), which has been shown to interact with Parkin, PINK1, and DJ-1.[13-16] Analysis of three PD-associated variants in the *mortalin* gene suggested a loss of protective mortalin function in human cells.[16] Another mitochondrial molecular chaperone is TRAP1 (TNF receptor-associated protein 1), also known as Hsp75, which has been shown to be phosphorylated by the mitochondrial serine/threonine kinase PINK1 to prevent oxidative stress-induced apoptosis.[17] Further work has suggested that in *Drosophila*, *Trap1* works downstream of *PINK1* and in parallel with *parkin*.[18] PINK1 has been shown to interact with several other chaperone proteins in mammalian cells, such as Hsp90 and Cdc37/p50, and these were suggested to regulate its stability.[19]

Unfolded proteins that can no longer be refolded by molecular chaperones must be degraded by mitochondrial proteases to prevent ROS production and mitochondrial stress. Clearance of dysfunctional proteins within the mitochondria is facilitated by an array of subcompartment-specific mitochondrial proteases,[12] such as the LON protease family[20] and the ClpXP protease[21] in the matrix, and the i-AAA and m-AAA proteases[22] embedded in the inner mitochondrial membrane (IMM) and facing the intermembrane space (IMS) and the matrix, respectively. Additional peptidases exist in the IMS, in particular the serine protease HtrA2/Omi, which was proposed to be involved in a protein quality control system within mitochondria in a similar manner to the homologous stress-adaptive proteins DegP and DegS in bacteria,[23,24] however, its precise function remains unclear.[25,26] Mutations in the *HtrA2* gene have been identified as high risk for developing PD,[27,28] although the pathogenicity of these variants remains controversial.[29] HtrA2 can be phosphorylated in a PINK1-dependent manner and HtrA2 phosphorylation is reduced in the brains of PD patients carrying mutations in *PINK1*. This model suggests that PINK1-dependent phosphorylation of HtrA2 might modulate its proteolytic activity, thereby contributing to an increased resistance of cells to mitochondrial stress.[30] Studies in mammals and in *Drosophila* have further suggested that PINK1 and HtrA2 work together in common pathways,[31,32] although in *Drosophila*, this is disputed.[33] Presenilin-associated rhomboid-like (PARL), a mitochondrial resident rhomboid serine protease, has been implicated in post-translational modification of both HtrA2[34] and PINK1.[35,36] All together, these studies suggest that PINK1 may play an important role in the first line of defense of MQC, monitoring the proper folding or degradation of unfolded and damaged proteins, respectively.

Although clearance of dysfunctional proteins is usually efficiently achieved in the mitochondria, clearance of defective outer mitochondrial membrane (OMM) proteins usually requires further support from the cytoplasmic UPS,[37,38] with some studies also reporting a role for the UPS in the regulation of IMM proteins.[37,39] Interestingly, the majority of OMM proteins degraded by the UPS play a role in apoptosis or in mitochondrial dynamics. Mitochondria-associated degradation (MAD) is also referred to as the OMM-associated degradation (OMMAD) system, by analogy with the endoplasmic-reticulum-associated protein degradation (ERAD) pathway. In OMMAD, OMM proteins are ubiquitinated and retro-translocated into the cytosol for degradation. OMMAD and ERAD share at least one component, the AAA-ATPase p97/VCP,[40,41] that extracts ubiquitinated proteins from the OMM (see Section 4.5 for details). Several E3 ubiquitin ligases, including MITOL/MARCH5,[42] MULAN,[43] and RNF185,[44] as well as deubiquitinating enzymes, such as Ubp16/USP30,[38] are embedded in the OMM and are known to play a role in a number of cellular processes including MQC. Furthermore, the cytosolic E3 ubiquitin ligase Parkin is involved in the widespread ubiquitination of the mitochondrial surface following mitochondrial dysfunction (see Section 4.5).

While mitochondrial chaperones and proteases are equipped to turn over a small subset of unfolded or damaged mitochondrial proteins, there are circumstances in which the proteolytic and proteosomal degradation capacity is overwhelmed by demand. As a response, a retrograde signal is sent from the mitochondria to the nucleus to increase expression of chaperones and proteases, hereafter referred as ^{mt}UPR.[45,46] Reduced mortalin function was shown to lead to activation of the ^{mt}UPR and increased autophagic degradation of fragmented mitochondria. This mitochondrial stress induced by loss of mortalin function was shown to be rescued by PINK1 and Parkin.[47]

2.2 Vesicular Transport to Lysosomes

The second line of defense involves the mitochondrial-derived vesicles (MDVs).[48] These vesicles sequester selected mitochondrial cargos, bud from mitochondria, and deliver them to lysosomes for degradation. This pathway, active under steady-state conditions, and further activated by oxidative stress, involves three PD-associated proteins, namely PINK1, Parkin, and VPS35, and is different from canonical mitophagy.[49–51]

2.3 Mitophagy

The pathways mentioned previously account for degradation of a subset of unfolded or damaged mitochondrial proteins. However, when mitochondria are too heavily damaged to be repaired by fusion with functional mitochondria (see Chapter 3), it is essential for the whole mitochondrion to be removed to avoid its toxic accumulation. This occurs through a selective autophagic process, hereafter called mitophagy, the third MQC line of defense. Autophagy is an evolutionary conserved process in which intracellular components are engulfed in a double membrane vesicle called autophagosome. Increasing evidence suggests autophagy is a protective mechanism in the brain as it maintains the optimal balance of protein synthesis, degradation, and recycling of cellular resources (Chapter 6). The presence of mitochondria within autophagosome structures was first observed in 1957.[52] Mitophagy is not only important for the removal of damaged mitochondria but also for ensuring optimal cellular energy production, maintaining the steady-state turnover of mitochondria, and fine-tuning their numbers. Studies in *S. cerevisiae* provided the first genetic evidence that mitochondrial degradation by autophagy could be a selective process. The identification of the AuTophaGy (*ATG*) genes in yeast has been instrumental in identifying mammalian *ATG* homologues through sequence comparison. A number of the ATG genes have been identified to play a role in mitophagy, however, they also play a number of additional roles in macroautophagy. Mitophagy is mediated by Atg32 in yeast, and by NIX/BNIP3L, a protein related to Bcl-2, during red blood cell differentiation in mammals.[36,53] The majority of what we know about the selective degradation of damaged mitochondria originates from the PD-associated genes *PINK1* and *Parkin*, with research over the past 10 years shedding light on the molecular mechanisms of the process (see later for details on the mitophagy process).

Oxidative stress generated by the accumulation of damaged mitochondria can spread to other mitochondria and eventually lead to the death of neurons.[54] It has recently been suggested that mitochondria can be unloaded by retinal ganglion axons, and that they are subsequently engulfed and degraded by the lysosomes in neighboring astrocytes. This process, called transcellular degradation of mitochondria or transmitophagy,[55] would challenge the general assumption that mitochondria can only be degraded cell-autonomously. Similarly, mesenchymal stem cells (MSCs) were recently shown to undergo mitophagy and use arrestin domain-containing protein 1–mediated vesicles to shed mitochondria, which are then internalized by macrophages to increase bioenergetics.[56] Whether transmitophagy is a widespread phenomenon in the CNS, and whether this process is selective for damaged mitochondria, remains to be firmly determined.

2.4 Apoptosis

The last level of quality control occurs at the cellular level, whereby extensive mitochondrial damage leads to programmed cell death.[57] Bioenergetic dysfunctions, calcium dyshomeostasis, aberrant generation of ROS, release of proapoptotic proteins, and altered stress signaling are among the leading causes of mitochondrial cell death.[10,54] Additionally, a novel pathway of mitochondrial cell death has been recently identified in yeast, termed mitochondrial precursor overaccumulation stress (mPOS). It was proposed that mitochondrial damage could affect protein import, leading to an exacerbated accumulation of mitochondrial precursors in the cytosol. The resulting mPOS triggers cell death in yeast.[58]

3 PINK1 AND PARKIN

Most of what we currently know about the mitophagy process comes from the study of the rare genetic forms of autosomal recessive PD. A number of the gene products associated with PD have been found to associate with mitochondria, and of particular interest, the two proteins PINK1 and Parkin.

3.1 Parkin

3.1.1 Parkin Structure

Loss of function mutations in the *Parkin* gene are the most common cause of autosomal recessive PD (see also Chapter 1). The gene is located within a highly unstable genomic region and over 120 mutations in the

Parkin gene have already been linked to familial forms of PD.[59] Soon after its discovery,[60] Parkin was shown to be a 465 amino acid E3 ubiquitin ligase[61] capable of mediating mono, multimono, and polyubiquitination of substrates with different chain topologies.[62,63] Based on the conserved function of E3 ubiquitin ligases in targeting substrates for degradation, it was hypothesized that loss of function mutations in the *Parkin* gene would lead to the toxic accumulation of proteins and subsequent death of DA neurons. However, research during the last 15 years has highlighted several other functions for Parkin, independent from the proteasomal degradation of toxic substrates. The primary Parkin function responsible for dopaminergic neurodegeneration remains to be confirmed, but the recent solving of Parkin crystal structure has hugely contributed to our understanding the regulation of its enzymatic activity and the impact of the pathogenic mutations. Several groups have reported on the crystal structure of Parkin either in its N-terminally truncated form or in its full-length form, resolved at a lower resolution, where they all come to similar conclusions.[64–67] Parkin belongs to the RING-Between-RING (RBR) class of E3 ubiquitin ligases, which share common features with both the RING and the homologous to E6-AP carboxyl terminus (HECT) families of E3 ubiquitin ligases.[68] Similar to RING E3 ligases, Parkin recruits E2 ubiquitin-conjugating enzymes via a really interesting new gene (RING) domain. The ubiquitin is then transferred from the E2 to a conserved catalytic cysteine residue in Parkin (Cys431) forming a transient thioester intermediate, similarly to HECT E3 ligases. Parkin structure comprises a ubiquitin-like (UBL) domain at the N terminus and a RBR domain at the C terminus. The RBR consists of two RING domains, RING1 and RING2, separated by an in-between RING (IBR) domain. An extra atypical RING domain, named RING0 or unique parkin domain (UPD), separates the UBL and the RBR domains.[60] Under basal conditions, Parkin is in a tightly folded nonactive auto-inhibited state. Some of the pathogenic mutations were shown to interfere with Parkin autoinhibition, resulting in increased autoubiquitination and turnover, which at least partially explains the loss of Parkin function in PD. How Parkin can switch from its inactive form to its active form is discussed later.

3.1.2 Parkin Localization

Parkin is expressed abundantly in the brain, heart, skeletal muscles, testis, and at lower levels in fibroblasts and peripheral leukocytes.[69,70] Although Parkin subcellular localization is mainly cytosolic,[61] Parkin expression was detected in the Golgi apparatus,[71,72] endoplasmic reticulum (ER),[73] mitochondria,[74,75] aggresomes,[76] neurites,[77] and synaptic vesicles.[72,78] Some pathogenic mutations in Parkin, in which the catalytic activity is retained, have been reported to alter Parkin localization, thereby forming aggresome-like structures.[79–81]

3.1.3 Parkin Protective Function

In a context-specific manner, Parkin displays a wide range of neuro-protective roles in cellular and animal models; these include promoting cell survival, preventing cell death, and mediating cell signaling and xenophagy.[82,83] Pathogenic mutations were shown to interfere with these cellular processes. In addition to its key role in the selective degradation of damaged mitochondria by mitophagy (described in detail in the Section "PINK1-Parkin-mediated Mitophagy"), Parkin is thought to mediate its neuroprotective function through two main molecular mechanisms: pro-teasomal degradation of toxic substrates and nondegradative ubiquitin-dependent signal transduction.

3.1.3.1 Proteasomal Degradation of Toxic Substrates

Based on the conserved function of E3 ubiquitin ligases in targeting substrates for degradation, a number of groups have searched for Parkin-interacting proteins and have identified an array of aggregation-prone protein substrates for Parkin.[84] Whether these are direct substrates for Parkin E3 ubiquitin ligase activity is often difficult to prove due to the poor enzymatic activity of Parkin in a variety of in vitro and in vivo cel-lular assays. Transgenic overexpression of a Parkin substrate, aminoacyl-tRNA synthetase complex interacting multifunctional protein-2 (AIMP2), leads to a selective, age-dependent, progressive DA neuronal death and impairment of motor coordination in mice.[85] Similarly, a new Parkin in-teracting substrate, PARIS (ZNF746), master transcriptional repressor of peroxisome proliferator-activated receptor gamma (PPARγ) coactivator-1α (PGC-1α), accumulates in Parkin knockout (KO) mice and leads to the progressive loss of DA neurons.[86] This suggests a role for Parkin in mitochondrial biogenesis. Parkin was shown to mediate polyubiquitina-tion of misfolded DJ-1, thus targeting DJ-1 to aggresomes via interaction with histone deacetylase 6 (HDAC6), an adaptor protein that binds the dynein-dynactin complex.[87] Parkin was also shown to ubiquitinate Tar-DNA binding protein 43 (TDP-43) and facilitate its cytosolic accumulation via interaction with HDAC6.[88] Parkin-mediated proteasomal degradation of the multisubunit E3 ubiquitin ligase SCF substrate adapter Fbw7β leads to stabilisation of the mitochondrial antiapoptotic factor Mcl-1.[89] Finally Parkin was shown to play a key role in proteasomal degradation of mito-fusins[90,91] and Miro,[92] mitochondrial GTPases leading to mitophagy and mitochondrial arrest, two processes which are intrinsically linked (see also Chapter 4).

3.1.3.2 Nondegradative Ubiquitin-Dependent Signal Transduction

Parkin is not only involved in proteasome-dependent degradation of substrates, it has also been shown to mediate its neuroprotective activity via nondegradative ubiquitin-dependent signaling pathways.[82,83] Parkin acts together with PINK1 to repair moderately damaged mitochondria through an autophagy-independent process via the formation of MDVs[48] (Section 2.2). Parkin promotes mitochondrial biogenesis[93,94] and protects mitochondrial genome integrity.[95,96] It activates prosurvival pathways by increasing the NF-kB signaling pathway[97] or decreasing activation of the c-jun N-terminal kinase.[98] Furthermore it can ubiquitinate Bax and suppress its proapoptotic function.[99] It modulates epidermal growth factor receptor (EGFR) internalization and downstream Akt signaling.[100] It represses p53 transcription in ubiquitin ligase-independent manner.[101] Conversely Parkin was shown to be a p53 target gene, mediating the Warburg effect of p53 on glucose metabolism (see also Chapter 2).[102] Parkin may play a key role in innate immunity by promoting xenophagy.[103,104] Finally Parkin has also been shown to be a tumor suppressor, and its inactivation has been reported in various human cancers.[105–107] Parkin's key roles in mitophagy, in proteasomal degradation, and in nondegradative signal transduction are detailed in the following reviews:.[82,83,108]

3.2 PINK1

3.2.1 PINK1 Structure

PINK1 was originally shown to be upregulated by the tumor suppressor gene phosphatase and tensin homolog (PTEN) in cancer cells.[109] Its cDNA was predicted to encode a 581-amino acid protein with a predictive C-terminal serine/threonine protein kinase catalytic domain containing three insertional loops and a mitochondrial targeting sequence (MTS) at the N-terminus. This MTS supported prior evidence for the key role of mitochondrial dysfunction in PD pathogenesis. Loss of function mutations in the PINK1 gene are the second most common genetic cause of autosomal recessive PD[110] (Chapter 1). Over 90 mutations have been identified so far, most of which lie within the kinase domain, or affect its kinase activity.[111,112] Clinically, patients carrying PINK1 mutations resemble those harboring mutations in the Parkin gene.[113]

3.2.2 PINK1 Processing

Under normal steady-state conditions, PINK1 is synthesized in the cytosol and imported to the mitochondria via its MTS, across the OMM via the translocase of the outer membrane (TOM), and subsequently across the IMM via the translocase of the inner membrane (TIM). PINK1 undergoes a series of proteolytic cleavages, with the 64-kDa full length form being sequentially cleaved into 60- and 52-kDa fragments. First, full length PINK1 MTS is reported to be cleaved in the IMM by the generic mitochondrial processing peptidase MPP, as mitochondrial proteins commonly are. However, those cell biology experiments suggesting that PINK1 MTS is processed by MPP have used MPP RNAi knockdown

(KD), therefore further studies are required to ascertain whether PINK1 is processed by MPP or by one of MPP's substrates. In fact a study using cell-free systems suggests that PINK1 MTS is not processed by MPP.[114] PINK1 is then cleaved in its hydrophobic domain by the rhomboid protease PARL.[34,115–117] It has been suggested that AGF3-like AAA ATPase 2 (AFG3L2) could facilitate PARL-dependent cleavage of PINK1.[118] Notably however, PINK1 cleavage is not completely abolished in PARL KO mouse embryonic fibroblasts (MEFs),[119] suggesting that other proteases can cleave PINK1. Studies in mammalian cells and in *Drosophila* identified the matrix-localized protease LON as a key player in the regulation of PINK1 degradation, however, the mechanism remains controversial.[120,121] The 52 kDa N-terminal-deleted PINK1 is then released in the cytosol, where it is degraded by the proteasome through the N-end rule pathway.[122] The constant import and degradation of PINK1 yields very low, almost undetectable levels of protein in healthy mitochondria, making its cellular detection at endogenous level in steady-state conditions difficult, at least with the currently available antibodies.

3.2.3 PINK1 Localization

PINK1 transcript is expressed in several tissues, with high levels in the heart, skeletal muscle, liver, pancreas, and testis.[109,123] PINK1 expression is detected in mitochondrial fractions, localized both in the OMM[124] and the IMM.[17,125] 52-kDa cleaved PINK1 is detected in the cytosol.[126] Finally, PINK1 has also been localized to the ER.[127]

3.2.4 PINK1 Protective Function

PINK1 deficiency has been associated with mitochondrial dysfunction, in particular reduced complex I activity[128–130] (see also Chapter 2), calcium dyshomeostasis,[131] oxidative stress-induced apoptosis,[110,132] mitochondrial fusion and fission[133] (Chapter 3), mitochondrial trafficking[92] (Chapter 4), and mitophagy,[134,135] in a number of cell and animal models. Although all these processes are intrinsically linked, their regulation by PINK1 appears to not always be codependent.[128,130] Pathogenic mutations in the PINK1 gene interfere with these cellular processes. PINK1 has also been shown to regulate a number of cellular processes of significance in cancer biology, such as cell survival, mitochondrial homeostasis, and cell cycle regulation.[106,107,136,137]

3.2.4.1 PINK1 Substrates

A number of proteins associated with mitochondria have been suggested to be directly phosphorylated by PINK1, in particular TRAP1,[17] a mitochondrial chaperone protecting against oxidative stress-induced apoptosis, and Miro, a mitochondrial GTPase regulating mitochondrial morphogenesis and trafficking along microtubules.[92] Other mitochondrial proteins have been shown to be phosphorylated in a PINK1-dependent manner, including HtrA2, mitofusin 2 (Mfn2), and NdufA10, a subunit of the ETC complex I. It was suggested that PINK1-dependent phosphorylation of HtrA2 may increase cell resistance to mitochondrial stress.[30] PINK1 was shown to mediate Phos-Tag phosphorylation of Mfn2, with phosphorylated Mfn2 acting as a mitochondrial receptor for Parkin.[138] However, these data have been contradicted by another study showing that Parkin translocates constitutively in Mfn1/Mfn2-KO cells.[139] Finally PINK1 has been shown to be important for phosphorylation of NdufA10, and subsequent ubiquinone reduction by complex I (see Chapter 2 for details).[128] Mammalian PINK1 low catalytic activity in vitro has challenged the identification of its direct specific substrates. However identification of catalytically active insect PINK1 orthologs, in particular *Tribolium castaneum* PINK1 (TcPINK1), has allowed the development of robust in vitro kinase assays and subsequent identification of PINK1 physiological substrates.[140] TcPINK1 only share 26% identity with the human protein, thus validating PINK1 substrates in mammalian systems is preferable. Notably in vitro kinase assays using recombinant human or insect PINK1 have failed to confirm reliable direct phosphorylation of the aforementioned substrates.[141,142] Finally, a recent quantitative phosphoproteomics study has shown that PINK1 is able to regulate the phosphorylation of Rab GTPases in an immortalized cell line.[143]

Using TcPINK1, it was shown that PINK1 can phosphorylate Parkin at Ser65, in its UBL domain, thereby increasing its E3 ubiquitin ligase activity and its recruitment to mitochondria.[141,144] Several E2 ubiquitin ligases can transfer ubiquitin to PINK1-activated Parkin in cell-free systems, including UBE2K, UBE2D1, UBE2D2, UBE2E1, UBE2L3, and UBE2C.[145,146] Recently, three groups have simultaneously and independently shown that PINK1 can also phosphorylate ubiquitin at Ser65 (at a motif homologous to the one contained in Parkin) signifying PINK1 as the first identified ubiquitin kinase, and that phospho-ubiquitin activates Parkin E3 ubiquitin ligase activity.[147–149] The discovery that ubiquitin can be phosphorylated by PINK1 has been instrumental in understanding optimal activation of Parkin E3 ubiquitin ligase activity. The possible mechanisms underlying phospho-ubiquitin-mediated Parkin activation are discussed in the section later.

4 PINK1-PARKIN-MEDIATED MITOPHAGY

The decisive advancement linking PINK1 and Parkin in the same pathway was the genetic epistasis analysis of the mutant flies. PINK1 and parkin KO flies share clear phenotypic similarities, including degeneration of dopaminergic neurons and indirect flight muscles, associated with mitochondrial morphology defects. The phenotypes caused by loss of *Drosophila* PINK1 could be complemented by wild-type human PINK1 or partly also by Parkin, but not by the pathogenic forms of PINK1, or wild-type DJ-1. Conversely the phenotype of the parkin mutant flies couldn't be rescued by overexpression of human PINK1. Altogether these data suggest that PINK1 and Parkin act in a common pathway, with PINK1 functioning upstream of Parkin, to regulate mitochondrial homeostasis.[150–153] These studies supported an earlier study in mammalian cells showing that Parkin could prevent mitochondrial damage.[74] Subsequent cell biology studies in mammalian immortalized cell lines have shown that while PINK1 and Parkin have different biological functions and reside in different subcellular compartments in steady-state conditions, they can act in concert to mediate mitophagy. Parkin can be recruited from the cytosol to mitochondria depolarized with high doses of the mitochondrial uncoupler carbonyl cyanide m-chlorophenyl hydrazone (CCCP), mediating their selective degradation by autophagy.[134,135,154,155] PINK1 accumulates on damaged mitochondria and its kinase activity is required for Parkin recruitment, thus linking the function of PINK1 and Parkin into a common signaling pathway.[134,156] Pathogenic mutations in both *PINK1* and *Parkin* genes interfere with the pathway, emphasizing the importance of mitophagy for PD pathogenesis.

4.1 PINK1 Accumulation at the Surface of Damaged Mitochondria

PINK1's rapid and constitutive degradation in healthy mitochondria is bypassed when PINK1 senses mitochondrial depolarization. Mitochondrial membrane potential ($\Delta\Psi_m$) is a key indicator of mitochondrial health and injury. Small molecules and ions equilibrate freely across the OMM, restricting $\Delta\Psi_m$ to the IMM. As a result PINK1 is imported as normal to the OMM of depolarized mitochondria, however the loss of $\Delta\Psi_m$ prevents its import through the TIM complex, and its subsequent cleavage in the IMM.[157] When mitochondrial import through the TIM complex is disrupted by depolarization, unprocessed PINK1 accumulates specifically at the OMM of dysfunctional mitochondria, discriminating the damaged mitochondria from their healthy counterparts and flagging them for degradation. While accumulating at the surface of mitochondria, it has been reported that PINK1 binds to the TOM complex. Its kinase domain is exposed outward, allowing access to potential cytosolic substrates.[158] When key residues in PINK1 MTS are mutated, PINK1 also accumulates at the OMM and recruits Parkin, further suggesting that PINK1 possesses a OMM targeting domain that interacts with the TOM import machinery.[157,159,160] However, the molecular identity of the TOM components incorporated in the complex with PINK1 remains controversial.[157,161] PINK1-mediated mitophagy can be initiated by mitochondrial depolarisation, but also by mitochondrial DNA (mtDNA) mutations, hepatitis C virus infection or an excessive amount of unfolded proteins in the mitochondrial matrix, independently of mitochondrial membrane potential.[121,162–165] This suggests that additional mechanisms, other than mitochondrial depolarization, are involved in PINK1 accumulation. In fact PINK1 has been shown to accumulate in cell and animal models deficient for the LON matrix protease, nonetheless the mechanism of LON-dependent PINK1 accumulation remains to be confirmed.[120,121]

4.2 PINK1-Dependent Parkin Recruitment to Damaged Mitochondria

Accumulation of intact PINK1 on the mitochondrial surface leads to Parkin translocation from the cytosol to damaged mitochondria through a mechanism as yet not fully characterized. In mammalian cells, expression of a PINK1 fusion protein stably expressed at the OMM is sufficient for Parkin recruitment, even in the absence of mitochondrial depolarization.[166] In addition artificial targeting of PINK1 to peroxisomes recruits Parkin to these organelles and mediates their autophagic degradation.[157] These experiments suggest that PINK1 and Parkin are sufficient for damaged mitochondria to be selectively degraded by mitophagy, at least in human cell lines and in dopaminergic neurons derived from induced pluripotent stem (iPS) cells overexpressing Parkin.[111,167]

PINK1 has been recently shown to directly phosphorylate Parkin in its UBL domain, at Ser65, thereby stimulating both its recruitment to damaged mitochondria and its E3 ubiquitin ligase activity (see Section 4.4).[141,168] However, a number of experiments suggest that other mechanisms are required for optimal PINK1-dependent Parkin activity. First, despite the fact that mutation of the Parkin Ser65 site into alanine (Ala) prevents its phosphorylation by PINK1, neither Ser65Ala mutation nor deletion of Parkin whole UBL domain completely prevent its relocation to damaged mitochondria.[144,148,169] Furthermore, PINK1 and parkin mutant flies can be partially rescued by Ser65Ala Parkin. This suggests that Parkin Ser65 phosphorylation-independent mechanisms, such as other Parkin phosphorylation events, additional Parkin posttranslational modifications (eg, autoubiquitination[170]), or unknown substrates for PINK1 are required for Parkin recruitment to damaged mitochondria, and its full activation.[168]

The discovery that PINK1 phosphorylates ubiquitin at Ser65, at a motif homologous to the one of Parkin Ser65, has been instrumental in starting to elucidate the molecular mechanism of full Parkin recruitment and full activation of its E3 ubiquitin ligase activity (see Section 4.4). It was shown that inhibition of ubiquitin phosphorylation resulted in inhibition of Parkin translocation to damaged mitochondria. Phospho-ubiquitin peptides were purified from OMM purified peptides, suggesting that mitochondrial phospho-ubiquitin could be the receptor for Parkin at the mitochondrial surface.[148,171]

4.3 Modifiers of Parkin Recruitment to Damaged Mitochondria

Other than PINK1, a number of regulators of Parkin translocation to depolarized mitochondria have been identified. Similar to PINK1 and Parkin, F-box only protein 7 (Fbxo7) is associated with an autosomal recessive early onset parkinsonian disorder. It encodes a member of the F-box protein family. The F-box proteins constitute one of the four subunits of the ubiquitin protein ligase complex called SCFs (SKP1-cullin-F-box), which function in phosphorylation-dependent ubiquitination. Fbxo7 has been shown to be important for Parkin recruitment to depolarized mitochondria and that pathogenic mutations in the *Fbxo7* gene interfere with this process, further emphasizing the importance of Parkin recruitment and subsequent mitophagy for PD pathogenesis.[172] Furthermore, upon stress, Fbxo7 expression levels are increased, forming aggregates at the mitochondria. These studies revealed that Fbxo7 expression levels are increased in PD patient fibroblasts and that Fbxo7 can form aggregates in the brains of LRRK2 transgenic mice and in the brain of PD patients.[173] A number of RNAi screens in mammalian and *Drosophila* cells have identified other regulators of Parkin recruitment, including hexokinases, HSPA1L, BAG4, TOMM7, SIAH3, ATPIF1, FBXW7, and SREBF1.[159,174–176] With the exception of PINK1, none of the regulators previously mentioned has been identified in two independent screens, possibly due to variability in the methods and cell models used. Some of these regulators were shown to be essential for PINK1 accumulation, prior to Parkin recruitment, in particular TOMM7 and SIAH3.[159] Notably genome-wide association studies (GWAS) have also identified SREBF1 as a risk locus for sporadic PD, thus giving more credence for this gene's association with disease pathogenesis.[176,177] The short mitochondrial isoform of ARF (smARF), previously identified as an alternate translation product of the tumor suppressor p19ARF, was shown to depolarize mitochondria (without the requirement of exogenous chemical stressors) and trigger Parkin-dependent mitophagy.[178] The protease PARL was suggested to be a negative regulator of PINK1 and Parkin-mediated mitophagy.[179] Activation of microphthalmia/transcription factor E (MiT/TFE) transcription factor was shown to facilitate Parkin-dependent mitophagy.[180] Other factors that have been identified to regulate Parkin translocation to depolarized mitochondria include components of the TOMM machinery, E2 ubiquitin-conjugating enzymes, deubiquitinases (DUBs), and Bcl2-associated proteins.[181–187]

4.4 PINK1-Dependent Activation of Parkin E3 Ubiquitin Ligase Activity

Following its recruitment, Parkin's catalytic activity is increased by PINK1-mediated phosphorylation of both Parkin Ser65 and ubiquitin Ser65.[141,145,147–149,156,188] Ser65Asp phospho-mimetic Parkin mutant display greater ubiquitin ligase activity than wild-type Parkin, suggesting that PINK1-dependent phosphorylation of Parkin Ser65 is important for its activation. Moreover, phospho-Ser65 ubiquitin is sufficient to activate Parkin E3 ubiquitin ligase activity, suggesting that PINK1-dependent phosphorylation of ubiquitin Ser65 is important for its activation. Parkin can interact with phospho-ubiquitin, and the artificial targeting of phospho-ubiquitin to mitochondria, or lysosomes, is sufficient to promote Parkin recruitment in a PINK1-independent manner.[168,189] All together these data suggest that PINK1-dependent phosphorylation of Parkin Ser65, PINK1-dependent phosphorylation of ubiquitin Ser65, and binding of phospho-ubiquitin to Parkin are required for switching Parkin from an inactive to an active form and its subsequent full activation. Although these mechanisms are only just beginning to be elucidated, evidence suggests that following its recruitment to damaged mitochondria and subsequent PINK1-dependent phosphorylation at Ser65, Parkin amplifies the mitophagy process by generating more ubiquitin chains at the surface of damaged mitochondria, thus providing more substrate for PINK1 kinase activity. This strong positive feedback cycle is thought to generate the specific clearance of damaged mitochondria.

PINK1 kinase activity is shown to be regulated by autophosphorylation, however the precise mechanism by which PINK1 phosphorylates itself and its substrates remains to be fully elucidated.[141,190,191] Several recent studies have highlighted the probable relevance of phospho-Ser65 ubiquitin for disease pathogenesis. First, pathogenic mutations in the domains of Parkin that preferably bind ubiquitin in a phosphorylation-dependent manner, namely IBR and RING1, block Parkin's interaction with phospho-Ser65 ubiquitin in mammalian cells.[192] Second, antiphospho-Ser65-ubiquitin immunoreactive granules partially colocalize with mitochondria and lysosomes in human postmortem brains. Finally, the number of the granules positive for phospho-Ser65 ubiquitin increase with age and with PD.[193]

4.5 Parkin-Dependent Ubiquitination of the Mitochondrial Surface

While PINK1 initiates mitophagy, Parkin is responsible for directing the proteasomal and autophagic degradation of damaged mitochondria. Once in its active conformation, Parkin mediates the ubiquitination of multiple substrates at the mitochondria, with a preference for proteins of the OMM.[194-197] The large number of substrates for Parkin E3 ubiquitin ligase activity suggests that its specificity is conferred by the phosphorylation of the ubiquitin chain, the type of ubiquitin chain linked to the substrates, and/or the density of the substrates rather than the molecular identity of the substrate. Parkin has been shown to form various ubiquitin conjugates on its substrates, in particular K6, K11, K48, and K63 ubiquitin linkages.[171,196] While the role of K6-linked chains is unknown, K48- and K11-linked ubiquitination were suggested to lead to OMMAD when K63-linked ubiquitination were proposed to recruit the autophagy machinery. The identification of K48 and K63 linkages wasn't a surprise as they had been previously shown for mitofusins and VDAC, respectively.[134,197,198]

K48-linked ubiquitin are proposed to be important for the recruitment of the proteasomal machinery. Parkin-mediated formation of K48-linked conjugates on mitofusins can recruit the AAA-ATPase p97/VCP, thus directing ubiquitinated mitofusins for proteasomal degradation after extracting them from mitochondria.[90,197,199] The damaged mitochondria that lack mitofusins are presumably less likely to fuse with undamaged mitochondria, and are instead targeted for mitophagy.[53,200] VCP mutations are the cause of inclusion body myopathy, Paget's disease of the bone, and frontotemporal dementia (IBMPFD), and they account for 1–2% of familial amyotrophic lateral sclerosis (ALS). In addition to its role in mitochondrial dysfunction and PINK1-Parkin-mediated mitophagy, VCP was shown to play a role in a wide variety of cellular activities.[201-203] Noteworthy, mitophagy could be prevented with a dominant-negative mutant of Drp1, further suggesting that fission is required for mitophagy.[204] The PINK1-dependent Ser156 phosphorylation of Miro, an outer mitochondrial membrane Rho GTPase that anchors kinesin to the mitochondrial surface, was suggested to activate its proteasomal degradation in a Parkin-dependent manner. Although Parkin-dependent Ser156 phosphorylation of Miro is contested,[169] the consensus is that the removal of Miro from the mitochondria results in kinesin detachment from the mitochondrial surface, thus arresting the mitochondria and facilitating their degradation by mitophagy. These functions of PINK1, Parkin, and Miro contribute to our understanding of the complex interplay between mitophagy and mitochondrial dynamics.

K63-linked ubiquitin chains are proposed to be important for the recruitment of the autophagy machinery. Parkin was suggested to form K63- but also K27-linked chains on VDAC,[134] and promote the subsequent degradation of depolarized mitochondria by autophagy. However, these findings were challenged by another study suggesting that VDAC isn't necessary for mitophagy.[205] Other substrates shown to be ubiquitinated by Parkin following CCCP treatment include Parkin itself,[170] TOM20, and Miro.[92,142] Like VDAC, Miro was suggested to be K27-ubiquitinated in a

PINK1 and Parkin-dependent manner. Yet no Parkin-mediated K27 ubiquitin linkage was identified by mass spectrometry,[171,196] thus the importance of K27-linked ubiquitination in PINK1-Parkin-mediated mitophagy remains to be verified.

While K6-, K11-, K48-, and K63-linked ubiquitination have been detected by mass spectrometry in depolarized cells,[171,196] their specific role in PINK1-Parkin-dependent mitophagy, if any, remains to be fully elucidated. Activity of E3 ubiquitin ligases is often regulated by DUBs, either by directly deubiquitinating them or by indirect deubiquitination of their substrates. Parkin is no exception as two DUBs, namely USP8 and USP15, have been shown to regulate Parkin-mediated mitophagy via deubiquitination of its substrates.[185,195,206] Pathogenic mutations in Parkin, but also in Fbxo7, have been shown to interfere with ubiquitination of OMM proteins, such as Mfn.[172,198]

4.6 Interaction of PINK1 and Parkin With the Core Autophagy Machinery

PINK1-Parkin-mediated mitophagy requires some ATG proteins and additional core components of the autophagy machinery. While genetic and pharmacological inhibition of the core autophagic machinery doesn't prevent neither PINK1 accumulation to depolarized mitochondria nor Parkin translocation, it prevents or delays the clearance of damaged mitochondria.[75,207] Several autophagy components have been suggested to be recruited to damaged mitochondria, including HDAC6, p62, optineurin (OPTN), NDP52, and AMBRA1.[134,208–210] Whether p62 is required only for Parkin-induced mitochondrial clustering or whether it is also required for mitophagy remains controversial.[205] It was also suggested that PINK1 and Parkin can directly interact with the Beclin-1 PI3K complex.[211,212] Parkin-dependent recruitment of AMBRA1, an activator of the Beclin-1 complex, to depolarized mitochondria, was shown to target damaged mitochondria to LC3-positive autophagosomes and promote their subsequent fusion with lysosomes. Proteins, such as OPTN and NDP52 were shown to be involved in linking the ubiquitin chains with autophagosomes via LC3. Two mitochondrial Rab GTPase-activating proteins (GAPs) bound to the OMM fission protein Fis1, namely TBC1D15 and TBC1D17, were suggested to mediate autophagosome encapsulation of damaged mitochondria by regulating Rab7 activity at the interface between mitochondria and isolation membranes.[213]

4.7 PINK1 and Parkin-Mediated Mitophagy: the Current Model

In summary, following PINK1 accumulation at the surface of damaged mitochondria and subsequent phosphorylation of the ubiquitin molecules already attached to the OMM, Parkin is recruited to the mitochondria, probably using phospho-ubiquitin as a receptor. Parkin is then phosphorylated by PINK1 in its UBL domain at Ser65. Mitochondrial phospho-Parkin then redecorates the mitochondrial surface with more ubiquitin molecules that are phosphorylated by PINK1 and further activates Parkin. The exact sequence of phosphorylation and ubiquitination events in this amplification loop remains to be fully elucidated.

5 PINK1 AND/OR PARKIN-INDEPENDENT SELECTIVE MITOPHAGY

While the rare genetic forms of autosomal recessive PD have been instrumental in understanding the mitophagy process, it has become clear that PINK1 and Parkin cannot be the only pathway for specific clearing of dysfunctional mitochondria. Several observations support this conclusion.

First, PINK1- and/or Parkin-independent mitophagy has been observed in mammalian cell lines. A recent study suggests that rather than being a receptor for Parkin, ubiquitin phosphorylated by PINK1 would be the receptor for two autophagy receptors (OPTN and NDF52) that recruits the upstream autophagy machinery to mitochondria and ultimately induces mitophagy. As opposed to being absolutely required for mitophagy, Parkin then amplifies the process by generating more ubiquitin chain substrates for PINK1, and subsequently recruiting more autophagy receptors.[209] Notably, most of the mitophagy studies so far have used Parkin-overexpression models (see Section 6), making the assumption that the mitophagy detected was Parkin-dependent. Whether Parkin was playing an indispensable function for mitophagy in all these studies may need to be confirmed. Other studies have shown that selective mitophagy may occur without PINK1 stabilization or Parkin activation, for example, it was shown that loss of iron can trigger PINK1 and Parkin-independent mitophagy.[214] Whether other E3 ubiquitin ligases can play the function of Parkin and whether kinases other than PINK1 can phosphorylate ubiquitin in these models remains to be determined.

Second, neither human neurons carrying mutations in the PINK1 and Parkin genes, nor PINK1/Parkin KO flies or mice, display any increase in mitochondrial mass. While PINK1 and Parkin patient-specific iPSC-derived dopaminergic neurons comprise abnormalities in mitochondrial homeostasis, no mitochondrial mass increase was observed.[215,216] Transgenic mice appear largely normal and only display very subtle phenotypes, even at a late onset, which would not be expected if these animals were accumulating a large amount of damaged, and thus toxic, mitochondria. PINK1 and Parkin transgenic flies, however, have a marked phenotype including degeneration of dopaminergic neurons and indirect flight muscles, associated with mitochondrial morphology defects.[150,151] In addition, a quantitative proteomic approach has found that parkin KO flies have a decreased rate of mitochondrial respiratory chain subunits turnover, suggesting that parkin can play a role in mitophagy in vivo.[217] Nonetheless there are some major discrepancies between *Drosophila* and mammalian models. First, while PINK1 mitochondrial accumulation appears to be both necessary and sufficient to promote Parkin recruitment and activation in mammalian immortalized cell lines, overexpression of human Parkin partly complements the mitochondrial defects observed in PINK1 null *Drosophila*, suggesting that for this function PINK1 isn't essential in flies. In addition, although in vitro studies and studies in mammalian cell lines indicate that Parkin phosphorylation at Ser65 is required for its activation, PINK1 and parkin KO flies can both be partially rescued by Ser65Ala Parkin mutant, suggesting that phosphorylation of Parkin displays Ser65 phosphorylation-independent functions in vivo.[168] Furthermore, ectopic expression of human Fbxo7 can rescue parkin KO but not PINK1 KO flies. Conversely, VCP can rescue PINK1 KO flies, but not parkin KO flies. Mitochondrial morphology defects and muscle degeneration displayed by PINK1-deficient flies could also be complemented by TRAP1 or by increasing the signaling mediated by Ret, the receptor for glial cell line-derived neurotrophic factor (GDNF), in a Parkin-independent manner.[18,218] However, Parkin and RET signaling converge to control mitochondrial integrity and maintain DA neurons in two different mouse models.[219] Some of the discrepancies observed in mammalian and in *Drosophila* models may be explained by differences in compensating for PINK1 and Parkin deficiency. Moreover, mitophagy-independent neuroprotective functions of PINK1 and Parkin may play a more or less important role in all these cell and animal models.

6 METHODS TO DETECT MITOPHAGY

In order to fully understand the role of mitophagy in disease, as well as identifying potential drug targets within the pathway, sensitive and reliable methods are required to measure mitophagic levels.

6.1 Inducers of Mitophagy

In healthy mammalian cells, mitophagy is an infrequent event; therefore, studies typically make use of cell stressors to initiate the process. Exposure of cells to the mitochondrial uncoupler CCCP causes a vast increase in membrane proton conductance and consequently severe loss of $\Delta\Psi m$.[220] However, CCCP is a nonspecific ionophore with several off-target effects, including the interference with autophagosomal degradation and lysosomal function.[221] Moreover, application of CCCP in neuronal cultures can trigger apoptotic processes before Parkin translocation is observed, therefore its use is not recommended in all cell types.[222]

Alternative mitophagy-inducing compounds include valinomycin, a K^+ ionophore, or the combination of antimycin A and oligomycin, inhibitors of complex III and ATP synthase respectively. Mitophagy can also be initiated by mtDNA mutations, hepatitis C virus infection, or an excessive amount of unfolded proteins in the mitochondrial matrix.[121,162–165]

A substantial disadvantage to the use of all these compounds as a model of mitophagy is that the entire mitochondrial network becomes depolarized. Cellular response to the immediate global loss of a functional mitochondrial network is likely to differ from the relatively sporadic occurrence of mitophagy, which hampers the validity of this model. Although biochemical studies benefit from the widespread induction of mitophagy, damage to subsets of mitochondria more closely resemble physiological conditions.[11] Spatiotemporally controlled ROS-mediated mitochondrial damage can be attained by photobleaching neurons containing mitochondrial KillerRed (mt-KR).[223] Increased ROS levels in the mitochondrial matrix, as opposed to a general increase in cellular ROS, has been shown to activate Parkin recruitment to mitochondria.[224] In order to only damage a subset of mitochondria while preserving the integrity of the mitochondrial network, microfluidic devices may be used. This technique has been shown to typically target three or four mitochondria, inducing mitophagy, while preserving the overall integrity of the network. An advantage of inducing targeted local damage to mitochondria is that mitophagy can be studied with fewer off-target effects caused by nonspecific potent compounds and without the activation of other damage response pathways when nonphysiological levels of damage are induced. Nevertheless, targeting few mitochondria at a time means that approaches to study the mitophagy pathway are largely limited to the use of imaging techniques.

6.2 Techniques for Detecting Mitophagy

Several approaches can be employed to monitor mitophagy but these vary in sensitivity and reliability when quantified. Traditional electron microscopy techniques provide absolute confirmation of the engulfment of dysfunctional mitochondria by the autophagosome, however quantification of mitophagy requires lengthy analysis. The use of this technique with statistically relevant sample sizes would be limited due to its time-consuming nature and consequently give rise to biases. Mitophagy is a transient process and as a result, challenging to reliably quantify. The approaches most commonly used, and that will be discussed in this section, are based on the measurement of mitochondrial protein expression, or the delivery of mitochondria to autophagic machinery.[225]

6.2.1 Expression of Mitochondrial Proteins

Mitophagy can be quantified by the loss of expression of mitochondrial proteins using immunoblot assays or immunostaining. This technique is used to conveniently confirm the elimination of mitochondria, however changes in protein levels can be heavily influenced, for example, by proteasomal degradation pathways and alterations in biogenesis.[226] Mitophagy may be underestimated using this method of quantification when substantial mitochondrial biosynthesis is occurring simultaneously.

Measurement of mitochondrial content typically utilizes markers of several mitochondrial proteins or quantification of mitochondrial DNA. Early studies of mitophagy used degradation of TOM20, an OMM protein, as a marker of mitochondrial mass.[75] However, Parkin has also been shown to mediate proteasome-dependent degradation of OMM proteins, in addition to mitophagy, therefore, loss of OMM proteins alone is not a sufficient measure.[227] Furthermore, it should be considered that intermembrane space proteins can be released after permeability transition[228] and that turnover rate of different inner membrane proteins can be highly variable, even within the same respiratory complex.[229] As a result, mitochondrial protein expression assays should utilize several proteins from different mitochondrial subcompartments to provide precise measurements of mitophagy. Similarly, loss of mitochondrial DNA can be measured using quantitative PCR or by immunostaining using DNA antibodies.[209]

Assessing mitophagy through the use of mitochondrial markers is a simple and quantitative method to determine complete degradation of mitochondria. However, this assay quantifies the net change in mitochondrial content and therefore is unable to distinguish between mitophagy and degradation by other mechanisms that may occur simultaneously.

6.2.2 Colocalization With Markers of Autophagic Machinery

Fluorescent microscopy can be used to monitor mitophagy by assessing the colocalization of mitochondria and autophagosomes. Most commonly this involves transfection of the autophagosome-specific marker GFP-LC3 and visualizing mitochondria using a marker independent of membrane potential.[225] For immunocytochemical experiments in fixed cells, antibodies against many mitochondrial proteins, such as VDAC, can be used to visualize mitochondria, however when imaging in live cells, fewer dyes are suitable. MitoTracker Green FM reacts with free thiol groups of cysteine residues, covalently binding mitochondrial proteins; therefore, unlike tetramethylrhodamine, methyl ester (TMRM), and certain other MitoTracker dyes, staining is not lost with mitochondrial membrane potential.[225,230]

Analysis of colocalization often involves merging images captured with two different channels and counting the areas of overlap, or assessing the intensity profile of each channel at these colocalized areas.[231] This information is used to quantify the mitochondria sequestered into the autophagosome, however, its accuracy relies on comparable gray-level (fluorescence intensity) dynamics between the two channels during image acquisition. A significant advantage of this technique is the ability to adapt its application to high-throughput systems, with improved sensitivity and specificity to selective mitophagy.[232,233]

The use of LC3-based probes however has some limitations. LC3 has been shown to aggregate in an autophagy-independent manner, leading to a high rate of false positives.[234] In reality, the interaction of mitochondria and autophagosomes is transient and colocalization is rare unless lysosomal degradation is blocked.[225]

6.2.3 Delivery to the Lysosome

Following engulfment of defective mitochondria, autophagosomes fuse with hydrolase-containing lysosomes.[235] Antibodies against lysosomal proteins, such as lysosomal-associated membrane protein 1 (LAMP1) can be used in conjunction with mitochondrial markers to assess mitophagy by immunocytochemistry. The use of LAMP1 as a late endosome/lysosome marker may be preferable to monitoring autophagosome colocalization with mitochondria, due to the high rate of false positives observed with the formation of GFP-LC3 aggregates. Analysis of the colocalization between mitochondrial proteins and lysosomes should be quantified by strength of correlation, rather than subjective methods of analysis, such as the counting of puncta. To visualize the delivery to the lysosome in live cell imaging, lysosomes can be labeled with LysoTracker dyes. LysoTracker is retained in acidic subcellular compartments, however, not exclusively autolysosomes. Moreover, live imaging over long time-courses may cause an increase in intracellular pH with the accumulation of LysoTracker, which can result in quenching of the fluorescent dye.[236] Colocalization of autophagic markers represents the interaction of mitochondria with the autophagosomes and subsequent fusion of mitochondria-containing autophagosomes with lysosomes, however, these methods do not quantify the complete and successful degradation of mitochondria.

An alternative method to measure mitophagy makes use of the acidic environment of lysosomes. Localization of mitochondria can be depicted by the different sensitivities of fluorescent proteins to the lysosomal environment.[237] One such assay consists of cells expressing a mitochondrial-targeted tandem mCherry-GFP tag, which under neutral conditions fluoresce red and green.[214] When mitochondria are delivered to lysosomes, the GFP signal, but not the mCherry, is quenched by the low pH and the mitochondria fluoresce red. Loss of GFP fluorescence in response to pH change occurs rapidly, <1 ms, as a result of protonation of the fluorophore.[238] In this assay only the GFP signal is quenched in lysosomes because its pKa, defined as the pH at which the fluorescence signal is at 50% of its maximum, is relatively high compared to mCherry. Mitophagy can be quantified using this technique by counting the red-alone puncta within fixed cells and determining a threshold for the minimum number of punta that constitutes a cell undergoing mitophagy.

In contrast to utilizing pH-dependent loss of fluorescence as a mitophagy marker, several proteins emit different color fluorescent signals at acidic and neutral pHs.[239]

In addition to its pH-dependent fluorescent properties, the coral-derived molecule, Keima, is resistant to lysosomal degradation, which is of particular benefit in mitophagy assays. This assay can be used to produce a cumulative readout without the requirement for complex time-lapse imaging.[239] A mitochondrial matrix-targeted form, mt-Keima, can be used to identify mitophagy through a shift of excitation spectrum peak from 440 to 586 nm upon delivery to lysosomes. The bimodal excitation spectrum of Keima corresponds to the neutral and ionized states of the chromophore's phenolic hydroxyl moiety.[240] Mitophagy can be quantified by finding the percentage of pixels that fluoresce red, or using fluorescence-activated cell sorting (FACS) and an arbitrary threshold to denote a cell undergoing mitophagy.[209,241] This technique has been used to provide a robust study of mitophagy in vivo, in which mouse embryonic fibroblasts derived from an mt-Keima reporter mouse were imaged and resultant red mt-Keima signal shown to be confined to lysosomes.[241]

6.3 Limitations of PINK1/Parkin-Dependent Mitophagy Studies

It has been shown that endogenous PINK1/Parkin-dependent mitophagy occurs in mouse primary hippocampal neurons over a period of days, without pharmacological intervention.[195] However, in studies using neuroblastoma cell lines, fibroblasts, and IPSC-derived neurons, endogenous Parkin is not sufficient to induce mitophagy despite the use of depolarizing agents.[242] As a result, overexpression of Parkin is usually required to study PINK1-dependent mitophagy, at least with the techniques currently available, and knowledge of the pathway to date is primarily based on Parkin-overexpressing nonneuronal cell models. Assays, such as the mCherry-GFP fluorescence assay, can be used as screens for inducers of mitophagy but are not restricted to those that recruit PINK1 and Parkin.[214] The assessment of mitochondrial protein expression and detection of colocalization between mitochondria and autophagic machinery also do not exclude the activation of Parkin-independent mitophagy.[243] It remains unclear how well the current model of mitophagy translates to the process in neurons and its physiological relevance in PD pathogenesis.

6.4 Animal Models to Study Mitophagy

Animal models are essential for studying cellular processes in the context of larger functional networks. Mouse models of disease are desirable due to the large similarities between the human and mouse genome, and the ease of genetic manipulation. Both Parkin- and PINK1-KO mice have been generated for use as models of Parkinson's disease, however, neither have shown neurodegenerative phenotype.

In an effort to generate a valid disease model, embryonic KO mice for Parkin have included deletions of exons 2, 3, and 7 of the *PARK2* gene; and exons 4–7 to generate PINK-KO mice.[244-247] PINK1- and Parkin-KO mice display mitochondrial dysfunction in the CNS; in particular, aged PINK1-KO mice show complex I deficiency that can be rescued by wild-type PINK1, but not PINK1 containing clinical mutations.[129] Nevertheless, regardless of disruptions to mitochondrial protein expression or function, dopaminergic neurons remain viable in all PINK1- and Parkin-KO mice models. Aged Parkin/DJ-1/ PINK1 triple knockout mice show no nigral degeneration, despite the requirement for each of these proteins for the survival of human DA neurons.[248]

Furthermore, PINK1- and Parkin-KO mice lack obvious motor phenotypes, which may be a consequence of using embryonic models that facilitate potential developmental compensation. In support of this, a conditional Cre-loxP exon 7 Parkin-KO mouse exhibits a progressive loss of dopaminergic neurons, however any behavioral alterations have not been reported.[86]

As discussed previously, PINK1- and Parkin-KO flies display mitochondrial morphology defects, dopaminergic degeneration, and impairment of flight muscles. The generation of these mutant flies has been instrumental in understanding the involvement of PINK1 and Parkin in the same pathway. In addition, studies in *Drosophila* have also identified other regulators of the PINK1/Parkin pathway, such as HtrA2. Through the use of double-mutant combinations, HtrA2 was identified to act in a parallel pathway to Parkin downstream of PINK1, since unlike PINK1/HtrA2 double mutants, Parkin/HtrA2 mutants exhibited a more severe phenotype than either alone.[31]

The use of *Drosophila* in models of disease has several advantages compared to higher organisms. *Drosophila* have a short life-cycle and can easily be maintained in large numbers, which makes them well suited for generating mutant stocks and performing genetic or chemical compound screens.

Importantly, the use of *Drosophila* has been essential for studying PINK1/Parkin-mediated mitophagy in vivo. Much of the PINK1/ Parkin-dependent mitophagy pathway has been characterized in cultured cells overexpressing Parkin, treated with nonspecific mitochondrial uncouplers. However, studies in *Drosophila* demonstrated the occurrence of this pathway under physiological conditions for the first time.[217,249] Parkin null *Drosophila* mutants were found to exhibit a significant slowing of mitochondrial protein turnover, comparable to that seen in autophagy-deficient Atg7 mutants.[217] Furthermore, dopaminergic neurons from *Drosophila* parkin mutants accumulate enlarged, depolarized mitochondria to a greater extent than found in cholinergic neurons, demonstrating the selective vulnerability of dopaminergic neurons observed in PD and further validating their use as a disease model.[249]

7　OUTSTANDING QUESTIONS

While the current body of evidence highlights the PINK1/Parkin mitophagy pathway as an important MQC mechanism, many questions remain unanswered.

7.1　Do Endogenous PINK1 and Parkin Modulate Mitophagy in Neurons?

Much of the work characterizing the PINK1-Parkin-mediated pathway has been carried out in immortalized cell lines, and whether exogenous Parkin can translocate to depolarized mitochondria in neurons remains controversial. Critically, some studies even suggest that endogenous Parkin doesn't mediate mitophagy in neurons and in cell lines.[167,222,242,250,251] The differences observed between the different studies may be due to a number of factors, including the neuronal type, the age of the cultures, the presence of apoptotic inhibitors, and B-27 supplement in the culture medium, the presence or not of a glial bed and the type of mitophagy-inducing agent.[252] Whether Parkin is recruited to depolarized mitochondria in neurons, and if so whether endogenous PINK1 and Parkin contribute to selective mitophagy in neurons will thus require more investigation.

7.2　Why do Different Cell Types Suffer Differently From Loss of PINK1 and Parkin?

OXPHOS in the mitochondria and glycolysis in the cytosol are the two main pathways to generate ATP in mammalian cells. Neurons divert glucose away from glycolysis to the pentose phosphate pathway—in order to maintain high levels of reduced glutathione—and instead rely mainly, although not exclusively, on mitochondrial OXPHOS for energy production.[10,253,254] As a result, neurons are particularly vulnerable to mitochondrial damage, and mutations in mitochondrial genes, such as *Mfn2* and

Opa1 cause neurodegenerative conditions, Charcot Marie Tooth disease type 2A and dominant optic atrophy, respectively.[255]

On the contrary to neurons, immortalized cell lines preferentially use aerobic glycolysis for ATP production.[256] By comparing neurons and myocytes, it was shown that PINK1 deficiency results in different responses in mitochondria depending on the energy metabolism background.[257] Furthermore several lines of evidence suggest that these bioenergetic differences between cell types explain, at least partially, the discrepancies observed with regards to mitophagy.[258] First, it was shown that yeast readily undergo mitophagy under starvation conditions. However, when grown in the presence of a media forcing cells into dependence on OXPHOS for energy production, mitophagy levels are considerably decreased, even under severe starvation conditions.[259] Second, it was shown that Parkin is recruited to depolarized mitochondria in a glycolytic, but not OXPHOS-dependent, Hela immortalized cell line.[174,250] Until now the mitophagy process has been primarily characterized in Hela and SHSY5Y cells. These immortalized cell lines preferentially generate ATP through glycolysis, and thus do not rely on OXPHOS so are much less dependent on mitochondria. In neurons, Parkin translocation to mitochondria depolarized by CCCP is considerably lower, if not absent, likely due to low glycolytic flux. However maintaining the ATP levels in neurons after mitochondrial depolarization increases Parkin recruitment to mitochondria. Altogether these data suggest that bioenergetic differences between neurons and cultured cell lines contribute to different levels of mitophagy in these cells.[244] Given the importance of mitochondria to neuronal function, it is likely that other regulatory mechanisms to prevent excessive degradation of mitochondria will be identified. These may include balancing ROS and antioxidant levels,[251,260] increasing mitochondrial fusion, or boosting mitochondrial biogenesis.[11]

In addition to mitophagy defects, PINK1 deficiency is associated with Parkin-independent functions, such as mitochondrial calcium overload, increased ROS production, and complex I deficiency.[129,131] Whether all these stresses happen in a sequential manner or whether they accumulate and reinforce each other up to a breaking point remains to be determined. The final straw is likely to depend on individual predispositions (eg, genetic background, age) and the specific vulnerabilities of various cell types to mitochondrial dysfunctions.[261]

7.3 Do Endogenous PINK1 and Parkin Modulate Mitophagy in Neurons In Vivo?

The lack of robust methods to assess mitophagy in vivo has hampered progress in determining the precise function of mitophagy in normal physiology and in disease. However, parkin null mutant flies show a significant slowing of mitochondrial protein turnover, similar to the slowing observed in autophagy-deficient Atg7 mutants. This suggests that Parkin acts upstream of Atg7 to promote mitophagy in vivo.[93] In addition, a recent study was able to measure mitophagy in mice in vivo. This study shows that mitophagy declines with age, suggesting that mitophagy is a contributor to neurodegenerative diseases, and aging itself.[241] Whether PINK1 and Parkin modulate mitophagy in neurons in vivo however remains to be confirmed.

7.4 Does Mitophagy Represent a Pathway to Treat or Prevent PD?

Temporary symptomatic relief remains the cornerstone of current treatments, with no disease-modifying therapies yet available. Identifying targets for disease-modifying therapies is a key aim of research into this disorder, an aim that has received a major boost by recent advances in our understanding of the genetic architecture of the disorder. Since the identification of *PINK1* mutations in 2004,[110] our understanding of the role of PINK1 and Parkin in mitochondrial biology has hugely increased. Modulating mitochondrial physiology using genetic or chemical strategies are able to overcome neuronal death secondary to PINK1 deficiency. However there has been a major challenge in translating compounds that improve mitochondrial health into successful clinical trials. This may be in part attributed to the way these compounds improve mitochondrial health (including normal mitochondria), for example by altering electron transport (coQ10, vitamin K_2), by scavenging ROS (vitamin E and mitoQ), or by enhancing nucleotide metabolism,[10,54,128,262] rather than selectively targeting the mitochondrial pathways affected in disease, such as mitophagy. Inhibiting DUBs could represent a therapeutic approach for increasing PINK1-Parkin mediated mitophagy and protect against neurodegeneration.[263] Furthermore, the recent identification of PINK1 as the first ubiquitin kinase is likely to be key in disentangling the complex relationship between dysfunctional mitochondria, mitophagy, and neuronal viability. Finally, the discovery of neosubstrates for PINK1, such as ATP analog kinetin triphosphate (KTP), could provide a modality for regulating PINK1 kinase activity.[264]

While intensive efforts are being made to dissect the mitophagy pathway downstream of PINK1 and Parkin, the regulatory molecular mechanisms upstream of this process remain mostly unknown. Identifying the upstream signaling mechanisms that trigger mitophagy will be of major importance to identify the insults that affect mitochondrial homeostasis in PD and will be key in identifying new therapeutic targets in PD. One such pathway is the Akt signaling pathway, which was shown to regulate Parkin recruitment to depolarized mitochondria and subsequent mitophagy.[174]

7.5 Future Perspectives

To open new therapeutic avenues, it will be essential to understand further the PINK1-Parkin biology (eg, resolving PINK1 crystal structure), the mitophagy process, and the upstream pathways regulating mitophagy. Improving methods to assess mitophagy will be crucial for unraveling the pathway in neurons and in vivo. Finally, combining the recent leap forward in our understanding of the genetic architecture of PD with advances in gene expression, cell biology, and bioinformatics will be key for investigating the functional links between genes implicated in the aetiology of PD and cellular degeneration in the human brain.

References

1. Langston JW, Ballard P, Tetrud JW, Irwin I. Chronic Parkinsonism in humans due to a product of meperidine-analog synthesis. *Science.* 1983;219:979–980.
2. Beal MF. Experimental models of Parkinson's disease. *Nat Rev Neurosci.* 2001;2:325–334.
3. Hirsch EC, et al. Animal models of Parkinson's disease in rodents induced by toxins: an update. *J Neural Transm Suppl.* 2003;65:89–100.
4. Dauer W, Przedborski S. Parkinson's disease: mechanisms and models. *Neuron.* 2003;39:889–909.
5. Parker WD, Boyson SJ, Parks JK. Abnormalities of the electron transport chain in idiopathic Parkinson's disease. *Ann Neurol.* 1989;26:719–723.
6. Swerdlow RH, et al. Origin and functional consequences of the complex I defect in Parkinson's disease. *Ann Neurol.* 1996;40:663–671.
7. Schapira AH, et al. Mitochondrial complex I deficiency in Parkinson's disease. *J Neurochem.* 1990;54:823–827.
8. Bender A, et al. High levels of mitochondrial DNA deletions in substantia nigra neurons in aging and Parkinson disease. *Nat Genet.* 2006;38:515–517.
9. Kraytsberg Y, et al. Mitochondrial DNA deletions are abundant and cause functional impairment in aged human substantia nigra neurons. *Nat Genet.* 2006;38:518–520.
10. Burchell VS, et al. Targeting mitochondrial dysfunction in neurodegenerative disease: part I. *Expert Opin Ther Targets.* 2010;14:369–385.
11. Ashrafi G, Schwarz TL. The pathways of mitophagy for quality control and clearance of mitochondria. *Cell Death Differ.* 2013;20:31–42.
12. Baker MJ, Tatsuta T, Langer T. Quality control of mitochondrial proteostasis. *Cold Spring Harb Perspect Biol.* 2011;3:1–19.
13. Jin J, et al. Identification of novel proteins associated with both alpha-synuclein and DJ-1. *Mol Cell Proteomics MCP.* 2007;6:845–859.
14. Davison EJ, et al. Proteomic analysis of increased Parkin expression and its interactants provides evidence for a role in modulation of mitochondrial function. *Proteomics.* 2009;9:4284–4297.
15. Rakovic A, et al. PINK1-interacting proteins: proteomic analysis of overexpressed PINK1. *Park Dis.* 2011;2011:153979.
16. Burbulla LF, et al. Dissecting the role of the mitochondrial chaperone mortalin in Parkinson's disease: functional impact of disease-related variants on mitochondrial homeostasis. *Hum Mol Genet.* 2010;19:4437–4452.
17. Pridgeon JW, Olzmann JA, Chin L-S, Li L. PINK1 protects against oxidative stress by phosphorylating mitochondrial chaperone TRAP1. *PLoS Biol.* 2007;5:e172.
18. Costa AC, Loh SHY, Martins LM. Drosophila Trap1 protects against mitochondrial dysfunction in a PINK1/parkin model of Parkinson's disease. *Cell Death Dis.* 2013;4:e467.

19. Moriwaki Y, et al. L347P PINK1 mutant that fails to bind to Hsp90/Cdc37 chaperones is rapidly degraded in a proteasome-dependent manner. *Neurosci Res.* 2008;61:43–48.
20. Venkatesh S, Lee J, Singh K, Lee I, Suzuki CK. Multitasking in the mitochondrion by the ATP-dependent Lon protease. *Biochim Biophys Acta.* 2012;1823:56–66.
21. Truscott KN, Bezawork-Geleta A, Dougan DA. Unfolded protein responses in bacteria and mitochondria: a central role for the ClpXP machine. *IUBMB Life.* 2011;63:955–963.
22. Tatsuta T, Langer T. AAA proteases in mitochondria: diverse functions of membrane-bound proteolytic machines. *Res Microbiol.* 2009;160:711–717.
23. Spiess C, Beil A, Ehrmann M. A temperature-dependent switch from chaperone to protease in a widely conserved heat shock protein. *Cell.* 1999;97:339–347.
24. Walsh NP, Alba BM, Bose B, Gross CA, Sauer RT. OMP peptide signals initiate the envelope-stress response by activating DegS protease via relief of inhibition mediated by its PDZ domain. *Cell.* 2003;113:61–71.
25. Moisoi N, et al. Mitochondrial dysfunction triggered by loss of HtrA2 results in the activation of a brain-specific transcriptional stress response. *Cell Death Differ.* 2009;16:449–464.
26. Martins LM, et al. Neuroprotective role of the Reaper-related serine protease HtrA2/Omi revealed by targeted deletion in mice. *Mol Cell Biol.* 2004;24:9848–9862.
27. Strauss KM, et al. Loss of function mutations in the gene encoding Omi/HtrA2 in Parkinson's disease. *Hum Mol Genet.* 2005;14:2099–2111.
28. Bogaerts V, et al. Genetic variability in the mitochondrial serine protease HTRA2 contributes to risk for Parkinson disease. *Hum Mutat.* 2008;29:832–840.
29. Simón-Sánchez J, Singleton AB. Sequencing analysis of OMI/HTRA2 shows previously reported pathogenic mutations in neurologically normal controls. *Hum Mol Genet.* 2008;17:1988–1993.
30. Plun-Favreau H, et al. The mitochondrial protease HtrA2 is regulated by Parkinson's disease-associated kinase PINK1. *Nat Cell Biol.* 2007;9:1243–1252.
31. Tain LS, et al. Drosophila HtrA2 is dispensable for apoptosis but acts downstream of PINK1 independently from Parkin. *Cell Death Differ.* 2009;16:1118–1125.
32. Dagda RK, Chu CT. Mitochondrial quality control: insights on how Parkinson's disease related genes PINK1, parkin, and Omi/HtrA2 interact to maintain mitochondrial homeostasis. *J Bioenerg Biomembr.* 2009;41:473–479.
33. Yun J, et al. Loss-of-function analysis suggests that Omi/HtrA2 is not an essential component of the PINK1/PARKIN pathway in vivo. *J Neurosci Off J Soc Neurosci.* 2008;28:14500–14510.
34. Whitworth AJ, et al. Rhomboid-7 and HtrA2/Omi act in a common pathway with the Parkinson's disease factors Pink1 and Parkin. *Dis Model Mech.* 2008;1:168–174.
35. Jin SM, et al. Mitochondrial membrane potential regulates PINK1 import and proteolytic destabilization by PARL. *J Cell Biol.* 2010;191:933–942.
36. Deas E, Wood NW, Plun-Favreau H. Mitophagy and Parkinson's disease: the PINK1-parkin link. *Biochim Biophys Acta.* 2011;1813:623–633.
37. Karbowski M, Youle RJ. Regulating mitochondrial outer membrane proteins by ubiquitination and proteasomal degradation. *Curr Opin Cell Biol.* 2011;23:476–482.
38. Livnat-Levanon N, Glickman MH. Ubiquitin-proteasome system and mitochondria - reciprocity. *Biochim Biophys Acta.* 2011;1809:80–87.
39. Ross JM, Olson L, Coppotelli G. Mitochondrial and ubiquitin proteasome system dysfunction in ageing and disease: two sides of the same coin? *Int J Mol Sci.* 2015;16:19458–19476.
40. Tanaka A, et al. Proteasome and p97 mediate mitophagy and degradation of mitofusins induced by Parkin. *J Cell Biol.* 2010;191:1367–1380.
41. Kim NC, et al. VCP is essential for mitochondrial quality control by PINK1/Parkin and this function is impaired by VCP mutations. *Neuron.* 2013;78:65–80.
42. Nagashima S, Tokuyama T, Yonashiro R, Inatome R, Yanagi S. Roles of mitochondrial ubiquitin ligase MITOL/MARCH5 in mitochondrial dynamics and diseases. *J Biochem.* 2014;155:273–279.

43. Li W, et al. Genome-wide and functional annotation of human E3 ubiquitin ligases identifies MULAN, a mitochondrial E3 that regulates the organelle's dynamics and signaling. *PloS One*. 2008;3:e1487.
44. Tang F, et al. RNF185, a novel mitochondrial ubiquitin E3 ligase, regulates autophagy through interaction with BNIP1. *PloS One*. 2011;6:e24367.
45. Martinus RD, et al. Selective induction of mitochondrial chaperones in response to loss of the mitochondrial genome. *Eur J Biochem FEBS*. 1996;240:98–103.
46. Zhao Q, et al. A mitochondrial specific stress response in mammalian cells. *EMBO J*. 2002;21:4411–4419.
47. Burbulla LF, et al. Mitochondrial proteolytic stress induced by loss of mortalin function is rescued by Parkin and PINK1. *Cell Death Dis*. 2014;5:e1180.
48. Soubannier V, et al. A vesicular transport pathway shuttles cargo from mitochondria to lysosomes. *Curr Biol CB*. 2012;22:135–141.
49. McLelland G-L, Soubannier V, Chen CX, McBride HM, Fon EA. Parkin and PINK1 function in a vesicular trafficking pathway regulating mitochondrial quality control. *EMBO J*. 2014;33:282–295.
50. Sugiura A, McLelland G-L, Fon EA, McBride HM. A new pathway for mitochondrial quality control: mitochondrial-derived vesicles. *EMBO J*. 2014;33:2142–2156.
51. Malik BR, Godena VK, Whitworth AJ. VPS35 pathogenic mutations confer no dominant toxicity but partial loss of function in Drosophila and genetically interact with parkin. *Hum Mol Genet*. 2015;24:6106–6117.
52. Clark SL. Cellular differentiation in the kidneys of newborn mice studies with the electron microscope. *J Biophys Biochem Cytol*. 1957;3:349–362.
53. Youle RJ, Narendra DP. Mechanisms of mitophagy. *Nat Rev Mol Cell Biol*. 2011;12:9–14.
54. Burchell VS, et al. Targeting mitochondrial dysfunction in neurodegenerative disease: Part II. *Expert Opin Ther Targets*. 2010;14:497–511.
55. Davis CO, et al. Transcellular degradation of axonal mitochondria. *Proc Natl Acad Sci USA*. 2014;111:9633–9638.
56. Phinney DG, et al. Mesenchymal stem cells use extracellular vesicles to outsource mitophagy and shuttle microRNAs. *Nat Commun*. 2015;6:8472.
57. Martin SJ. Cell biology. Opening the cellular poison cabinet. *Science*. 2010;330:1330–1331.
58. Wang X, Chen XJ. A cytosolic network suppressing mitochondria-mediated proteostatic stress and cell death. *Nature*. 2015;524:481–484.
59. Durcan TM, Fon EA. The three 'P's of mitophagy: PARKIN, PINK1, and post-translational modifications. *Genes Dev*. 2015;29:989–999.
60. Kitada T, et al. Mutations in the parkin gene cause autosomal recessive juvenile parkinsonism. *Nature*. 1998;392:605–608.
61. Shimura H, et al. Familial Parkinson disease gene product, parkin, is a ubiquitin-protein ligase. *Nat Genet*. 2000;25:302–305.
62. Walden H, Martinez-Torres RJ. Regulation of Parkin E3 ubiquitin ligase activity. *Cell Mol Life Sci CMLS*. 2012;69:3053–3067.
63. Spratt DE, Walden H, Shaw GS. RBR E3 ubiquitin ligases: new structures, new insights, new questions. *Biochem J*. 2014;458:421–437.
64. Trempe J-F, et al. Structure of parkin reveals mechanisms for ubiquitin ligase activation. *Science*. 2013;340:1451–1455.
65. Riley BE, et al. Structure and function of Parkin E3 ubiquitin ligase reveals aspects of RING and HECT ligases. *Nat Commun*. 2013;4:1982.
66. Wauer T, Komander D. Structure of the human Parkin ligase domain in an autoinhibited state. *EMBO J*. 2013;32:2099–2112.
67. Spratt DE, Walden H, Shaw GS. RBR E3 ubiquitin ligases: new structures, new insights, new questions. *Biochem J*. 2014;458:421–437.
68. Spratt DE, Walden H, Shaw GS. RBR E3 ubiquitin ligases: new structures, new insights, new questions. *Biochem J*. 2014;458:421–437.

69. Kasap M, Akpinar G, Sazci A, Idrisoglu HA, Vahaboğlu H. Evidence for the presence of full-length PARK2 mRNA and Parkin protein in human blood. *Neurosci Lett.* 2009;460:196–200.

70. Nakaso K, Adachi Y, Yasui K, Sakuma K, Nakashima K. Detection of compound heterozygous deletions in the parkin gene of fibroblasts in patients with autosomal recessive hereditary parkinsonism (PARK2). *Neurosci Lett.* 2006;400:44–47.

71. Huynh DP, Nguyen DT, Pulst-Korenberg JB, Brice A, Pulst S-M. Parkin is an E3 ubiquitin-ligase for normal and mutant ataxin-2 and prevents ataxin-2-induced cell death. *Exp Neurol.* 2007;203:531–541.

72. Kubo SI, et al. Parkin is associated with cellular vesicles. *J Neurochem.* 2001;78:42–54.

73. Imai Y, et al. An unfolded putative transmembrane polypeptide, which can lead to endoplasmic reticulum stress, is a substrate of Parkin. *Cell.* 2001;105:891–902.

74. Darios F, et al. Parkin prevents mitochondrial swelling and cytochrome c release in mitochondria-dependent cell death. *Hum Mol Genet.* 2003;12:517–526.

75. Narendra D, Tanaka A, Suen D-F, Youle RJ. Parkin is recruited selectively to impaired mitochondria and promotes their autophagy. *J Cell Biol.* 2008;183:795–803.

76. Muqit MMK, et al. Parkin is recruited into aggresomes in a stress-specific manner: over-expression of parkin reduces aggresome formation but can be dissociated from parkin's effect on neuronal survival. *Hum Mol Genet.* 2004;13:117–135.

77. Huynh DP, Dy M, Nguyen D, Kiehl TR, Pulst SM. Differential expression and tissue distribution of parkin isoforms during mouse development. *Brain Res Dev Brain Res.* 2001;130:173–181.

78. Zhang Y, et al. Parkin functions as an E2-dependent ubiquitin-protein ligase and promotes the degradation of the synaptic vesicle-associated protein, CDCrel-1. *Proc Natl Acad Sci USA.* 2000;97:13354–13359.

79. Cookson MR, et al. RING finger 1 mutations in Parkin produce altered localization of the protein. *Hum Mol Genet.* 2003;12:2957–2965.

80. Sriram SR, et al. Familial-associated mutations differentially disrupt the solubility, localization, binding and ubiquitination properties of parkin. *Hum Mol Genet.* 2005;14:2571–2586.

81. Wang C, et al. Alterations in the solubility and intracellular localization of parkin by several familial Parkinson's disease-linked point mutations. *J Neurochem.* 2005;93:422–431.

82. Seirafi M, Kozlov G, Gehring K. Parkin structure and function. *F EBS J.* 2015;282:2076–2088.

83. Winklhofer KF. Parkin and mitochondrial quality control: toward assembling the puzzle. *Trends Cell Biol.* 2014;24:332–341.

84. Kahle PJ, Haass C. How does parkin ligate ubiquitin to Parkinson's disease? *EMBO Rep.* 2004;5:681–685.

85. Lee Y, et al. Parthanatos mediates AIMP2-activated age-dependent dopaminergic neuronal loss. *Nat Neurosci.* 2013;16:1392–1400.

86. Shin J-H, et al. PARIS (ZNF746) repression of PGC-1α contributes to neurodegeneration in Parkinson's disease. *Cell.* 2011;144:689–702.

87. Olzmann JA, et al. Parkin-mediated K63-linked polyubiquitination targets misfolded DJ-1 to aggresomes via binding to HDAC6. *J Cell Biol.* 2007;178:1025–1038.

88. Hebron ML, et al. Parkin ubiquitinates Tar-DNA binding protein-43 (TDP-43) and promotes its cytosolic accumulation via interaction with histone deacetylase 6 (HDAC6). *J Biol ChemV 288.* 2013;4103–4115.

89. Ekholm-Reed S, Goldberg MS, Schlossmacher MG, Reed SI. Parkin-dependent degradation of the F-box protein Fbw7β promotes neuronal survival in response to oxidative stress by stabilizing Mcl-1. *Mol Cell Biol.* 2013;33:3627–3643.

90. Ziviani E, Tao RN, Whitworth AJ. Drosophila parkin requires PINK1 for mitochondrial translocation and ubiquitinates mitofusin. *Proc Natl Acad Sci USA.* 2010;107:5018–5023.

91. Gegg ME, et al. Mitofusin 1 and mitofusin 2 are ubiquitinated in a PINK1/parkin-dependent manner upon induction of mitophagy. *Hum Mol Genet.* 2010;19:4861–4870.

92. Wang X, et al. PINK1 and Parkin target Miro for phosphorylation and degradation to arrest mitochondrial motility. *Cell.* 2011;147:893–906.

93. Vincow ES, et al. The PINK1-Parkin pathway promotes both mitophagy and selective respiratory chain turnover in vivo. *Proc Natl Acad Sci USA*. 2013;110:6400–6405.

94. Kuroda Y, et al. Parkin enhances mitochondrial biogenesis in proliferating cells. *Hum Mol Genet*. 2006;15:883–895.

95. Rothfuss O, et al. Parkin protects mitochondrial genome integrity and supports mitochondrial DNA repair. *Hum Mol Genet*. 2009;18:3832–3850.

96. Pickrell AM, et al. Endogenous Parkin preserves dopaminergic substantia nigral neurons following mitochondrial DNA mutagenic stress. *Neuron*. 2015;87:371–381.

97. Müller-Rischart AK, et al. The E3 ligase parkin maintains mitochondrial integrity by increasing linear ubiquitination of NEMO. *Mol Cell*. 2013;49:908–921.

98. Hwang S, et al. Parkin suppresses c-Jun N-terminal kinase-induced cell death via transcriptional regulation in Drosophila. *Mol Cells*. 2010;29:575–580.

99. Johnson BN, Berger AK, Cortese GP, Lavoie MJ. The ubiquitin E3 ligase parkin regulates the proapoptotic function of Bax. *Proc Natl Acad Sci USA*. 2012;109:6283–6288.

100. Fallon L, et al. A regulated interaction with the UIM protein Eps15 implicates parkin in EGF receptor trafficking and PI(3)K-Akt signalling. *Nat Cell Biol*. 2006;8:834–842.

101. da Costa CA, et al. Transcriptional repression of p53 by parkin and impairment by mutations associated with autosomal recessive juvenile Parkinson's disease. *Nat Cell Biol*. 2009;11:1370–1375.

102. Zhang C, et al. Parkin, a p53 target gene, mediates the role of p53 in glucose metabolism and the Warburg effect. *Proc Natl Acad Sci USA*. 2011;108:16259–16264.

103. Mira MT, et al. Susceptibility to leprosy is associated with PARK2 and PACRG. *Nature*. 2004;427:636–640.

104. Manzanillo PS, et al. The ubiquitin ligase parkin mediates resistance to intracellular pathogens. *Nature*. 2013;501:512–516.

105. Veeriah S, et al. Somatic mutations of the Parkinson's disease-associated gene PARK2 in glioblastoma and other human malignancies. *Nat Genet*. 2010;42:77–82.

106. Plun-Favreau H, Lewis PA, Hardy J, Martins LM, Wood NW. Cancer and neurodegeneration: between the devil and the deep blue sea. *PLoS Genet*. 2010;6:e1001257.

107. Devine MJ, Plun-Favreau H, Wood NW. Parkinson's disease and cancer: two wars, one front. *Nat Rev Cancer*. 2011;11:812–823.

108. Scarffe LA, Stevens DA, Dawson VL, Dawson TM. Parkin and PINK1: much more than mitophagy. *Trends Neurosci*. 2014;37:315–324.

109. Unoki M, Nakamura Y. Growth-suppressive effects of BPOZ and EGR2, two genes involved in the PTEN signaling pathway. *Oncogene*. 2001;20:4457–4465.

110. Valente EM, et al. Hereditary early-onset Parkinson's disease caused by mutations in PINK1. *Science*. 2004;304:1158–1160.

111. Pickrell AM, Youle RJ. The roles of PINK1, parkin, and mitochondrial fidelity in Parkinson's disease. *Neuron*. 2015;85:257–273.

112. Deas E, Plun-Favreau H, Wood NW. PINK1 function in health and disease. *EMBO Mol Med*. 2009;1:152–165.

113. Khan NL, et al. Clinical and subclinical dopaminergic dysfunction in PARK6-linked parkinsonism: an 18F-dopa PET study. *Ann Neurol*. 2002;52:849–853.

114. Kato H, Lu Q, Rapaport D, Kozjak-Pavlovic V. Tom70 is essential for PINK1 import into mitochondria. *PloS One*. 2013;8:e58435.

115. Jin SM, et al. Mitochondrial membrane potential regulates PINK1 import and proteolytic destabilization by PARL. *J Cell Biol*. 2010;191:933–942.

116. Meissner C, Lorenz H, Weihofen A, Selkoe DJ, Lemberg MK. The mitochondrial intramembrane protease PARL cleaves human Pink1 to regulate Pink1 trafficking. *J Neurochem*. 2011;117:856–867.

117. Deas E, et al. PINK1 cleavage at position A103 by the mitochondrial protease PARL. *Hum Mol Genet*. 2011;20:867–879.

118. Greene AW, et al. Mitochondrial processing peptidase regulates PINK1 processing, import and Parkin recruitment. *EMBO Rep.* 2012;13:378–385.

119. Deas E, et al. PINK1 cleavage at position A103 by the mitochondrial protease PARL. *Hum Mol Genet.* 2011;20:867–879.

120. Thomas RE, Andrews LA, Burman JL, Lin W-Y, Pallanck LJ. PINK1-Parkin pathway activity is regulated by degradation of PINK1 in the mitochondrial matrix. *PLoS Genet.* 2014;10:e1004279.

121. Jin SM, Youle RJ. The accumulation of misfolded proteins in the mitochondrial matrix is sensed by PINK1 to induce PARK2/Parkin-mediated mitophagy of polarized mitochondria. *Autophagy.* 2013;9:1750–1757.

122. Yamano K, Youle RJ. PINK1 is degraded through the N-end rule pathway. *Autophagy.* 2013;9:1758–1769.

123. Nakajima A, Kataoka K, Hong M, Sakaguchi M, Huh N. BRPK, a novel protein kinase showing increased expression in mouse cancer cell lines with higher metastatic potential. *Cancer Lett.* 2003;201:195–201.

124. Gandhi S, et al. PINK1 protein in normal human brain and Parkinson's disease. *Brain J Neurol.* 2006;129:1720–1731.

125. Silvestri L, et al. Mitochondrial import and enzymatic activity of PINK1 mutants associated to recessive parkinsonism. *Hum Mol Genet.* 2005;14:3477–3492.

126. Fedorowicz MA, et al. Cytosolic cleaved PINK1 represses Parkin translocation to mitochondria and mitophagy. *EMBO Rep.* 2014;15:86–93.

127. Weihofen A, Ostaszewski B, Minami Y, Selkoe DJ. Pink1 Parkinson mutations, the Cdc37/Hsp90 chaperones and Parkin all influence the maturation or subcellular distribution of Pink1. *Hum Mol Genet.* 2008;17:602–616.

128. Morais VA, et al. PINK1 loss-of-function mutations affect mitochondrial complex I activity via NdufA10 ubiquinone uncoupling. *Science.* 2014;344:203–207.

129. Morais VA, et al. Parkinson's disease mutations in PINK1 result in decreased complex I activity and deficient synaptic function. *EMBO Mol Med.* 2009;1:99–111.

130. Pogson JH, et al. The complex I subunit NDUFA10 selectively rescues Drosophila pink1 mutants through a mechanism independent of mitophagy. *PLoS Genet.* 2014;10:e1004815.

131. Gandhi S, et al. PINK1-associated Parkinson's disease is caused by neuronal vulnerability to calcium-induced cell death. *Mol Cell.* 2009;33:627–638.

132. Gautier CA, Kitada T, Shen J. Loss of PINK1 causes mitochondrial functional defects and increased sensitivity to oxidative stress. *Proc Natl Acad Sci USA.* 2008;105:11364–11369.

133. Poole AC, et al. The PINK1/Parkin pathway regulates mitochondrial morphology. *Proc Natl Acad Sci USA.* 2008;105:1638–1643.

134. Geisler S, et al. PINK1/Parkin-mediated mitophagy is dependent on VDAC1 and p62/SQSTM1. *Nat Cell Biol.* 2010;12:119–131.

135. Narendra DP, et al. PINK1 is selectively stabilized on impaired mitochondria to activate Parkin. *PLoS Biol.* 2010;8:e1000298.

136. O'Flanagan CH, Morais VA, Wurst W, De Strooper B, O'Neill C. The Parkinson's gene PINK1 regulates cell cycle progression and promotes cancer-associated phenotypes. *Oncogene.* 2015;34:1363–1374.

137. O'Flanagan CH, O'Neill C. PINK1 signalling in cancer biology. *Biochim Biophys Acta.* 2014;1846:590–598.

138. Chen Y, Dorn GW. PINK1-phosphorylated mitofusin 2 is a Parkin receptor for culling damaged mitochondria. *Science.* 2013;340:471–475.

139. Narendra D, Tanaka A, Suen D-F, Youle RJ. Parkin is recruited selectively to impaired mitochondria and promotes their autophagy. *J Cell Biol.* 2008;183:795–803.

140. Woodroof HI, et al. Discovery of catalytically active orthologues of the Parkinson's disease kinase PINK1: analysis of substrate specificity and impact of mutations. *Open Biol.* 2011;1:110012.

141. Kondapalli C, et al. PINK1 is activated by mitochondrial membrane potential depolarization and stimulates Parkin E3 ligase activity by phosphorylating Serine 65. *Open Biol.* 2012;2:120080.

142. Liu S, et al. Parkinson's disease-associated kinase PINK1 regulates Miro protein level and axonal transport of mitochondria. *PLoS Genet.* 2012;8:e1002537.

143. Lai Y-C, et al. Phosphoproteomic screening identifies Rab GTPases as novel downstream targets of PINK1. *EMBO J.* 2015;34:2840–2861.

144. Shiba-Fukushima K, et al. PINK1-mediated phosphorylation of the Parkin ubiquitin-like domain primes mitochondrial translocation of Parkin and regulates mitophagy. *Sci Rep.* 2012;2:1002.

145. Kazlauskaite A, et al. Phosphorylation of Parkin at Serine65 is essential for activation: elaboration of a Miro1 substrate-based assay of Parkin E3 ligase activity. *Open Biol.* 2014;4:130213.

146. Lazarou M, et al. PINK1 drives Parkin self-association and HECT-like E3 activity upstream of mitochondrial binding. *J Cell Biol.* 2013;200:163–172.

147. Kazlauskaite A, et al. Parkin is activated by PINK1-dependent phosphorylation of ubiquitin at Ser65. *Biochem J.* 2014;460:127–139.

148. Kane LA, et al. PINK1 phosphorylates ubiquitin to activate Parkin E3 ubiquitin ligase activity. *J Cell Biol.* 2014;205:143–153.

149. Koyano F, et al. Ubiquitin is phosphorylated by PINK1 to activate parkin. *Nature.* 2014;510:162–166.

150. Clark IE, et al. Drosophila pink1 is required for mitochondrial function and interacts genetically with parkin. *Nature.* 2006;441:1162–1166.

151. Park J, et al. Mitochondrial dysfunction in Drosophila PINK1 mutants is complemented by parkin. *Nature.* 2006;441:1157–1161.

152. Yang Y, et al. Mitochondrial pathology and muscle and dopaminergic neuron degeneration caused by inactivation of Drosophila Pink1 is rescued by Parkin. *Proc Natl Acad Sci USA.* 2006;103:10793–10798.

153. Greene JC, et al. Mitochondrial pathology and apoptotic muscle degeneration in Drosophila parkin mutants. *Proc Natl Acad Sci USA.* 2003;100:4078–4083.

154. Narendra D, Tanaka A, Suen D-F, Youle RJ. Parkin is recruited selectively to impaired mitochondria and promotes their autophagy. *J Cell Biol.* 2008;183:795–803.

155. Vives-Bauza C, et al. PINK1-dependent recruitment of Parkin to mitochondria in mitophagy. *Proc Natl Acad Sci USA.* 2010;107:378–383.

156. Matsuda N, et al. PINK1 stabilized by mitochondrial depolarization recruits Parkin to damaged mitochondria and activates latent Parkin for mitophagy. *J Cell Biol.* 2010;189:211–221.

157. Lazarou M, Jin SM, Kane LA, Youle RJ. Role of PINK1 binding to the TOM complex and alternate intracellular membranes in recruitment and activation of the E3 ligase Parkin. *Dev Cell.* 2012;22:320–333.

158. Zhou C, et al. The kinase domain of mitochondrial PINK1 faces the cytoplasm. *Proc Natl Acad Sci USA.* 2008;105:12022–12027.

159. Hasson SA, et al. High-content genome-wide RNAi screens identify regulators of parkin upstream of mitophagy. *Nature.* 2013;504:291–295.

160. Okatsu K, Kimura M, Oka T, Tanaka K, Matsuda N. Unconventional PINK1 localization to the outer membrane of depolarized mitochondria drives Parkin recruitment. *J Cell Sci.* 2015;128:964–978.

161. Becker D, Richter J, Tocilescu MA, Przedborski S, Voos W. Pink1 kinase and its membrane potential ($\Delta\psi$ -dependent cleavage product both localize to outer mitochondrial membrane by unique targeting mode. *J Biol Chem.* 2012;287:22969–22987.

162. Suen D-F, Narendra DP, Tanaka A, Manfredi G, Youle RJ. Parkin overexpression selects against a deleterious mtDNA mutation in heteroplasmic cybrid cells. *Proc Natl Acad Sci USA.* 2010;107:11835–11840.

163. Hämäläinen RH, et al. Tissue- and cell-type-specific manifestations of heteroplasmic mtDNA 3243A>G mutation in human induced pluripotent stem cell-derived disease model. *Proc Natl Acad Sci USA*. 2013;110:E3622–E3630.
164. Gilkerson RW, et al. Mitochondrial autophagy in cells with mtDNA mutations results from synergistic loss of transmembrane potential and mTORC1 inhibition. *Hum Mol Genet*. 2012;21:978–990.
165. Kim S-J, Syed GH, Siddiqui A. Hepatitis C virus induces the mitochondrial translocation of Parkin and subsequent mitophagy. *PLoS Pathog*. 2013;9:e1003285.
166. Narendra DP, et al. PINK1 is selectively stabilized on impaired mitochondria to activate Parkin. *PLoS Biol*. 2010;8:e1000298.
167. Seibler P, et al. Mitochondrial Parkin recruitment is impaired in neurons derived from mutant PINK1 induced pluripotent stem cells. *J Neurosci Off J Soc Neurosci*. 2011;31:5970–5976.
168. Shiba-Fukushima K, Inoshita T, Hattori N, Imai Y. PINK1-mediated phosphorylation of Parkin boosts Parkin activity in Drosophila. *PLoS Genet*. 2014;10:e1004391.
169. Birsa N, et al. Lysine 27 ubiquitination of the mitochondrial transport protein Miro is dependent on serine 65 of the Parkin ubiquitin ligase. *J Biol Chem*. 2014;289:14569–14582.
170. Durcan TM, et al. USP8 regulates mitophagy by removing K6-linked ubiquitin conjugates from parkin. *EMBO J*. 2014;33:2473–2491.
171. Ordureau A, et al. Quantitative proteomics reveal a feedforward mechanism for mitochondrial PARKIN translocation and ubiquitin chain synthesis. *Mol Cell*. 2014;56:360–375.
172. Burchell VS, et al. The Parkinson's disease-linked proteins Fbxo7 and Parkin interact to mediate mitophagy. *Nat Neurosci*. 2013;16:1257–1265.
173. Zhou ZD, et al. F-box protein 7 mutations promote protein aggregation in mitochondria and inhibit mitophagy. *Hum Mol Genet*. 2015;24:6314–6330.
174. McCoy MK, Kaganovich A, Rudenko IN, Ding J, Cookson MR. Hexokinase activity is required for recruitment of parkin to depolarized mitochondria. *Hum Mol Genet*. 2014;23:145–156.
175. Lefebvre V, et al. Genome-wide RNAi screen identifies ATPase inhibitory factor 1 (ATPIF1) as essential for PARK2 recruitment and mitophagy. *Autophagy*. 2013;9:1770–1779.
176. Ivatt RM, et al. Genome-wide RNAi screen identifies the Parkinson disease GWAS risk locus SREBF1 as a regulator of mitophagy. *Proc Natl Acad Sci USA*. 2014;111:8494–8499.
177. Do CB, et al. Web-based genome-wide association study identifies two novel loci and a substantial genetic component for Parkinson's disease. *PLoS Genet*. 2011;7:e1002141.
178. Grenier K, Kontogiannea M, Fon EA. Short mitochondrial ARF triggers Parkin/PINK1-dependent mitophagy. *J Biol Chem*. 2014;289:29519–29530.
179. Meissner C, Lorenz H, Hehn B, Lemberg MK. Intramembrane protease PARL defines a negative regulator of PINK1- and PARK2/Parkin-dependent mitophagy. *Autophagy*. 2015;11:1484–1498.
180. Nezich CL, Wang C, Fogel AI, Youle RJ. MiT/TFE transcription factors are activated during mitophagy downstream of Parkin and Atg5. *J Cell Biol*. 2015;210:435–450.
181. Bertolin G, et al. The TOMM machinery is a molecular switch in PINK1 and PARK2/PARKIN-dependent mitochondrial clearance. *Autophagy*. 2013;9:1801–1817.
182. Haddad DM, et al. Mutations in the intellectual disability gene Ube2a cause neuronal dysfunction and impair parkin-dependent mitophagy. *Mol Cell*. 2013;50:831–843.
183. Geisler S, Vollmer S, Golombek S, Kahle PJ. The ubiquitin-conjugating enzymes UBE2N, UBE2L3 and UBE2D2/3 are essential for Parkin-dependent mitophagy. *J Cell Sci*. 2014;127:3280–3293.
184. Bingol B, et al. The mitochondrial deubiquitinase USP30 opposes parkin-mediated mitophagy. *Nature*. 2014;510:370–375.
185. Cornelissen T, et al. The deubiquitinase USP15 antagonizes Parkin-mediated mitochondrial ubiquitination and mitophagy. *Hum Mol Genet*. 2014;23:5227–5242.

186. Hollville E, Carroll RG, Cullen SP, Martin SJ. Bcl-2 family proteins participate in mitochondrial quality control by regulating Parkin/PINK1-dependent mitophagy. *Mol Cell*. 2014;55:451–466.

187. Qu, D., et al. Bag2 mediated regulation of Pink1 is critical for mitochondrial translocation of Parkin and neuronal survival. *J Biol Chem*. 2015. doi:10.1074/jbc.M115.677815.

188. Wauer T, Simicek M, Schubert A, Komander D. Mechanism of phospho-ubiquitin-induced PARKIN activation. *Nature*. 2015;524:370–374.

189. Okatsu K, et al. Phosphorylated ubiquitin chain is the genuine Parkin receptor. *J Cell Biol*. 2015;209:111–128.

190. Okatsu K, et al. PINK1 autophosphorylation upon membrane potential dissipation is essential for Parkin recruitment to damaged mitochondria. *Nat Commun*. 2012;3:1016.

191. Aerts L, Craessaerts K, De Strooper B, Morais VA. PINK1 kinase catalytic activity is regulated by phosphorylation on serines 228 and 402. *J Biol Chem*. 2015;290:2798–2811.

192. Yamano K, et al. Site-specific interaction mapping of phosphorylated ubiquitin to uncover parkin activation. *J Biol Chem*. 2015;290:25199–25211.

193. Fiesel FC, Springer W. Disease relevance of phosphorylated ubiquitin (p-S65-Ub). *Autophagy*. 2015;11:2125–2126.

194. Sarraf SA, et al. Landscape of the PARKIN-dependent ubiquitylome in response to mitochondrial depolarization. *Nature*. 2013;496:372–376.

195. Bingol B, et al. The mitochondrial deubiquitinase USP30 opposes parkin-mediated mitophagy. *Nature*. 2014;510:370–375.

196. Cunningham CN, et al. USP30 and parkin homeostatically regulate atypical ubiquitin chains on mitochondria. *Nat Cell Biol*. 2015;17:160–169.

197. Chan NC, et al. Broad activation of the ubiquitin-proteasome system by Parkin is critical for mitophagy. *Hum Mol Genet*. 2011;20:1726–1737.

198. Rakovic A, et al. Mutations in PINK1 and Parkin impair ubiquitination of Mitofusins in human fibroblasts. *PloS One*. 2011;6:e16746.

199. Gegg ME, et al. Mitofusin 1 and mitofusin 2 are ubiquitinated in a PINK1/parkin-dependent manner upon induction of mitophagy. *Hum Mol Genet*. 2010;19:4861–4870.

200. Pallanck LJ. Culling sick mitochondria from the herd. *J Cell Biol*. 2010;191:1225–1227.

201. Bartolome F, et al. Pathogenic VCP mutations induce mitochondrial uncoupling and reduced ATP levels. *Neuron*. 2013;78:57–64.

202. Meyer H, Bug M, Bremer S. Emerging functions of the VCP/p97 AAA-ATPase in the ubiquitin system. *Nat Cell Biol*. 2012;14:117–123.

203. Yamanaka K, Sasagawa Y, Ogura T. Recent advances in p97/VCP/Cdc48 cellular functions. *Biochim Biophys Acta*. 2012;1823:130–137.

204. Twig G, et al. Fission and selective fusion govern mitochondrial segregation and elimination by autophagy. *EMBO J*. 2008;27:433–446.

205. Narendra D, Kane LA, Hauser DN, Fearnley IM, Youle RJ. p62/SQSTM1 is required for Parkin-induced mitochondrial clustering but not mitophagy; VDAC1 is dispensable for both. *Autophagy*. 2010;6:1090–1106.

206. Dikic I, Bremm A. DUBs counteract parkin for efficient mitophagy. *EMBO J*. 2014;33:2442–2443.

207. Fogel AI, et al. Role of membrane association and Atg14-dependent phosphorylation in beclin-1-mediated autophagy. *Mol Cell Biol*. 2013;33:3675–3688.

208. Lee J-Y, Nagano Y, Taylor JP, Lim KL, Yao T-P. Disease-causing mutations in parkin impair mitochondrial ubiquitination, aggregation, and HDAC6-dependent mitophagy. *J Cell Biol*. 2010;189:671–679.

209. Lazarou M, et al. The ubiquitin kinase PINK1 recruits autophagy receptors to induce mitophagy. *Nature*. 2015;524:309–314.

210. Strappazzon F, et al. AMBRA1 is able to induce mitophagy via LC3 binding, regardless of Parkin and p62/SQSTM1. *Cell Death Differ*. 2015;22:419–432.

211. Michiorri S, et al. The Parkinson-associated protein PINK1 interacts with Beclin1 and promotes autophagy. *Cell Death Differ*. 2010;17:962–974.

212. Van Humbeeck C, et al. Parkin interacts with Ambra1 to induce mitophagy. *J Neurosci Off J Soc Neurosci.* 2011;31:10249–10261.

213. Yamano K, Fogel AI, Wang C, van der Bliek AM, Youle RJ. Mitochondrial Rab GAPs govern autophagosome biogenesis during mitophagy. *eLife.* 2014;3:e01612.

214. Allen GFG, Toth R, James J, Ganley IG. Loss of iron triggers PINK1/Parkin-independent mitophagy. *EMBO Rep.* 2013;14:1127–1135.

215. Rakovic A, Seibler P, Klein C. iPS models of Parkin and PINK1. *Biochem Soc Trans.* 2015;43:302–307.

216. Shaltouki A, et al. Mitochondrial alterations by Parkin in dopaminergic neurons using PARK2 patient-specific and PARK2 knockout isogenic iPSC lines. *Stem Cell Rep.* 2015;4:847–859.

217. Vincow ES, et al. The PINK1-Parkin pathway promotes both mitophagy and selective respiratory chain turnover in vivo. *Proc Natl Acad Sci USA.* 2013;110:6400–6405.

218. Klein P, et al. Ret rescues mitochondrial morphology and muscle degeneration of Drosophila Pink1 mutants. *EMBO J.* 2014;33:341–355.

219. Meka DP, et al. Parkin cooperates with GDNF/RET signaling to prevent dopaminergic neuron degeneration. *J Clin Invest.* 2015;125:1873–1885.

220. Kasianowicz J, Benz R, McLaughlin S. The kinetic mechanism by which CCCP (carbonyl cyanide m-chlorophenylhydrazone) transports protons across membranes. *J Membr Biol.* 1984;82:179–190.

221. Padman BS, Bach M, Lucarelli G, Prescott M, Ramm G. The protonophore CCCP interferes with lysosomal degradation of autophagic cargo in yeast and mammalian cells. *Autophagy.* 2013;9:1862–1875.

222. Cai Q, Zakaria HM, Simone A, Sheng Z-H. Spatial parkin translocation and degradation of damaged mitochondria via mitophagy in live cortical neurons. *Curr Biol CB.* 2012;22:545–552.

223. Ashrafi G, Schlehe JS, LaVoie MJ, Schwarz TL. Mitophagy of damaged mitochondria occurs locally in distal neuronal axons and requires PINK1 and Parkin. *J Cell Biol.* 2014;206:655–670.

224. Wang Y, Nartiss Y, Steipe B, McQuibban GA, Kim PK. ROS-induced mitochondrial depolarization initiates PARK2/PARKIN-dependent mitochondrial degradation by autophagy. *Autophagy.* 2012;8:1462–1476.

225. Klionsky DJ, et al. Guidelines for the use and interpretation of assays for monitoring autophagy. *Autophagy.* 2012;8:445–544.

226. Tolkovsky AM, Xue L, Fletcher GC, Borutaite V. Mitochondrial disappearance from cells: a clue to the role of autophagy in programmed cell death and disease? *Biochimie.* 2002;84:233–240.

227. Yoshii SR, Kishi C, Ishihara N, Mizushima N. Parkin mediates proteasome-dependent protein degradation and rupture of the outer mitochondrial membrane. *J Biol Chem.* 2011;286:19630–19640.

228. Scarlett JL, Murphy MP. Release of apoptogenic proteins from the mitochondrial intermembrane space during the mitochondrial permeability transition. *FEBS Lett.* 1997;418:282–286.

229. Hare JF, Hodges R. Turnover of mitochondrial inner membrane proteins in hepatoma monolayer cultures. *J Biol Chem.* 1982;257:3575–3580.

230. Elmore SP, Qian T, Grissom SF, Lemasters JJ. The mitochondrial permeability transition initiates autophagy in rat hepatocytes. *FASEB J Off Publ Fed Am Soc Exp Biol.* 2001;15:2286–2287.

231. Patergnani S, Pinton P. Mitophagy and mitochondrial balance. *Methods Mol Biol Clifton NJ.* 2015;1241:181–194.

232. Diot A, et al. A novel quantitative assay of mitophagy: Combining high content fluorescence microscopy and mitochondrial DNA load to quantify mitophagy and identify novel pharmacological tools against pathogenic heteroplasmic mtDNA. *Pharmacol Res.* 2015;100:24–35.

233. Sargsyan A, et al. Rapid parallel measurements of macroautophagy and mitophagy in mammalian cells using a single fluorescent biosensor. *Sci Rep*. 2015;5:12397.

234. Kuma A, Matsui M, Mizushima N. LC3, an autophagosome marker, can be incorporated into protein aggregates independent of autophagy: caution in the interpretation of LC3 localization. *Autophagy*. 2007;3:323–328.

235. Xie Z, Klionsky DJ. Autophagosome formation: core machinery and adaptations. *Nat Cell Biol*. 2007;9:1102–1109.

236. Chen X, et al. Lysosomal targeting with stable and sensitive fluorescent probes (Superior LysoProbes): applications for lysosome labeling and tracking during apoptosis. *Sci Rep*. 2015;5:9004.

237. Kimura S, Noda T, Yoshimori T. Dissection of the autophagosome maturation process by a novel reporter protein, tandem fluorescent-tagged LC3. *Autophagy*. 2007;3:452–460.

238. Kneen M, Farinas J, Li Y, Verkman AS. Green fluorescent protein as a noninvasive intracellular pH indicator. *Biophys J*. 1998;74:1591–1599.

239. Katayama H, Kogure T, Mizushima N, Yoshimori T, Miyawaki A. A sensitive and quantitative technique for detecting autophagic events based on lysosomal delivery. *Chem Biol*. 2011;18:1042–1052.

240. Violot S, Carpentier P, Blanchoin L, Bourgeois D. Reverse pH-dependence of chromophore protonation explains the large Stokes shift of the red fluorescent protein mKeima. *J Am Chem Soc*. 2009;131:10356–10357.

241. Sun N, et al. Measuring In Vivo Mitophagy. *Mol Cell*. 2015;60:685–696.

242. Rakovic A, et al. Phosphatase and tensin homolog (PTEN)-induced putative kinase 1 (PINK1)-dependent ubiquitination of endogenous Parkin attenuates mitophagy: study in human primary fibroblasts and induced pluripotent stem cell-derived neurons. *J Biol Chem*. 2013;288:2223–2237.

243. Strappazzon F, et al. AMBRA1 is able to induce mitophagy via LC3 binding, regardless of PARKIN and p62/SQSTM1. *Cell Death Differ*. 2015;22:419–432.

244. Goldberg MS, et al. Parkin-deficient mice exhibit nigrostriatal deficits but not loss of dopaminergic neurons. *J Biol Chem*. 2003;278:43628–43635.

245. Von Coelln R, et al. Loss of locus coeruleus neurons and reduced startle in parkin null mice. *Proc Natl Acad Sci USA*. 2004;101:10744–10749.

246. Perez FA, Palmiter RD. Parkin-deficient mice are not a robust model of parkinsonism. *Proc Natl Acad Sci USA*. 2005;102:2174–2179.

247. Kitada T, et al. Impaired dopamine release and synaptic plasticity in the striatum of PINK1-deficient mice. *Proc Natl Acad Sci USA*. 2007;104:11441–11446.

248. Kitada T, Tong Y, Gautier CA, Shen J. Absence of nigral degeneration in aged parkin/DJ-1/PINK1 triple knockout mice. *J Neurochem*. 2009;111:696–702.

249. Burman JL, Yu S, Poole AC, Decal RB, Pallanck L. Analysis of neural subtypes reveals selective mitochondrial dysfunction in dopaminergic neurons from parkin mutants. *Proc Natl Acad Sci USA*. 2012;109:10438–10443.

250. Van Laar VS, et al. Bioenergetics of neurons inhibit the translocation response of Parkin following rapid mitochondrial depolarization. *Hum Mol Genet*. 2011;20:927–940.

251. Joselin AP, et al. ROS-dependent regulation of Parkin and DJ-1 localization during oxidative stress in neurons. *Hum Mol Genet*. 2012;21:4888–4903.

252. Grenier K, McLelland G-L, Fon EA. Parkin- and PINK1-dependent mitophagy in neurons: will the real pathway please stand up? *Front Neurol*. 2013;4:100.

253. Rangaraju V, Calloway N, Ryan TA. Activity-driven local ATP synthesis is required for synaptic function. *Cell*. 2014;156:825–835.

254. Zala D, et al. Vesicular glycolysis provides on-board energy for fast axonal transport. *Cell*. 2013;152:479–491.

255. Youle RJ, van der Bliek AM. Mitochondrial fission, fusion, and stress. *Science*. 2012;337:1062–1065.

256. Warburg O. On the origin of cancer cells. *Science*. 1956;123:309–314.

257. Yao Z, et al. Cell metabolism affects selective vulnerability in PINK1-associated Parkinson's disease. *J Cell Sci*. 2011;124:4194–4202.

258. Melser S, Lavie J, Bénard G. Mitochondrial degradation and energy metabolism. *Biochim Biophys Acta*. 2015;1853:2812–2821.

259. Kanki T, Klionsky DJ. Mitophagy in yeast occurs through a selective mechanism. *J Biol Chem*. 2008;283:32386–32393.

260. Gatliff J, Campanella M. TSPO is a REDOX regulator of cell mitophagy. *Biochem Soc Trans*. 2015;43:543–552.

261. Saxena S, Caroni P. Selective neuronal vulnerability in neurodegenerative diseases: from stressor thresholds to degeneration. *Neuron*. 2011;71:35–48.

262. Tufi R, et al. Enhancing nucleotide metabolism protects against mitochondrial dysfunction and neurodegeneration in a PINK1 model of Parkinson's disease. *Nat Cell Biol*. 2014;16:157–166.

263. Nardin A, Schrepfer E, Ziviani E. Counteracting PINK/Parkin deficiency in the activation of mitophagy: a potential therapeutic intervention for Parkinson's Disease. *Curr Neuropharmacol*. 2016;14:250–259.

264. Hertz NT, et al. A neo-substrate that amplifies catalytic activity of Parkinson's-disease-related kinase PINK1. *Cell*. 2013;154:737–747.

Autophagy

P.A. Lewis*,**, M. Perez-Carrion†, G. Piccoli†,‡

*School of Pharmacy, University of Reading, Reading, United Kingdom;
**Department of Molecular Neuroscience, UCL Institute of Neurology,
London, United Kingdom; †Centre for Integrative Biology (CIBIO),
Università degli Studi di Trento, Trento, Italy; ‡Dulbecco Telethon Institute

1 INTRODUCTION

The accumulation of protein aggregates and dysfunctional mitochondria within the brain have long been recognized as critical components of the molecular etiology of Parkinsons disease. The catabolic pathways collectively known as autophagy represent a key determinant of whether protein aggregates persist, or whether dysfunctional mitochondria are disposed of correctly, and as such are likely to play an important, if not central, role in the molecular events that lead to nigral degeneration. These

Parkinson's Disease. http://dx.doi.org/10.1016/B978-0-12-803783-6.00006-7

pathways, whose name derives from the Greek to self-eat (αὐτός Φαγεῖν), are related cellular processes that identify, gather up, and degrade cellular material. In this chapter, the reported links between the various forms of autophagy and Parkinsons disease will be discussed, with a particular focus on the molecular mechanisms binding the two together, and the therapeutic potential of modulating autophagy in Parkinsons disease.

1.1 What Is Autophagy?

As noted earlier, autophagy is a broad church of closely connected catabolic cellular processes.[1] These processes can be divided up into three interrelated systems—all of which terminate in the lysosomes (Fig. 6.1). Macroautophagy, a bulk degradation process that allows for the large-scale trafficking of waste organelles and material from the cytoplasm to the lysosomes, is the most intensely studied form of autophagy and is often referred to as simply autophagy (although in this chapter the distinction between macroautophagy and autophagy will be maintained). It relies on the formation of membrane-bound sacks, or autophagosomes, that first encapsulate, and then transport targeted waste to be recycled or degraded. Chaperone-mediated autophagy (CMA) is a more focused and smaller scale form of autophagy that tags particular proteins for degradation, with these proteins trafficked to the lysosomes and aided into

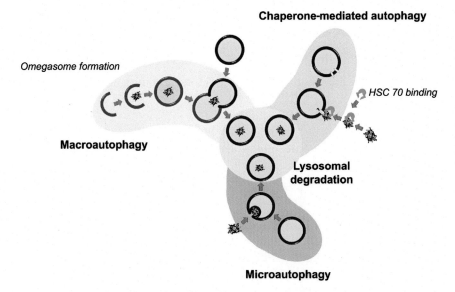

FIGURE 6.1 **The tripartite nature of autophagy**. The different pathways to lysosomal degradation are shown here in a simplified form. Lysosomes are displayed in red, with the formation and maturation of autophagosomes is shown in blue.

the lysosomal vacuole by a molecular assembly of heat shock cognate 70 (HSC70), heat shock protein 90 (HSP90), and Lamp2a. Finally, microautophagy relies on the direct invagination of the lysosomal vacuole membrane for the targeted sequestration of material for degradation. This relatively simple autophagic landscape is complicated somewhat by the existence of multiple specialized forms of autophagy. These range from xenophagy (the selective encapsulation and degradation of bacterial and parasitical invaders) to mitophagy, a process that governs the selection and removal of dysfunctional mitochondria.[2] The latter form of autophagy, essentially a specialized category of macroautophagy, has multiple lines of evidence linking it to cell death in the *substantia nigra pars compacta* and parkinsonism, and is discussed in greater detail in Chapter 5. A further complication, and one that is intrinsic to all homeostatic cellular systems, is that autophagy in all its guises is a highly dynamic process. Turning to the example of macroautophagy, this is a process that is constantly ongoing in all eukaryotic cells, with new autophagic vesicles being formed, fusing with the lysosomes, delivering their cargo, and being recycled all the time. In certain situations, for example, when nutrients are removed in starvation conditions or in certain disease conditions, the rate at which this occurs—and the flux of material through the system—may be upregulated. Quantifying this, however, is challenging, especially with the static snapshots of cellular behavior provided by immunoblotting and immunocytochemistry. As such, researchers have developed a series of tools and specific protocols to examine autophagy, examples of which are provided in Box 6.1.[3]

Returning to Parkinsons disease, what evidence is there that disruption or alteration of these various autophagic pathways can result in nigral degeneration?

2 LESSONS FROM THE GENETICS OF PARKINSONS DISEASE

As is the case with many neurodegenerative disorders, genetics has provided an essential source of information relating to the molecular mechanisms underpinning neuronal cell death (see also Chapter 1). In the Parkinson's field, this was instigated by the 1997 identification of mutations linked to a rare, autosomal dominant form of Parkinsons in the *SNCA* gene located on chromosome 4.[4] The first mutation to be discovered, in a large Italian pedigree originally from the town of Contursi, caused an amino acid substitution (alanine to threonine) at position 53 in the alpha synuclein protein product of the *SNCA* gene, and there are now a number of other point mutations and gene multiplications described in this gene that are causative for Parkinson's.[5] Since 1997, an increasing

BOX 6.1

ASSAYS TO INVESTIGATE AUTOPHAGY

Characterization of Autophagic Vesicles

The most powerful tool to distinguish between different autophagic vesicles is electron microscopy. Morphologically, autophagosomes are identified as double-membrane cytoplasmic structures that include organelles or undigested cellular material inside. Instead, late autophagic vesicles, such as autophagolysosomes, are single-membrane structures as a consequence of the fusion of the autophagosome and the lysosome, with partially digested content. Although it is a useful instrument to study autophagy, it is a limited method that requires expertise to distinguish autophagic vesicles from other cytoplasmic vacuoles.

Methods to Analyze Autophagosomes Amount

The number of autophagosomes is an indicator of the level of autophagic activity running. During autophagy, the LC3 protein (microtubule-associated protein light chain 3) is first modified in the C-terminal part, becoming the cytosolic form LC3B-I, which is then conjugated with phospholipids and converted into LC3BII. LC3BII is considered a specific marker of the autophagosome, localized around either inner or outer membrane and subsequently degraded after fusion of the autophagosome with the lysosome. Two different approaches can be successfully used to assess the autophagosome number in physiological and pathological conditions.

Quantification of Number or Autophagosomes

Autophagosomes can be monitored by fluorescence microscopy following endogenous LC3 levels with specific antibodies (Novus Biological, NB 100-2331) as well as analyzing punctate structures after GFP-LC3 overexpression. Dots can be counted automatically using a specific software (MetaMorph version 7.0 by Molecular Devices and G-Count by G-Angstrom) or manually, avoiding personal subjectivity. Quantification of dots is accepted as an accurate measure of autophagosome number.

LC3B-II Levels

As LC3B-II is a specific marker of the autophagosome, an increment on LC3B-II expression usually correlates with an increased number of autophagosomes. Conversion of LC3B-I into LC3B-II can be measured by immunoblotting, analyzing endogenous levels of the protein as well as conversion of GFP-LC3-I–GFP-GFPII tracking GFP signal. A good interpretation requires to have in consideration that LC3B-II migrates faster in SDS-PAGE gels than LC3BI, showing an apparently small molecular weight, although it is a larger form of the protein.

Methods to Monitor Autophagic Flux

Considering that autophagy is a dynamic process, accumulation of autophagosomes could indicate either activation of early stages of the pathway or inhibition of late steps concerning autophagosomes maturation. As a consequence, it is convenient to monitor autophagy evaluating several phases of the process performing autophagic flux analyses. Autophagic flux is defined as a measure of completion of the pathway through amount of autophagic content effectively delivered to the lysosome. This approach can distinguish whether an increased number of autophagosomes is a consequence of autophagy induction or is due to the blockage of fusion of autophagosome to lysosome.

LC3B Turnover Assay

After fusion of autophagosome and lysosome, LC3B-II is degraded into the autophagolysosome. Drugs that block this step avoid LC3B-II degradation and are used to and incorporate to the lysosome for degradation. Between the most common lysosomal inhibitors are included pH modifiers of lysosome such as ammonium chloride, chloroquine, bafilomycin A1, and inhibitors of lysosomal proteases such as E64 and pepstatin A.

Degradation of Selective Substrates

There are some proteins preferably eliminated by autophagy that are considered appropriate candidates to monitor autophagic flux. LC3B, which is finally degraded in the autophagolysosome, can be measured after long periods of autophagy induction as a good indicator of autophagic state. LC3B protein expression can be followed by Western blot assay, as well as changes on GFP-LC3 puncta through fluorescence microscopy or flow cytometry. P62 is degraded by autophagy after binding to LC3 during autophagosome formation and is accepted as an autophagic substrate. However, it is not totally clear if p62 is exclusively eliminated by autophagy or if ubiquitin–proteasome pathway could also be involved.

Autophagic flux can also be analyzed through measurement of long-lived protein degradation where amino acids have been specifically labeled with different isotopes, such as ^{14}C or ^{3}H.

Analyses of GFP-mCherry-LC3 or GFP-RPF-LC3 Reporter

One of the most useful approaches to follow autophagic flux, step by step, involves overexpression of LC3 with a double-tag construct: GFP-mCherry-LC3 or GFP-RFP-LC3. This vector consists of a pH-sensitive reporter that expresses mCherry or RFP tag and GFP fused to LC3. The GFP signal is quenched by acidic lysosomal pH of the autophagolysosome, whereas mCherry or RFP tag is acid-insensitive (Pankiv et al., 2007). Autophagosomes correspond to yellow dots derived from the green and red merged signal; instead, autophagolysosomes are labeled just in red.

number of genes have been linked to Parkinsons either as Mendelian loci, where there is a clear mode of inheritance, or as risk factors.[6] With regard to the former, there now exists a long list of genes that are unambiguously associated with familial forms of parkinsonism (Table 6.1) as also outlined in Chapter 1. The use of the term *parkinsonism*, the movement disorder symptoms associated with disruption of dopaminergic neuronal circuits originating in the midbrain, as distinct from the disease entity of Parkinsons disease, is quite deliberate as many of these mutated genes result in complex neurological, and sometimes multisystem, disorders that share some cellular and clinical phenotypes with Parkinsons, but are clearly not Parkinsons disease in the classic clinical sense of the term. In addition to the Mendelian forms of PD, the advent of genome-wide analyses, and in particular genome-wide association (GWA) studies, has provided researchers with an even more extensive list of loci significantly linked to small increases in the life time risk of developing PD.[7]

2.1 Familial Parkinsons and Autophagy

Concentrating on the genes linked to Mendelian inheritance of Parkinsons and parkinsonism, studies in a wide range of experimental systems have implicated the function of PD genes in the various forms of autophagy. These range from cellular studies, through in vivo model systems to human pathological studies—directly examining the brain tissue most affected in Parkinsons.

Alpha synuclein, expressed from the *SNCA* locus, is undoubtedly at the center of the pathological events that lead to Parkinsons disease. Soon after the discovery of mutations in this gene, it was realized that alpha synuclein was the main protein component of Lewy bodies, the pathognomic hallmark of Parkinsons disease.[8] Based on biochemical and genetic studies, a clear line of causation can be drawn between increased aggregation (dependent on point mutations or increased absolute levels of expression) of alpha synuclein and the death of neurons in the *substantia nigra*.[9] Despite a huge amount of research effort, the molecular details of this line of causation are still a matter of debate. These experimental investigations have, however, uncovered several links to the various forms of autophagy. Alpha synuclein can be degraded by both macroautophagy and CMA, immediately suggesting that a decrease in degradation via either of these pathways could result in increased cellular levels of alpha synuclein.[10,11] Indeed, recent data from a mouse model in which macroautophagy is specifically downregulated in the brain suggest that this is the case.[12] An increase in alpha synuclein levels in the cell is, in turn, likely to favor increased aggregation of the protein and lead to cytotoxicity.[13] Conversely, a number of studies have reported that aggregated alpha synuclein can inhibit autophagy—supporting a model in which alterations in autophagy

TABLE 6.1   Genes Linked to Inherited Forms of Parkinsons Disease

Gene	Chromosome	Protein	Phenotype	Inheritance	Links to autophagy
PINK1	1	PINK1	Young onset/juvenile parkinsonism	Autosomal recessive	Mitophagy
DJ1	1	DJ1	Young onset/juvenile parkinsonism	Autosomal recessive	Mitophagy
ATP13a2	1	ATP13a2	Kufor-Rakeb syndrome	Autosomal recessive	Lysosomal ATPase
GBA	1	GBA	Parkinsons (heterozygote state)	Risk factor	Lysosomal hydrolase
DNAJC13	3	DNAJC13	Parkinsons	Autosomal dominant	Endosomal trafficking
SNCA	4	Alpha synuclein	Parkinsons, Lewy body dementia	Autosomal dominant	CMA, macroautophagy, lysosomal function
PARK2	6	Parkin	Young onset/juvenile parkinsonism	Autosomal recessive	Mitophagy
LRRK2	12	LRRK2	Parkinsons	Autosomal dominant	Macroautophagy, CMA, endosomal trafficking
VPS35	16	VPS35	Parkinsons	Autosomal dominant	Endosomal trafficking, macroautophagy
FBXO7	22	FBXO7	Young onset/juvenile parkinsonism	Autosomal recessive	Mitophagy
RAB39B	X	RAB39	Early onset parkinsonism and intellectual disability	X-linked	Endosomal trafficking
WDR45	X	WDR45/WIPI4	Neurodegeneration with brain iron accumulation	X-linked	Macroautophagy

could increase aggregation of alpha synuclein, which then inhibits autophagic activity, resulting in a vicious circle.[14] Importantly, this implies that both de novo aggregation of alpha synuclein (caused by, eg, points mutations such as A53T) and disruptions in autophagy can initiate the sequence of events that lead to nigral degeneration. It should be noted here that there also exists robust evidence linking alpha synuclein to proteasomal degradation, another key catabolic system within the cell, and it is likely that the rate of degradation (and, based upon increased or decreased cellular levels of alpha synuclein, the propensity of this protein to aggregate) is a product of multiple, interconnected expression and degradation events within the cell.[15] Notwithstanding this, the close link between the degradation of alpha synuclein by autophagy, the aggregation of this protein into oligomeric assemblies and cytotoxicity means that perturbation of autophagy because of other genetic alterations can also push neuronal populations over the precipice toward nigral cell death. A key and well-studied example of where this is likely to be the case is presented by mutations in the *GBA* gene on chromosome 1.[16] Homozygous mutations in this gene cause Gaucher's disease, a serious and often early onset lysosomal storage disorder. Carrying mutations in this gene in a heterozygous state, however, greatly increases the lifetime risk of developing Parkinsons. While not strictly speaking a Mendelian form of Parkinsons, this is a clear familial and inherited risk factor for disease. Moreover, the disruption in protein function as a result of these mutations—resulting in a decrease in activity of the glucocerebrosidase (GCase) enzyme product of the *GBA* gene—directly impacts on lysosomal function. As the final destination of proteins tagged for destruction by macro-, micro-, and chaperone-mediated forms of autophagy this has important consequences for the clearance of alpha synuclein. Multiple lines of experimental and clinical evidence support this, from the presence of Lewy bodies in the brains of individuals who develop Parkinsons as a result of mutation in *GBA*, to cellular and in vivo analysis of reduced GCase function that demonstrates the accumulation of alpha synuclein as a consequence of this.[17-19] Intriguingly, aggregation of alpha synuclein appears to have an inhibitory impact on GCase activity.[20] Similar to the reciprocal links between autophagy and alpha synuclein aggregation, these data prompted the proposal of a model in which alpha synuclein aggregation and GCase activity are bound into a vicious circle, gyrating around the autophagy/lysosomal system. Genetic studies have uncovered other less common genetic-derived disruptions of the autophagy/lysosomal system resulting in Parkinsons disease or parkinsonism. Another fascinating example is provided by mutations in *ATP13a2*, the protein product of which is a lysosomal ATPase thought to be involved in regulating lysosomal acidification. Homozygous mutations in this gene cause a complicated neurological disorder called Kufor–Rakeb syndrome (KRS), one of the clinical symptoms of which is parkinsonism.[21]

Although there are important clinical differences between KRS and PD, the overlap in symptoms, and, critically, pathology, led to the designation of the *ATP13a2* locus as a *PARK* gene.[22] Here again, experimental evidence supports a causative link between mutations in this gene, lysosomal (and thence autophagic) dysfunction, and aggregation of alpha synuclein.[23]

A number of other genes causatively linked to either Parkinsons (eg, mutations in *VPS35* and *LRRK2*) or disorders in which parkinsonism is a central clinical component (eg, mutations in *WDR45*) have also been implicated or are directly involved in the regulation of autophagy. The protein product of the *VPS35* gene, mutations in which are a rare cause of Parkinsons, is a key component of the retromer complex involved in endosomal trafficking.[24] Recent studies suggested that alterations in endosomal trafficking because of mutations in VPS35 specifically alter the regulation of the induction of macroautophagy.[25,26] Similarly, multiple studies have linked the activity of LRRK2, the protein produced from the *LRRK2* locus on chromosome 12, to the regulation of macroautophagy[27–30] and of CMA.[31] As autosomal dominant mutations in *LRRK2* are the most common genetic cause of Parkinsons, this is an important connection between autophagy and PD. Notably, LRRK2 has also been reported to accumulate upon reduction of macroautophagy in the brain.[12]

Less easy to interpret, mainly because of lack of experimental evidence and a complicated clinical phenotype, are mutations in WDR45. These cause a very rare X-linked childhood disorder called neurodegeneration with brain iron accumulation (NBIA), which similar to KRS includes parkinsonism as part of its clinical presentation.[32] Very little is known, however, about the cellular consequence of mutations in this gene. A link to autophagy is provided by the putative function of this gene as a WIPI protein (the protein product of *WDR45* is variously known as WIPI4 or WDR45), involved in the formation and function of autophagosomes.[33] Similar to the impact of mutations, there is a dearth of information regarding the cellular function of WDR45, and so a more detailed understanding of a putative role for this gene in autophagy awaits further experimental investigations.

Likewise, mutations or risk variants in the genes producing the Rab proteins RAB39B and RAB7L1 have been associated with Parkinsons disease.[34,35] These proteins are important for vesicle trafficking within the cell, and RAB7L1 in particular has been linked to endosomal trafficking and sorting (along with LRRK2 and VPS35).[35,36] This, in turn, has the potential to disrupt lysosomal and autophagic pathways within the cell.

Feeding in to a model in which disruption of autophagy and accumulation of alpha synuclein is a (or even the) central event in the etiopathogenesis of PD, there is an abundance of genetic evidence connecting autophagy and PD. A cautionary note, however, should be struck when interpreting these data, as our knowledge of the systems regulating proteostasis in human cells, as well as the consequences of accumulation of alpha synuclein, remains incomplete. A good example of how human disease is often more complicated than our somewhat simplistic categorization and understanding of it is provided by Parkinsons linked to mutations in *LRRK2*. According to the Queen Square Brain Bank diagnostic criteria for Parkinsons disease, a full diagnosis of PD is dependent upon the presence of a set of clinical symptoms and final pathological confirmation of the presence of nigral degeneration and Lewy bodies in the *substantia nigra*. In *LRRK2* mutation cases, the former (clinical) criteria are met; however, the latter (pathological) criteria are more variable.[37] A small but significant proportion of *LRRK2* mutation carriers developing disease do not present with Lewy body pathology, instead presenting with either neurofibrillary tangle pathology or deposition of TDP43 in the brain.[38] This represents a break, therefore, with the dogma that connects Parkinsons disease and Lewy body pathology, and thence the model for alpha synuclein and the autophagy/lysosomal pathway as a determinant of disease state.

2.2 Risk Variants and Parkinsons

An important area of advancement in our understanding of the genetics of Parkinsons has been in the field of genetic risk variants (in contrast with Mendelian forms of PD). GWA studies, analyzing data from tens of thousands of cases and controls, have identified a long list of risk loci for PD.[7] Intriguingly, many of these are familiar from previous studies of inherited Parkinsons—for example, *GBA, SNCA,* and *LRRK2* are all associated with risk factors for PD, distinct from the coding changes that are associated with inherited disease. Although the impact of these risk variants is small, often a fraction of a percent increase in lifetime risk, the GWA studies have been critical in forming a bridge between Mendelian and idiopathic forms of Parkinsons. With reference to autophagy, it is noteworthy that the genes and mechanisms described earlier are part of the genetic burden that increases the risk of PD in the general population—and so it is likely, although not certain, that much of what holds true regarding the contribution of autophagy to the aetiology of inherited PD is also pertinent to idiopathic disease.

3 AGEING AND AUTOPHAGY

The single greatest risk factor for Parkinsons is age: the older you get, the more likely it is that you will develop PD. During the natural ageing process, organisms exhibit a specific set of biochemical and physiological alterations that can either result as a direct consequence of the ageing itself or be a secondary, compensatory answer. Alterations in cellular function may arise in part by the progressive accumulation of altered molecular species, such as DNA, protein, and lipids. A key component of this is the gradual decrease in activity of the processes devoted to protein catabolism. Changes in the catalytic activities of the proteasome, its supramolecular organization, and in expression of the enzymes involved in proteasomal degradation all occur with ageing.[39] Importantly, accumulating studies clearly describe the role of autophagy in ageing. An age-related downregulation decline of autophagy genes (Atg) transcription has been demonstrated in *Drosophila melanogaster*, and Atg8 overexpression correlates with an increase in the lifespan of several animal models.[40] Accordingly, RNAi-mediated acute downregulation of different autophagy genes in *Caenorhabditis elegans* resulted in a reduced survival.[41] Notably, studies in humans revealed that autophagy genes such as *ATG5*, *ATG7*, and *BECN1* are downregulated in the brains of aged individuals.[42] Indeed, both macroautophagy and CMA activity decline with age.[43] Interestingly, an experimental model of accelerated senescence, the SAMP8 mouse, shows Alzheimer's-like phenotype, accumulates ubiquitin-positive protein aggregates, and is characterized by a robustly impaired decreased autophagy activity.[44] The age-dependent alteration in the functionality of macroautophagy was addressed in an early study that correlated not only ageing but also food intake to autophagic-proteolytic response.[45] Further experimental data provided evidence that *ad libitum* feeding worsens the ageing process and suppressed macroautophagy.[46] Strikingly, a diet providing a reduced calorie intake ameliorated the ageing-dependent reduction in autophagic activity, and correlated with an increase in animal life span.[47] A significant dysregulation in the insulin signaling that regulates macroautophagy has been suggested as causative.[48] Furthermore, it has been reported that clearance of autophagosomes suffers a cumulative impairment during ageing, and it has been suggested that the accumulation of undigested proteins within lysosomes might negatively influence autophagosomes cargo degradation.[49] The severe reduction of lysosomal CMA receptors reported in older animals may also account for the reduction of CMA activity.[50] In fact, the cargo proteins shuffled by the cytosolic chaperone cannot enter the lysosome without binding to a receptor. Congruent with this, long-lasting pharmacological impairment of autophagy speeds up the ageing process: data obtained in several species and organ systems brought to the formulation of the "protease-inhibitor model of ageing."[51] Oxidative damage together with reduced function housekeeping machinery may result in a progressive accumulation of toxic molecular agents.[52] Focusing in on brain function, protein clearance plays a pivotal role in maintaining the homoeostasis of the nervous system: neurons have increased propensity to accumulate malfunctional proteins as they do not regenerate. Therefore the progressive deterioration of the cellular machinery handling protein degradation and refolding is gaining relevance as a critical mechanism underlying late onset degenerative processes in diseases such as Parkinsons. Although the genes and proteins implicated in

PD are present in the cell from birth, symptoms typically appear in old age, and so even in genetic forms of disease ageing is indeed the main risk factor for the onset of PD. Protein misfolding and aggregation have been suggested as the final trigger of the neurodegeneration occurring in the late phase of PD. The first indications suggesting degradation impairment in PD neurons were in fact the detection of protein inclusions in post-mortem specimen from PD patients (reviewed in Ref. [53]). Among brain areas, the *substantia nigra* is particularly susceptible: the cellular metabolism of dopamine induces by itself the formation of free radical and protein misfolding. Furthermore, dopamine by-products can precipitate the formation of toxic alpha-synuclein protofibrils and aggregates.[54] Further confirmation for the linkage between protein clearance and PD came from the observation that the systemic administration of proteasomal inhibitors causes neuropathological changes similar to those observed in PD, such as Lewy body–like inclusions.[55]

4 AUTOPHAGY IN THE NERVOUS SYSTEM: FRIEND OR FOE?

Research in the autophagy field started with the investigation of liver cells from rats, where autophagy was described as a response to nutrient deprivation aiming to catabolize and thus recycle nonessential biological molecules.[56] It is now clear that in higher organisms autophagy plays a crucial role during embryonic development, survival, and cell differentiation.[57] Notwithstanding the fact that autophagy is activated at very low level in neurons when compared to other cell types upon acute starvation, the survival and the proper function of the CNS requires basal autophagy. Basal autophagy in neurons may not be required to sustain metabolic needs—in basal conditions the number of autophagosomes as well as the levels of LC3-II protein, a molecular marker of the autophagic flux (Box 6.1), are quite low in comparison to other tissues.[58] Instead, it is becoming more and more evident that autophagy is required for housekeeping functions crucial for normal neuron physiology. Neurons need a constitutive basal level of autophagy: upon acute blockage of lysosomal degradation, autophagosome number quickly increases, supporting the hypothesis that a fast basal flux exists.[59] Electron microscopy investigation of human and animal model brain specimens has confirmed the presence of both lysosomes and autophagosomes in basal condition, thus suggesting the relevance of autophagy in physiological neuronal function.[60] Autophagy participates in neuronal plasticity, allowing neurite and growth cone deconstruction,[61] and assists activity-driven neuronal plasticity.[62] Early studies revealed how, upon axonal lesion, structures closely resembling autophagosomes are found near to the axonal terminal.[63]

Later, the existence of active bidirectional trafficking of autophagasomes along the microtubular cytoskeleton supporting neuronal survival was reported,[64] facilitating the somatic degradation of presynaptic aggregates.[65] Some evidence suggested that autophagosomes can transport trophic factors such as NGF back into the cell body from the peripheral parts of the neuron. If this was the case, perturbed autophagy might contribute to neuronal death via impaired NGF signalling.[66] The final confirmation that autophagy plays a protective role in neurons came from the investigation of a conditional rodent model lacking autophagy-related proteins in CNS.[67] Neural-specific knockout ablation of autophagic genes results in massive neurodegeneration, progressive impairment of motor performance, ataxia, and developmental delay. At the molecular level, neurons lacking functional autophagic machinery accumulate poly-ubiquitinated aggregates and damaged organelles. Such outcome is particularly deleterious in neurons, where the trafficking of membrane-engulfed cargo plays a critical role. The huge accumulation of defective organelles and/or protein aggregates may impair proper axonal traffic of vesicles and organelles, thus exacerbating cytotoxicity.[68] A paradigmatic model is the *Atg5*flox/flox;nestin-Cre mice. Being *Atg5* and *Atg7* necessary for embryonic development,[69] scientists generated conditional KO models where in which *Atg5* and *Atg7* expression was ablated only in neuronal lineage.[70,71] The *Atg5 and Atg7* flox/flox, nestin-Cre mice are characterized by growth retardation, progressive motor and behavioral impairment, and robust neurodegeneration. This model strongly supports intimate functional cross-talk between autophagy and the ubiquitin proteasome system (UPS). Strikingly, histological examination revealed the presence of large numbers of ubiquitin-positive inclusions in neurons. Given that proteasomal function was unaffected in Atg5 conditional mice, it is clear that autophagy plays a crucial role in clearance of poly-ubiquitinated proteins. The recently developed *Ambra1* knockout mice resulted in a useful animal model to dissect the role of autophagy in neuron development. Autophagy/beclin 1 regulator 1 (Ambra1) protein modulates autophagy via interaction with beclin 1 (BECN1).[72] Ambra1 mice are characterized by poly-ubiquitinated aggregates and massive neuronal death. Furthermore, biochemical and immunological analysis has revealed that ubiquitin-positive aggregates are strongly decorated by the p62/SQSTM1 protein.[73] Noteworthy, adaptor such as p62 demonstrates more affinity for Lys63-linked polyubiquitin chains, whereas Lys48-linked polyubiquitin proteins are preferential substrates for the UPS.[74] p62 protein binding can induce the formation of large protein aggregates.[75] Thus, p62 might exacerbate the deposition of large poly-ubiquitinated aggregates. Furthermore, p62 binds LC3,[76] and thus it can direct ubiquitinated protein aggregates toward autophagic degradation.[77] Furthermore, while pharmacological blockade of autophagy correlates with an increase in p62 level and aggregation of poly-ubiquitinated aggregates, UPS inhibition results in autophagy induction.[78] In conclusion, autophagy plays a crucial homeostatic function in tight accordance with the UPS. Thus, it does not surprise that dysfunctional autophagy correlates with several degenerative diseases. With specific reference to PD, one interesting mechanistic insight came from in vitro experiments taking advantage of PC12 cell clones overexpressing alpha synuclein.[79] Webb et al. found not only that alpha synuclein aggregates can be cleared by either/both UPS and autophagy, but also that pharmacological blockade of either of the two mechanisms causes the accumulation of alpha synuclein. Interestingly, the metabolism of wild-type alpha synuclein turnover is not reduced upon pharmacological blockade of autophagy. Instead, any alteration of proper autophagic flux resulted in the severe accumulation of mutant synuclein.[10,80] Such outcome may derive from the limited capability of the UPS to handle large oligomeric protein aggregates. In fact, in UPS-mediated degradation, the substrate must pass through the tight channel of the proteasomal structure. Thus, autophagy arises as the main, or maybe unique, solution to achieve efficient clearance of protein aggregates, as the one formed by mutant alpha-synuclein. Overall, it can thus be concluded that autophagy supports normal neuronal homeostasis, and that blockage of autophagy results in neuronal dysfunction and cell death. However, postmortem examination of specimens from patients suffering from Alzheimer's,[59,81,82] Parkinsons,[60] and Huntington's disease (HD)[83] often demonstrates the accumulation of autophagosomes and endosomes or lysosomes. But it must be taken into account that the correlation between autophagosome accumulation and disease is complex and nonlinear, given that an increased amount of autophagic organelles may be a consequence of an increase in autophagy or instead originates from a blockage in the autophagic flux. In particular, defects at the stage of autophagosome and lysosome fusion can result in the accumulation of abnormal autophagosome. Furthermore, autophagic flux moves on a track formed by microtubules.[84] Thus, any alteration of the microtubulin cytoskeleton could result in autophagosome accumulation. Accordingly, mutations affecting dynactin or kinesin subunits, the engine of any transport on tubulin tracks, cause neurodegeneration.[85] Finally, dysfunction of the lysosomal system can result in a failure to clear mature autophagosome and induce an increase in immature autophagosomes. Not surprisingly, lysosomal storage disorders are often characterized by neurodegeneration, accumulation of polyubiquitinated protein aggregates, damaged mitochondria, and early stage autophagosomes,[86] and link into the genetic forms of PD, for example, with mutations in *GBA*. Thus, the accumulation of autophagosomes revealed in postmortem specimens of neurodegenerative patients may be a result of dysfunctional autophagy rather than a *bona fide* increase in activity. Conversely, excessive autophagy can be detrimental for neurons. Macroautophagy at its most active allows cell self-destruction in a form of cannibalization. Initial data supporting this in a neurodegenerative paradigm came from the investigation of a HD mouse model. In particular, primary striatal culture overexpressing mutant human showed an increase in autophagy and oxidative stress upon exposure to an increased concentration of dopamine.[87] This evidence demonstrated that high intracellular amount of dopamine may induce oxidative stress, stimulate autophagy, and result in toxicity. Further support to a potential deleterious role of autophagy came from the study of models of neuronal injury. Acute pharmacological insult or deprivation of trophic factors results in massive cytotoxicity that requires autophagy. In particular, the inhibition of autophagy via treatment with 3-methyladenine (3-MA) partially protected neurons exposed to different insults, such as kainate, oxidative stress, and the lysosomal inhibitor chloroquine.[88] Similarly, 3-MA treatment as well as acute downregulation of ATG5 or BECN1 expression ameliorated photoreceptor survival upon oxidative stress.[89] Finally, the very same mouse model demonstrating the crucial role played by autophagy in supporting long-term neuronal survival, the Atg7flox/flox;nestin-Cre mice, shows almost no neurodegeneration upon neonatal or adult hypoxic brain injury.[90] Thus, depending on the experimental setup, one can draw quite opposite conclusions, describing autophagy either as neuroprotective or neurotoxic. Indeed, it should be taken into consideration the different strategies chosen to monitor autophagy in each study and the timing of the analysis. In particular, the investigation of rats exposed to a controlled cortical impact revealed that trauma induces autophagy within an hour, and that the activation lasts for excess of 1 month in the surviving cells. Interestingly, while early phase LC-3-positive cells were spared apoptotic cell death, already after 24 hours after injury, the number of caspase-3 and LC-3-positive cells markedly increased. Thus, the induction of autophagy may be protective at an early stage, but then contributes to cell death, supporting the elimination of damaged cells. This opposite double impact of autophagy on cell survival is consequent to the articulate functional cross-talk between the autophagy and cell death programs. Experimental models abrogating apoptotic response factors such as $Bax^{-/-}Bak^{-/-}$ knockout mice are resistant to various noxious insults, but die because of a delayed autophagic cellular response.[91] The block of apoptosis via pharmacological inhibition of caspase processing induces autophagic cell death.[92] Despite some exceptions,[93] it is becoming clear that autophagy and apoptosis are mutually exclusive cell death mechanisms.[94] Autophagy and apoptosis can be triggered upon exposure to similar noxious stimuli, such as oxidative stress, ceramide, and increased concentration of cytosolic calcium.[95–97] The mammalian target of rapamycin (mTOR) kinase senses extracellular signals such as a reduced concentration of nutrients, growth factors, energy, and stress, and negatively regulates autophagy. Upon starvation, mTOR is inhibited and consequently autophagy is activated. Recent evidence suggested that mTOR has tissue-specific impact on cellular death via apoptosis.[98] In fact, two recently reported mTOR interactors, the proline-rich AKT substrate (PRAS40) and the protein Q6MZQ0/FLJ14213/CAE45978, have been robustly implicated in the apoptotic pathway.[99] Beclin 1, the mammalian ortholog of yeast Atg6, controls one of the first step allowing autophagosome formation. Beclin 1 interacts with the antiapoptotic proteins BCL-2 and BCL-XL.[100] This interaction is conserved across different species and supports a role for beclin 1 as a key modulator of both autophagy and apoptosis. Growth factor depletion causes the proteolytic cleavage of beclin 1 and PI3K via caspase and results in inhibition of autophagy. Strikingly, the C-terminal fragment of beclin 1 released by the caspase processing binds a mitochondrial receptor and sensitizes cells to proapoptotic insults.[101] Furthermore, the proapoptotic protein BAX increases caspase cleavage of beclin 1, thus impairing autophagic flux.[102] It has been now shown that several proteins involved in autophagy are substrates for caspases: for example, caspase-3 cleaves ATG4D, a cysteine protease that is responsible for the correct processing of LC-3.[103] Similarly to that discussed in relation to apoptosis, the functional cross-talk between necrosis and autophagy is quite complex. The ability of autophagy to abolish necrosis is recognized as one of the main prosurvival roles played by autophagy. Accumulating evidence shows that upon exposure to TNF alpha and starvation, autophagy is activated and necrosis is impaired in different cell lines.[104] From a molecular point of view, it has been shown that zVAD prevents apoptosis and autophagy but induces necrosis.[105] This comes from the inhibition of caspases as well as lysosomal cathepsins enzymatic activity. Thus, the three major routes deciding cellular fate—apoptosis, necrosis, and autophagy—are tightly linked together. Indeed, the investigation of the functional cross-talk between autophagy and necrotic and apoptotic cell death deserves further effort.

5 AUTOPHAGY INDUCTION TO PROMOTE NEURONAL SURVIVAL

Chronic inhibition of autophagy accelerates the rate of ageing; it is therefore tempting to speculate that sustained autophagy may preserve from cellular senescence. Not surprisingly, accumulating evidence suggests that pharmacological induction of autophagy may be a potential therapeutic approach relevant to a number of neurodegenerative disorders.[106] The possibility that long-lasting activation of autophagy may result in deleterious digestion of organelles, critical cellular components, and cell self-eating should, however, be taken into account. These doubts have been somewhat laid to rest by recent data from a range of experimental systems. In fact, it has been found that autophagy is quite selective toward the clearance of protein aggregates rather than healthy cellular elements. Protein aggregates as well as the entire machinery necessary to sustain autophagy are confined to the pericentriolar region in a process requiring the microtubule cytoskeleton and the deacetylase HDAC6.[78] The achievement of therapeutic activation of autophagy relies on a deep understanding and precise manipulation of the upstream signaling pathways regulating this process. The master regulator of autophagy is the intracellular and extracellular concentration of amino acids. The protein kinase EIF2AK4 is activated by empty tRNAs and thus monitors the level of available amino acids. Activated EIF2AK4 initiates autophagy catalyzing the phosphorylation of eukaryotic initiation factor eIFα2.[107] The extracellular content of essential nutrients such as amino acids is instead monitored via a not-yet-defined transmembrane receptor[108] that may control mTOR. mTOR integrates a number of signaling pathways, such as insulin and growth factors, intracellular calcium, and inositol.[109] Energy status, in particular, glucose and lipids level, is monitored also by AMPK. AMPK is a serine and/or threonine kinase that is triggered by increasing AMP/decreasing ATP level. Specifically, AMPK is activated by AMP binding as well as by phosphorylation on Thr172 by LKB1.[110] AMPK inhibits mTOR, thus inducing autophagy. p53 proteins are also involved in the modulation of autophagy: whereas cytosolic p53 blocks autophagy,[111] nuclear translocated p53 triggers it.[112] In particular, p53 inhibits mTOR activity via induction of AMPK. Finally, the presence of intracellular protein aggregates can directly induce autophagy.[113] It is still unresolved which of these signaling cascades are centrally implicated in neurodegenerative disease. Rapamycin and the derived analogues, the rapalogues, induce autophagy via the inhibition of mTORC1.[114] In particular, in mammalian systems, rapamycin specifically inhibits mTORC1, thus abolishing the kinase activity of mTOR.[115] Several independent experimental findings demonstrate that rapalogues protect cells from the cytotoxic impact of huntingtin and alpha synuclein species.[116] Furthermore, rapalogues impaired neurodegeneration and increase survival in animal models of HD and spinocerebellar ataxia.[117,118] To further support the protective role of autophagy, the prosurvival activity of rapamycin and rapalogues was lost in cellular model of disease upon inhibition of autophagy via 3-MA.[119,120] Autophagy induction has been demonstrated to promote the survival of dopaminergic neurons in several PD models. Overexpression of beclin 1 reduced neuronal death in a rodent model of PD. Specifically, lentiviral-mediated overexpression of beclin 1 by stereotaxic injection into the temporal cortex and hippocampus of alpha synuclein–overexpressing mice activated autophagy, reduced the accumulation of alpha synuclein, and ameliorated neuronal pathology.[121] By supporting the clearance of protein aggregates and damaged mitochondria, autophagy might mitigate oxidative stress. For example, rapamycin reduced robustly not only the accumulation of polyubiquitinated aggregates, but also neuronal death in rats exposed to rotenone.[122] Although rapamycin is already in clinical use (eg, in oncology), it has deleterious side effects on lung function and the immune system. To limit the eventual side effects of rapamycin on B and T cells and to reduce the oxidative stress in dopaminergic neurons, it has been proposed to combine low doses of rapamycin or analogues with antioxidants. Strong evidence, however, points toward the necessity of basal levels of oxidative stress (in particular superoxide species), to induce an autophagic response. In particular, the coadministration with thiol or non-thiol antioxidants such as N-acetyl cysteine or vitamin E not only blocked the proautophagic action of rapamycin in primary neurons, but also impaired basal autophagy and precipitated the formation of toxic alpha synuclein deposits.[123] Comparable outcomes were reported upon treatment with different antioxidant molecules such as tocopherol and lipoic acid.[124,125] Such findings have clinical relevance: in fact, several compounds with putative or predicted antioxidant capabilities are in common use in the healthy population. Considering that over 60% of patients affected by neurological disease, specifically Huntington's and Parkinson's, take such supplements,[126,127] a deeper knowledge of the cross-talk between autophagy and oxidative stress is urgently required. Efforts to develop clinically relevant compounds would benefit from agents capable of buffering oxidative stress without impairing autophagy. Actually, some naturally occurring compounds, such as resveratrol, kaempferol, quercetin, and curcumin,[128] demonstrate these properties. For example, curcumin, a compound obtained from the spice cumin (derived from *Cuminum cyminum*), inhibits mTOR, activates autophagy, and supports the clearance of A53T aggregates. Furthermore, independent studies have demonstrated that certain functions of mTORC1 escape from rapamycin inhibition. Given these limitations, together with suboptimal pharmacokinetic properties such as a poor bioavailability, alternative pharmacological strategies have been sought. A number of small molecules have now been identified that abolish mTOR kinase activity by competing with ATP for the binding pocket of mTOR.[129] Compounds such as PP242, Torin1, WYE-354, and Ku-0063794 inhibit both mTORC1 and mTORC2 activity and are more efficient and selective than rapamycin in terms of autophagy induction.[130,131] Furthermore, a macromolecular complex composed of beclin 1, PI3K, UVRAG, and AMBRA1 is required to allow autophagosome formation and subsequent binding and cleavage of LC3 to phagosome membrane. Several molecules regulate both mTORC1 and PI3K such as PI-103 and NVPBEZ235. These compounds are more effective than agents targeting either mTOR or PI3K.[132] Recently, it has been found that metformin (N, N-dimethylbiguanide), a biguanide used as a therapy in diabetes impairs lymphoma cell proliferation via activation of AMPK.[133] Thus, metformin may emerge as a promising autophagy-inducer agent. Other drugs already in clinical use have been found to be capable of inducing autophagy. Lithium, a mood stabilizer agent acting on several pathways, including inositol and GSK3B, supports the clearance of huntigtin aggregates in models of HD via mTOR-independent autophagy induction.[134,135] Interestingly, a pharmacological regimen including valproate and lithium protected neurons from MPTP insult in a rodent PD model via autophagy.[136–138] Finally, accumulating evidence suggests the potential therapeutic application of trehalose. Trehalose is a disaccharide found in a number of different organisms, from bacteria to plants, but it is not synthesized in mammals. For example, it is the main sugar present in the hemolymph of invertebrates. Different human food sources contain significant amount of trehalose (mushrooms and honey among others). Trehalose has many interesting properties: it protects cells from high or low temperature, absence of water, and oxidative stress.[139] Trehalose is an important molecule for industrial and medical applications. These applications include use as a food additive to increase sweetness and promote freeze-dry preservation. Trehalose is also included in antibody preparations for stabilization during freezing or desiccation.[140] Many beneficial effects of trehalose arise from its chaperone action on proteins.[141] For instance, trehalose prevents amyloid deposition in a model of Alzheimer's disease,[142] as well as polyglutamine-mediated protein aggregation in vitro and in mouse models of HD.[143] Trehalose successfully reduced protein aggregation in mouse models of PD[144] and abolished alpha synuclein toxicity.[145–147] Recent studies have shown that trehalose increases the number of autophagosome and promotes mitophagy by an as-yet-unresolved mechanism. Seemingly, trehalose acts independently from mTOR pathway and increases the amount of proautophagic mediators such as beclin 1, Atg12, Atg7, or Atg5.[146] Trehalose does not pass freely through cell membranes but is uptaken into mammalian cells via endocytosis and pinocytosis.[106] A crucial issue is represented by the pharmacokinetics of trehalose in the human body. Intriguingly, although vertebrates lack the trehalose biosynthetic genes, they do express trehalase, an enzyme that efficiently hydrolyses trehalose into two glucose monomers.[148] Thus, upon ingestion, trehalose is quickly degraded. Trehalase is expressed as a GPI-anchored glycoprotein on the membrane of the microvillus intestinal mucosa and renal brush border.[149] In humans, it is believed that intestinal trehalase is the main enzyme metabolizing trehalose. Intestinal trehalase catalyzes the rapid degradation of the dietary trehalose to such an extent that trehalose is not found in the blood stream, not even in transitory or low levels. Despite this, trehalose in drinking water induced autophagy and reduced protein accumulation in the CNS and in ameliorated motor phenotype in several different mouse models of PD.[144,146,150] This evidence suggests that trehalose metabolism is less efficient in mice. No side effects were reported in phase I studies in which patients were exposed to increasing doses of trehalose.[151] However, it should be noted that alterations of trehalase activity result in digestive issues. In fact, subjects deficient for trehalase activity show strong trehalose intolerance and manifest abdominal pain diarrhea.[152] In any case, taking into account the complementary protective effects of trehalose in several in vivo and in vitro models of proteinopathies, trehalose is a promising treatment in neurodegenerative diseases, including Parkinsons.

6 CONCLUSIONS

In summary, there is a large and growing body of evidence linking the various forms of autophagy to Parkinsons disease—both from genetic studies of PD and more generally in ageing. This has already had a major impact on our understanding of Parkinsons, and more importantly provided much-needed insights relevant to drug development. As our knowledge of how autophagic pathways function, a knowledge that is still in its infancy, improves, it is likely that more advances will be made that will have even greater impact on efforts to develop novel treatments for Parkinsons.

References

1. Parzych KR, Klionsky DJ. An overview of autophagy: morphology, mechanism, and regulation. *Antioxid. Redox Signal.* 2014;20(3):460–473.
2. Sorbara MT, Girardin SE. Emerging themes in bacterial autophagy. *Curr Opin Microbiol.* 2015;23:163–170.
3. Klionsky DJ, Abdalla FC, Abeliovich H, Abraham S R.T., Acevedo-Arozena A, Adeli K, Agholme Lotta, et al. Guidelines for the use and interpretation of assays for monitoring autophagy. *Autophagy.* 2012;8(4):445–544.
4. Polymeropoulos MH, Lavedan C, Leroy E, Ide SE, Dehejia A, Dutra A, Pike B, et al. Mutation in the alpha-synuclein gene identified in families with Parkinsons disease. *Science.* 1997;276(5321):2045–2047.
5. Kara E, Lewis PA, Ling H, Proukakis Christos, Houlden Henry, Hardy J. α-Synuclein mutations cluster around a putative protein loop. *Neurosci Lett.* 2013;546:67–70. doi: 10.1016/j.neulet.2013.04.058.
6. Lill CM, Roehr JT, McQueen MB, Kavvoura FK, Bagade S, Schjeide B-MM, Schjeide LM, et al. Comprehensive research synopsis and systematic meta-analyses in Parkinsons disease genetics: the PDGene Database. *PLoS Genet.* 2012;8(3):e1002548.
7. Nalls MA, Pankratz N, Lill CM, Do CB, Hernandez DG, Saad M, DeStefano AL, et al. Large-scale meta-analysis of genome-wide association data identifies six new risk loci for Parkinsons disease. *Nature Genet.* 2014;46(9):989–993.
8. Spillantini MG, Schmidt ML, Lee VM, Trojanowski JQ, Jakes R, Goedert M. Alpha-synuclein in Lewy bodies. *Nature.* 1997;388(6645):839–840.
9. Lashuel HA, Overk CR, Oueslati A, Masliah E. The many faces of α-synuclein: from structure and toxicity to therapeutic target. *Nature Rev Neurosci.* 2013;14(1):38–48.
10. Vogiatzi T, Xilouri M, Vekrellis K, Stefanis L. Wild type alpha-synuclein is degraded by chaperone-mediated autophagy and macroautophagy in neuronal cells. *J Biol Chem.* 2008;283(35):23542–23556.

11. Cuervo AM, Stefanis L, Fredenburg R, Lansbury PT, Sulzer D. Impaired degradation of mutant alpha-synuclein by chaperone-mediated autophagy. *Science.* 2004;305(5688):1292–1295.

12. Friedman LG, Lachenmayer ML, Wang J, He L, Poulose SM, Komatsu M, Holstein GR, Yue Z. Disrupted autophagy leads to dopaminergic axon and dendrite degeneration and promotes presynaptic accumulation of α-synuclein and LRRK2 in the brain. *J Neurosci.* 2012;32(22):7585–7593.

13. Singleton A, Myers A, Hardy J. The law of mass action applied to neurodegenerative disease: a hypothesis concerning the etiology and pathogenesis of complex diseases. *Hum Mol Genet.* 2004;13(Spec No 1):R123–R126.

14. Winslow AR, Chen C-W, Corrochano S, Acevedo-Arozena A, Gordon DE, Peden AA, Lichtenberg M, et al. α-Synuclein impairs macroautophagy: implications for Parkinsons disease. *J Cell Biol.* 2010;190(6):1023–1037.

15. Lopes da Fonseca Tomás, Villar-Piqué Anna, Outeiro Tiago Fleming. The interplay between alpha-synuclein clearance and spreading. *Biomolecules.* 2015;5(2):435–471.

16. Schapira AHV. Glucocerebrosidase and Parkinson disease: recent advances. *Mol Cell Neurosci.* 2015;66(Pt A):37–42.

17. Neumann J, Bras J, Deas E, O'Sullivan SS, Parkkinen L, Lachmann RH, Abi L, et al. Glucocerebrosidase mutations in clinical and pathologically proven Parkinsons disease. *Brain J Neurol.* 2009;132(Pt 7):1783–1794.

18. Cullen VS, Sardi P, Ng J, Xu You-Hai, Sun Ying, Tomlinson Julianna J, Kolodziej Piotr, et al. Acid β-glucosidase mutants linked to Gaucher disease, Parkinson disease, and Lewy body dementia alter α-synuclein processing. *Ann Neurol.* 2011;69(6):940–953.

19. Yap TL, Gruschus JM, Velayati A, Westbroek W, Goldin E, Moaven N, Sidransky E, Lee JC. Alpha-synuclein interacts with glucocerebrosidase providing a molecular link between Parkinson and Gaucher diseases. *J Biol Chem.* 2011;286(32):28080–28088.

20. Mazzulli JR, Xu Y-H, Sun Y, Knight AL, McLean PJ, Caldwell GA, Sidransky E, Grabowski GA, Krainc D. Gaucher disease glucocerebrosidase and α-synuclein form a bidirectional pathogenic loop in synucleinopathies. *Cell.* 2011;146(1):37–52.

21. Ramirez A, Heimbach A, Gründemann J, Stiller B, Dan Hampshire L, Cid P, Goebel I, et al. Hereditary Parkinsonism with dementia is caused by mutations in ATP13A2, encoding a lysosomal type 5 P-type ATPase. *Nature Genet.* 2006;38(10):1184–1191.

22. Lees AJ, Singleton AB. Clinical heterogeneity of ATP13A2 linked disease (Kufor-Rakeb) justifies a PARK designation. *Neurology.* 2007;68(19):1553–1554.

23. Usenovic M, Tresse E, Mazzulli JR, Taylor JP, Krainc D. Deficiency of ATP13A2 leads to lysosomal dysfunction, α-synuclein accumulation, and neurotoxicity. *J Neurosci.* 2012;32(12):4240–4246.

24. Vilariño-Güell C, Wider C, Ross OA, Dachsel JC, Kachergus JM, Lincoln SJ, Soto-Ortolaza AI, et al. VPS35 mutations in Parkinson disease. *Am J Hum Genet.* 2011;89(1):162–167.

25. Zavodszky E, Seaman MNJ, Moreau K, Jimenez-Sanchez M, Breusegem SY, Harbour ME, Rubinsztein DC. Mutation in VPS35 associated with Parkinsons disease impairs WASH complex association and inhibits autophagy. *Nat Commun.* 2014;5:3828.

26. Tang F-L, Erion JR, Tian Y, Liu W, Yin D-M, Ye J, Tang B, Mei L, Xiong W-C. VPS35 in dopamine neurons is required for endosome-to-golgi retrieval of Lamp2a, a receptor of chaperone-mediated autophagy that is critical for α-synuclein degradation and prevention of pathogenesis of Parkinsons disease. *J Neurosci.* 2015;35(29):10613–10628.

27. Plowey ED, Cherra SJ, Liu Y-J, Chu CT. Role of autophagy in G2019S-LRRK2-associated neurite shortening in differentiated SH-SY5Y cells. *J Neurochem.* 2008;105(3):1048–1056.

28. Alegre-Abarrategui J, Christian H, Lufino MM, Mutihac R, Venda LL, Ansorge O, Wade-Martins R. LRRK2 regulates autophagic activity and localizes to specific membrane microdomains in a novel human genomic reporter cellular model. *Hum Mol Genet.* 2009;18(21):4022–4034.

29. Tong Y, Yamaguchi H, Giaime E, Boyle S, Kopan R, Kelleher RJ, Shen J. Loss of leucine-rich repeat kinase 2 causes impairment of protein degradation pathways, accumulation of alpha-synuclein, and apoptotic cell death in aged mice. *Proc Natl Acad Sci USA*. 2010;107(21):9879–9884.
30. Manzoni C, Mamais A, Dihanich S, Abeti R, Soutar MPM, Plun-Favreau H, Giunti P, Tooze SA, Bandopadhyay R, Lewis PA. Inhibition of LRRK2 kinase activity stimulates macroautophagy. *Biochimica Et Biophysica Acta*. 2013;1833(12):2900–2910.
31. Orenstein SJ, Kuo S-H, Tasset I, Arias E, Koga H, Fernandez-Carasa I, Cortes E, et al. Interplay of LRRK2 with chaperone-mediated autophagy. *Nature Neurosci*. 2013;16(4):394–406.
32. Haack TB, Hogarth P, Kruer MC, Gregory A, Wieland T, Schwarzmayr T, Graf E, et al. Exome sequencing reveals de novo WDR45 mutations causing a phenotypically distinct, X-linked dominant form of NBIA. *Am J Hum Genet*. 2012;91(6):1144–1149.
33. Proikas-Cezanne T, Takacs Z, Dönnes P, Kohlbacher O. WIPI proteins: essential PtdIns3P effectors at the nascent autophagosome. *J Cell Sci*. 2015;128(2):207–217.
34. Wilson GR, Sim JCH, McLean C, Giannandrea M, Galea CA, Riseley JR, Stephenson SEM, et al. Mutations in RAB39B cause X-linked intellectual disability and early-onset Parkinson disease with α-synuclein pathology. *Am J Hum Genet*. 2014;95(6):729–735.
35. MacLeod DA, Rhinn H, Kuwahara T, Zolin A, Di Paolo G, McCabe BD, MacCabe BD, et al. RAB7L1 Interacts with LRRK2 to modify intraneuronal protein sorting and Parkinsons disease risk. *Neuron*. 2013;77(3):425–439.
36. Beilina A, Rudenko IN, Kaganovich A, Civiero L, Chau H, Kalia SK, Kalia LV, et al. Unbiased screen for interactors of leucine-rich repeat kinase 2 supports a common pathway for sporadic and familial parkinson disease. *Proc Natl Acad Sci USA*. 2014;111(7):2626–2631.
37. Zimprich A, Biskup S, Leitner P, Lichtner P, Farrer M, Lincoln S, Kachergus J, et al. Mutations in LRRK2 cause autosomal-dominant parkinsonism with pleomorphic pathology. *Neuron*. 2004;44(4):601–607.
38. Wider C, Dickson DW, Wszolek ZK. Leucine-Rich Repeat kinase 2 gene-associated disease: redefining genotype-phenotype correlation. *Neuro-Degen Dis*. 2010;7(1–3):175–179.
39. Gray DA, Tsirigotis M, Woulfe J. Ubiquitin, proteasomes, and the aging brain. *Sci Aging Knowledge Environ SAGE KE*. 2003;2003(34):RE6.
40. Simonsen Anne, Cumming RC, Brech A, Isakson P, Schubert DR, Finley KD. Promoting basal levels of autophagy in the nervous system enhances longevity and oxidant resistance in adult drosophila. *Autophagy*. 2008;4(2):176–184.
41. Hars ES, Haiyan S Q, Ryazanov AG, Jin S, Cai L, Hu C, Liu LF. Autophagy regulates ageing in C. elegans. *Autophagy*. 2007;3(2):93–95.
42. Lipinski MM, Zheng B, Lu T, Yan Z, Py BF, Ng A, Xavier RJ, et al. Genome-wide analysis reveals mechanisms modulating autophagy in normal brain aging and in Alzheimer's disease. *Proc Natl Acad Sci USA*. 2010;107(32):14164–14169.
43. Martinez-Vicente M, Sovak G, Cuervo AM. Protein degradation and aging. *Experim Gerontol*. 2005;40(8-9):622–633.
44. Ma Qinying, Qiang Jing, Gu Ping, Wang Yanyong, Geng Yuan, Wang Mingwei. Age-related autophagy alterations in the brain of senescence accelerated mouse prone 8 (SAMP8) mice. *Experim Gerontol*. 2011;46(7):533–541.
45. Del Roso A, Bombara M, Fierabracci V, Fosella PV, Gori Z, Masini M, Masiello P, Pollera M, Bergamini E. Exploring the mechanisms of in vivo regulation of liver protein breakdown in the rat by the use of the new antilipolytic-agent model. *Acta Biologica Hungarica*. 1991;42(1-3):87–99.
46. Del Roso A, Vittorini S, Cavallini G, Donati A, Gori Z, Masini M, Pollera M, Bergamini E. Ageing-related changes in the in vivo function of rat liver macroautophagy and proteolysis. *Experim Gerontol*. 2003;38(5):519–527.
47. Donati A, Cavallini G, Paradiso C, Vittorini S, Pollera M, Gori Z, Bergamini E. Age-related changes in the autophagic proteolysis of rat isolated liver cells: effects of antiaging dietary restrictions. *J Gerontol*. 2001;56(9):B375–B383.

48. Bergamini E, Bizzarri R, Cavallini G, Cerbai B, Chiellini E, Donati A, Gori Z, et al. Ageing and oxidative stress: a role for dolichol in the antioxidant machinery of cell membranes? *J Alzheimer Dis.* 2004;6(2):129–135.

49. Terman A, Brunk UT. The aging myocardium: roles of mitochondrial damage and lysosomal degradation. *Heart Lung Circul.* 2005;14(2):107–114.

50. Cuervo AM, Dice JF. Age-related decline in chaperone-mediated autophagy. *J Biol Chem.* 2000;275(40):31505–31513.

51. Kitani K, Ohta M, Kanai S, Nokubo M, Sato Y, Otsubo K, Ivy GO. Morphological, physiological and biochemical alterations in livers of rodents induced by protease inhibitors: a comparison with old livers. *Adv Exp Med Biol.* 1989;266:75–92.

52. Troen BR. The biology of aging. *Mt Sinai J Medicine.* 2003;70(1):3–22.

53. Cookson MR. The biochemistry of Parkinsons disease. *Annu Rev Biochem.* 2005;74:29–52.

54. Olanow CW, McNaught K St P. Ubiquitin-proteasome system and Parkinsons disease. *Movem Disord.* 2006;21(11):1806–1823.

55. McNaught K St P, Björklund LM, Belizaire R, Isacson O, Jenner P, Olanow CW. Proteasome inhibition causes nigral degeneration with inclusion bodies in rats. *Neuroreport.* 2002;13(11):1437–1441.

56. Deter RL, Baudhuin P, De Duve C. Participation of lysosomes in cellular autophagy induced in rat liver by glucagon. *J Cell Biol.* 1967;35(2):C11–C16.

57. Mizushima Noboru, Levine Beth. Autophagy in mammalian development and differentiation. *Nature Cell Biol.* 2010;12(9):823–830.

58. Mizushima N, Levine B, Cuervo AM, Klionsky DJ. Autophagy fights disease through cellular self-digestion. *Nature.* 2008;451(7182):1069–1075.

59. Boland B, Kumar A, Lee S, Platt FM, Wegiel J, Yu WH, Nixon RA. Autophagy induction and autophagosome clearance in neurons: relationship to autophagic pathology in Alzheimer's disease. *J Neurosci.* 2008;28(27):6926–6937.

60. Anglade P, Vyas S, Javoy-Agid F, Herrero MT, Michel PP, Marquez J, Mouatt-Prigent A, Ruberg M, Hirsch EC, Agid Y. Apoptosis and autophagy in nigral neurons of patients with Parkinsons disease. *Histol Histopathol.* 1997;12(1):25–31.

61. Yue Zhenyu, Friedman Lauren, Komatsu Masaaki, Tanaka Keiji. The cellular pathways of neuronal autophagy and their implication in neurodegenerative diseases. *Biochimica Et Biophysica Acta.* 2009;1793(9):1496–1507.

62. Komatsu Masaaki, Wang S Qing Jun, Holstein Gay R, Friedrich Victor L, Iwata Jun-ichi, Kominami Eiki, Chait Brian T, Tanaka Keiji, Yue Zhenyu. Essential role for autophagy protein Atg7 in the maintenance of axonal homeostasis and the prevention of axonal degeneration. *Proc Natl Acad Sci USA.* 2007;104(36):14489–14494.

63. Dixon JS. "Phagocytic" lysosomes in chromatolytic neurones. *Nature.* 1967;215(5101): 657–658.

64. Reichardt LF, Mobley WC. Going the distance, or not, with neurotrophin signals. *Cell.* 2004;118(2):141–143.

65. Ravikumar B, Acevedo-Arozena A, Imarisio S, Berger Z, Vacher C, O'Kane CJ, Brown SDM, Rubinsztein DC. Dynein mutations impair autophagic clearance of aggregate-prone proteins. *Nature Genet.* 2005;37(7):771–776.

66. Kaasinen SK, Harvey L, Reynolds AJ, Hendry IA. Autophagy generates retrogradely transported organelles: a hypothesis. *Int J Dev Neurosci.* 2008;26(6):625–634.

67. Mariño G, Madeo F, Kroemer G. Autophagy for tissue homeostasis and neuroprotection. *Curr Opin Cell Biol.* 2011;23(2):198–206.

68. Nixon RA, Yang D-S, Lee J-H. Neurodegenerative lysosomal disorders: a continuum from development to late age. *Autophagy.* 2008;4(5):590–599.

69. Kuma A, Hatano M, Matsui M, Yamamoto A, Nakaya H, Yoshimori T, Ohsumi Y, Tokuhisa T, Mizushima N. The role of autophagy during the early neonatal starvation period. *Nature.* 2004;432(7020):1032–1036.

70. Komatsu M, Waguri S, Chiba T, Murata S, Iwata J-i, Tanida I, Ueno T, et al. Loss of autophagy in the central nervous system causes neurodegeneration in mice. *Nature.* 2006;441(7095):880–884.
71. Hara T, Nakamura K, Matsui M, Yamamoto A, Nakahara Y, Suzuki-Migishima R, Yokoyama M, et al. Suppression of basal autophagy in neural cells causes neurodegenerative disease in mice. *Nature.* 2006;441(7095):885–889.
72. Fimia GM, Stoykova A, Romagnoli A, Giunta L, Di Bartolomeo S, Nardacci R, Corazzari M, et al. Ambra1 regulates autophagy and development of the nervous system. *Nature.* 2007;447(7148):1121–1125.
73. Komatsu M, Waguri S, Koike M, Sou Y-S, Ueno T, Hara T, Mizushima N, et al. Homeostatic levels of p62 control cytoplasmic inclusion body formation in autophagy-deficient mice. *Cell.* 2007;131(6):1149–1163.
74. Tan JMM, Wong ESP, Kirkpatrick DS, Pletnikova O, Ko HS, Tay S-P, Ho Michelle WL, et al. Lysine 63-linked ubiquitination promotes the formation and autophagic clearance of protein inclusions associated with neurodegenerative diseases. *Hum Mol Genet.* 2008;17(3):431–439.
75. Lamark T, Perander M, Outzen H, Kristiansen K, Øvervatn A, Michaelsen E, Bjørkøy G, Johansen T. Interaction codes within the family of mammalian Phox and Bem1p domain-containing proteins. *J Biol Chem.* 2003;278(36):34568–34581.
76. Ichimura Y, Kumanomidou T, Sou Y-s, Mizushima T, Ezaki J, Ueno T, Kominami E, Yamane T, Tanaka K, Komatsu M. Structural basis for sorting mechanism of p62 in selective autophagy. *J Biol Chem.* 2008;283(33):22847–22857.
77. Bjørkøy G, Lamark T, Brech A, Outzen H, Perander M, Overvatn A, Stenmark H, Johansen T. p62/SQSTM1 forms protein aggregates degraded by autophagy and has a protective effect on Huntingtin-induced cell death. *J Cell Biol.* 2005;171(4):603–614.
78. Iwata A, Riley BE, Johnston JA, Kopito RR. HDAC6 and microtubules are required for autophagic degradation of aggregated Huntingtin. *J Biol Chem.* 2005;280(48):40282–40292.
79. Webb JL, Ravikumar B, Atkins J, Skepper JN, Rubinsztein DC. Alpha-synuclein is degraded by both autophagy and the proteasome. *J Biol Chem.* 2003;278(27):25009–25013.
80. Sarkar S, Ravikumar B, Rubinsztein DC. Autophagic clearance of aggregate-prone proteins associated with neurodegeneration. *Meth Enzymol.* 2009;453:83–110.
81. Nixon RA, Wegiel J, Kumar A, Yu WH, Peterhoff C, Cataldo A, Cuervo AM. Extensive involvement of autophagy in Alzheimer disease: an immuno-electron microscopy study. *J Neuropathol Exp Neurol.* 2005;64(2):113–122.
82. Cataldo AM, Hamilton DJ, Barnett JL, Paskevich PA, Nixon RA. Properties of the endosomal-lysosomal system in the human central nervous system: disturbances mark most neurons in populations at risk to degenerate in Alzheimer's disease. *J Neurosci.* 1996;16(1):186–199.
83. Rudnicki DD, Pletnikova O, Vonsattel J-PG, Ross CA, Margolis RL. A comparison of Huntington disease and Huntington disease-like 2 neuropathology. *J Neuropathol Exp Neurol.* 2008;67(4):366–374.
84. Fass E, Shvets E, Degani I, Hirschberg K, Elazar Z. Microtubules support production of starvation-induced autophagosomes but not their targeting and fusion with lysosomes. *J Biol Chem.* 2006;281(47):36303–36316.
85. Fujiwara T, Morimoto K, Kakita A, Takahashi H. Dynein and dynactin components modulate neurodegeneration induced by excitotoxicity. *J Neurochem.* 2012;122(1):162–174.
86. Bosch ME, Kielian T. Neuroinflammatory paradigms in lysosomal storage diseases. *Front Neurosci.* 2015;9:417.
87. Petersén A, null KE, Larsen GG, Behr N, Romero S, Przedborski P, Sulzer BD. Expanded CAG repeats in exon 1 of the Huntington's disease gene stimulate dopamine-mediated striatal neuron autophagy and degeneration. *Hum Mol Genet.* 2001;10(12):1243–1254.

88. Zaidi AU, McDonough JS, Klocke BJ, Latham CB, Korsmeyer SJ, Flavell RA, Schmidt RE, Roth F KA. Chloroquine-induced neuronal cell death is p53 and Bcl-2 family-dependent but caspase-independent. *J Neuropathol Exp Neurol.* 2001;60(10):937–945.

89. Kunchithapautham K, Rohrer B. Apoptosis and autophagy in photoreceptors exposed to oxidative stress. *Autophagy.* 2007;3(5):433–441.

90. Koike M, Shibata M, Tadakoshi M, Gotoh K, Komatsu M, Waguri S, Kawahara N, et al. Inhibition of autophagy prevents hippocampal pyramidal neuron death after hypoxic-ischemic injury. *Am J Pathol.* 2008;172(2):454–469.

91. Shimizu S, Kanaseki T, Mizushima N, Mizuta T, Arakawa-Kobayashi S, Thompson CB, Tsujimoto Y. Role of Bcl-2 family proteins in a non-apoptotic programmed cell death dependent on autophagy genes. *Nat Cell Biol.* 2004;6(12):1221–1228.

92. Madden DT, Egger L, Bredesen DE. A calpain-like protease inhibits autophagic cell death. *Autophagy.* 2007;3(5):519–522.

93. Christensen ST, Chemnitz J, Straarup EM, Kristiansen K, Wheatley DN, Rasmussen L. Staurosporine-induced cell death in tetrahymena thermophila has mixed characteristics of both apoptotic and autophagic degeneration. *Cell Biol Int.* 1998;22(7–8):591–598.

94. Boya P, González-Polo R-A, Casares N, Perfettini J-L, Dessen P, Larochette N, Métivier D, et al. Inhibition of macroautophagy triggers apoptosis. *Mol Cell Biol.* 2005;25(3):1025–1040.

95. Høyer-Hansen M, Bastholm L, Szyniarowski P, Campanella M, Szabadkai G, Farkas T, Bianchi K, et al. Control of macroautophagy by calcium, calmodulin-dependent kinase kinase-beta, and Bcl-2. *Mol Cell.* 2007;25(2):193–205.

96. Scherz-Shouval R, Shvets E, Fass E, Shorer H, Gil L, Elazar Z. Reactive oxygen species are essential for autophagy and specifically regulate the activity of Atg4. *EMBO J.* 2007;26(7):1749–1760.

97. Lavieu G, Scarlatti F, Sala G, Carpentier S, Levade T, Ghidoni R, Botti J, Codogno P. Regulation of autophagy by sphingosine kinase 1 and its role in cell survival during nutrient starvation. *J Biol Chem.* 2006;281(13):8518–8527.

98. Castedo M, Ferri KF, Kroemer G. Mammalian target of rapamycin (mTOR): pro- and anti-apoptotic. *Cell Death Diff.* 2002;9(2):99–100.

99. Thedieck K, Polak P, Kim ML, Molle KD, Cohen A, Jenö P, Arrieumerlou C, Hall MN. PRAS40 and PRR5-like protein are new mTOR interactors that regulate apoptosis. *PloS One.* 2007;2(11):e1217.

100. Pattingre S, Tassa A, Qu X, Garuti R, Liang XH, Mizushima N, Packer M, Schneider MD, Levine B. Bcl-2 antiapoptotic proteins inhibit Beclin 1-dependent autophagy. *Cell.* 2005;122(6):927–939.

101. Wirawan E, Vande Walle L, Kersse K, Cornelis S, Claerhout S, Vanoverberghe I, Roelandt R, et al. Caspase-mediated cleavage of Beclin-1 inactivates Beclin-1-induced autophagy and enhances apoptosis by promoting the release of proapoptotic factors from mitochondria. *Cell Death Dis.* 2010;1:e18.

102. Luo S, Rubinsztein DC. Apoptosis blocks beclin 1-dependent autophagosome synthesis: an effect rescued by Bcl-xL. *Cell Death Diff.* 2010;17(2):268–277.

103. Betin VMS, Lane JD. Caspase cleavage of Atg4D stimulates GABARAP-L1 processing and triggers mitochondrial targeting and apoptosis. *J Cell Sci.* 2009;122(Pt 14):2554–2566.

104. Bell BD, Leverrier S, Weist BM, Newton RH, Arechiga AF, Luhrs KA, Morrissette NS, Walsh CM. FADD and caspase-8 control the outcome of autophagic signaling in proliferating T cells. *Proc Natl Acad Sci USA.* 2008;105(43):16677–16682.

105. Wu Y-T, Tan H-L, Huang Q, Sun X-J, Zhu X, Shen H-M. zVAD-induced necroptosis in L929 cells depends on autocrine production of TNFα mediated by the PKC-MAPKs-AP-1 pathway. *Cell Death Diff.* 2011;18(1):26–37.

106. Emanuele E. Can trehalose prevent neurodegeneration? Insights from experimental studies. *Curr Drug Targ.* 2014;15(5):551–557.

107. Tallóczy Z, Jiang W, Virgin HW, Leib DA, Scheuner D, Kaufman RJ, Eskelinen E-L, Levine B. Regulation of starvation- and virus-Induced autophagy by the eIF2alpha kinase signaling pathway. *Proc Natl Acad Sci USA*. 2002;99(1):190–195.

108. Kanazawa Takumi, Taneike I, Akaishi R, Yoshizawa F, Furuya N, Fujimura S, Kadowaki M. Amino acids and insulin control autophagic proteolysis through different signaling pathways in relation to mTOR in isolated rat hepatocytes. *J Biol Chem*. 2004;279(9):8452–8459.

109. Corradetti MN, Guan K-L. Upstream of the mammalian target of rapamycin: do all roads pass through mTOR? *Oncogene*. 2006;25(48):6347–6360.

110. Imai K, Inukai K, Ikegami Y, Awata T, Katayama S. LKB1, an upstream AMPK kinase, regulates glucose and lipid metabolism in cultured liver and muscle cells. *Biochem Biophys Res Commun*. 2006;351(3):595–601.

111. Tasdemir EM, Maiuri C, Galluzzi L, Vitale I, Djavaheri-Mergny M, D'Amelio M, Criollo A, et al. Regulation of autophagy by cytoplasmic p53. *Nat Cell Biol*. 2008;10(6):676–687.

112. Crighton D, Wilkinson S, O'Prey J, Syed N, Smith P, Harrison PR, Gasco M, Garrone O, Crook T, Ryan KM. DRAM, a p53-induced modulator of autophagy, is critical for apoptosis. *Cell*. 2006;126(1):121–134.

113. Kopito RR. Aggresomes, inclusion bodies and protein aggregation. *Trends Cell Biol*. 2000;10(12):524–530.

114. Tekirdag KA, Korkmaz G, Ozturk DG, Agami R, Gozuacik D. MIR181A regulates starvation- and rapamycin-induced autophagy through targeting of ATG5. *Autophagy*. 2013;9(3):374–385.

115. Sarbassov DD, Ali SM, Sengupta S, Sheen J-H, Hsu PP, Bagley AF, Markhard AL, Sabatini DM. Prolonged rapamycin treatment inhibits mTORC2 assembly and Akt/PKB. *Mol Cell*. 2006;22(2):159–168.

116. Ravikumar B, Vacher C, Berger Z, Davies JE, Luo S, Oroz LG, Scaravilli F, et al. Inhibition of mTOR induces autophagy and reduces toxicity of polyglutamine expansions in fly and mouse models of Huntington disease. *Nature Genet*. 2004;36(6):585–595.

117. Ravikumar B, Duden R, Rubinsztein DC. Aggregate-prone proteins with polyglutamine and polyalanine expansions are degraded by autophagy. *Hum Mol Genet*. 2002;11(9):1107–1117.

118. Menzies FM, Huebener J, Renna M, Bonin M, Riess O, Rubinsztein DC. Autophagy induction reduces mutant ataxin-3 levels and toxicity in a mouse model of spinocerebellar ataxia type 3. *Brain J Neurol*. 2010;133(Pt 1):93–104.

119. Berger Z, Ravikumar B, Menzies FM, Oroz LG, Underwood BR, Pangalos MN, Schmitt I, et al. Rapamycin alleviates toxicity of different aggregate-prone proteins. *Hum Mol Genet*. 2006;15(3):433–442.

120. Wang T, Lao U, Edgar BA. TOR-mediated autophagy regulates cell death in drosophila neurodegenerative disease. *J Cell Biol*. 2009;186(5):703–711.

121. Spencer B, Potkar R, Trejo M, Rockenstein E, Patrick C, Gindi R, Adame A, Wyss-Coray T, Masliah E. Beclin 1 gene transfer activates autophagy and ameliorates the neurodegenerative pathology in alpha-synuclein models of Parkinsons and Lewy body diseases. *J Neurosci*. 2009;29(43):13578–13588.

122. Xiong N, Xiong J, Jia M, Liu L, Zhang X, Chen Z, Huang J, et al. The role of autophagy in Parkinsons Disease: rotenone-based modeling. *Behav Brain Func*. 2013;9:13.

123. Underwood BR, Imarisio S, Fleming A, Rose C, Krishna G, Heard P, Quick M, et al. Antioxidants can inhibit basal autophagy and enhance neurodegeneration in models of polyglutamine disease. *Hum Mol Genet*. 2010;19(17):3413–3429.

124. Persson HL, Svensson AI, Brunk UT. Alpha-lipoic acid and alpha-lipoamide prevent oxidant-induced lysosomal rupture and apoptosis. *Redox Rep Commun Free Radical Res*. 2001;6(5):327–334.

125. Persson HL, Nilsson KJ, Brunk UT. Novel cellular defenses against iron and oxidation: ferritin and autophagocytosis preserve lysosomal stability in airway epithelium. *Redox Rep Commun Free Radical Res*. 2001;6(1):57–63.

126. Underwood BR, Broadhurst D, Dunn WB, Ellis DI, Michell AW, Vacher C, Mosedale DE, et al. Huntington disease patients and transgenic mice have similar pro-catabolic serum metabolite profiles. *Brain J Neurol*. 2006;129(Pt 4):877–886.

127. Möller JC, Oertel WH. Use of nutritional supplements in patients with Parkinsons disease: a cause for concern? *Nat Clin Pract Neurol*. 2006;2(11):598–599.

128. Di Zanni E, Bachetti T, Parodi S, Bocca P, Prigione I, Di Lascio S, Fornasari D, Ravazzolo R, Ceccherini I. In vitro drug treatments reduce the deleterious effects of aggregates containing polyAla expanded PHOX2B proteins. *Neurobiol Dis*. 2012;45(1):508–518.

129. Liu Y-Q, Cheng X, Guo L-X, Mao C, Chen Y-J, Liu H-X, Xiao Q-C, Jiang S, Yao Z-J, Zhou G-B. Identification of an annonaceous acetogenin mimetic, AA005, as an AMPK activator and autophagy inducer in colon cancer cells. *PloS One*. 2012;7(10):e47049.

130. Feldman ME, Shokat KM. New inhibitors of the PI3K-Akt-mTOR pathway: insights into mTOR signaling from a new generation of tor kinase domain inhibitors (TORKinibs). *Curr Topics Microbiol Immunol*. 2010;347:241–262.

131. Feldman ME, Apsel B, Uotila A, Loewith R, Knight ZA, Ruggero D, Shokat KM. Active-site inhibitors of mTOR target rapamycin-resistant outputs of mTORC1 and mTORC2. *PLoS Biol*. 2009;7(2):e38.

132. Fokas E, Yoshimura M, Prevo R, Higgins G, Hackl W, Maira S-M, Bernhard EJ, McKenna GW, Muschel RJ. NVP-BEZ235 and NVP-BGT226, dual phosphatidylinositol 3-kinase/ mammalian target of rapamycin inhibitors, enhance tumor and endothelial cell radiosensitivity. *Rad Oncol*. 2012;7:48.

133. Rosilio C, Ben-Sahra I, Bost F, Peyron J-F. Metformin: a metabolic disruptor and antidiabetic drug to target human leukemia. *Cancer Lett*. 2014;346(2):188–196.

134. Sarkar S, Krishna G, Imarisio S, Saiki S, O'Kane CJ, Rubinsztein DC. A rational mechanism for combination treatment of Huntington's disease using lithium and rapamycin. *Hum Mol Genet*. 2008;17(2):170–178.

135. Sarkar SR, Floto A, Berger Z, Imarisio S, Cordenier A, Pasco M, Cook LJ, Rubinsztein DC. Lithium induces autophagy by inhibiting inositol monophosphatase. *J Cell Biol*. 2005;170(7):1101–1111.

136. Li X-Z, Chen X-P, Zhao K, Bai L-M, Zhang H, Zhou X-P. Therapeutic effects of valproate combined with lithium carbonate on MPTP-induced Parkinsonism in mice: possible mediation through enhanced autophagy. *Int J Neurosci*. 2013;123(2):73–79.

137. Castro AA, Ghisoni K, Latini A, Quevedo J, Tasca CI, Prediger RDS. Lithium and valproate prevent olfactory discrimination and short-term memory impairments in the intranasal 1-methyl-4-phenyl-1,2,3,6-tetrahydropyridine (MPTP) rat model of Parkinsons disease. *Behav Brain Res*. 2012;229(1):208–215.

138. Lauterbach EC, Fontenelle LF, Teixeira AL. The neuroprotective disease-modifying potential of psychotropics in Parkinsons disease. *Parkinsons Dis*. 2012;2012:753548.

139. Chen T, Acker JP, Eroglu A, Cheley S, Bayley H, Fowler A, Toner M. Beneficial effect of intracellular trehalose on the membrane integrity of dried mammalian cells. *Cryobiology*. 2001;43(2):168–181.

140. Wolkers WF, Tablin F, Crowe JH. From anhydrobiosis to freeze-drying of eukaryotic cells. *Comp Biochem Physiol Mol Integr Physiol*. 2002;131(3):535–543.

141. Singer MA, Lindquist S. Multiple effects of trehalose on protein folding in vitro and in vivo. *Mol Cell*. 1998;1(5):639–648.

142. Liu R, Barkhordarian H, Emadi S, Park CB, Sierks MR. Trehalose differentially inhibits aggregation and neurotoxicity of beta-amyloid 40 and 42. *Neurobiol Dis*. 2005;20(1):74–81.

143. Tanaka M, Machida Y, Nukina N. A novel therapeutic strategy for polyglutamine diseases by stabilizing aggregation-prone proteins with small molecules. *J Mol Med*. 2005;83(5):343–352. doi: 10.1007/s00109-004-0632-2.

144. Rodríguez-Navarro JA, Rodríguez L, Casarejos MJ, Solano RM, Gómez A, Perucho J, Cuervo AM, Yébenes JGde, Mena MA. Trehalose ameliorates dopaminergic and tau

pathology in parkin deleted/tau overexpressing mice through autophagy activation. *Neurobiol Dis.* 2010;39(3):423–438.

145. Lan D-M, Liu F-T, Zhao J, Chen Y, Wu J-J, Ding Z-T, Yue Z-Y, Ren H-M, Jiang Y-P, Wang J. Effect of trehalose on PC12 cells overexpressing wild-type or A53T mutant α-synuclein. *Neurochem Res.* 2012;37(9):2025–2032.

146. Tanji K, Miki Y, Maruyama A, Mimura J, Matsumiya T, Mori F, Imaizumi T, Itoh K, Wakabayashi K. Trehalosei induces chaperone molecules along with autophagy in a mouse model of Lewy body disease. *Biochem Biophys Res Commun.* 2015;465(4):746–752.

147. He Q, Koprich JB, Wang Y, Yu W-B, Xiao B-G, Brotchie JM, Wang J. Treatment with trehalose prevents behavioral and neurochemical deficits produced in an AAV α-synuclein rat model of Parkinsons disease. *Mol Neurobiol.* 2015.

148. Argüelles J-C. Why can't vertebrates synthesize trehalose? *J Mol Evol.* 2014;79(3–4):111–116.

149. Ishihara R, Taketani S, Sasai-Takedatsu M, Kino M, Tokunaga R, Kobayashi Y. Molecular cloning, sequencing and expression of cDNA encoding human trehalase. *Gene.* 1997;202(1-2):69–74.

150. Tanaka M, Machida Y, Niu S, Ikeda T, Jana NR, Doi H, Kurosawa M, Nekooki M, Nukina N. Trehalose alleviates polyglutamine-mediated pathology in a mouse model of Huntington disease. *Nat Med.* 2004;10(2):148–154.

151. Richards AB, Krakowka S, Dexter LB, Schmid H, Wolterbeek APM, Waalkens-Berendsen DH, Shigoyuki A, Kurimoto M. Trehalose: a review of properties, history of use and human tolerance, and results of multiple safety studies. *Food Chem Toxicol.* 2002;40(7):871–898.

152. Montalto M, Gallo A, Ojetti V, Gasbarrini A. Fructose, trehalose and sorbitol malabsorption. *Eur Rev Med Pharmacol Sci.* 2013;17(Suppl. 2):26–29.

CHAPTER

7

Endocytosis and Synaptic Function

S.F. Soukup, P. Verstreken, S. Vilain

Center for the Biology of Disease, Flemish Institute for Biotechnology (VIB), KU Leuven Department of Human Genetics, Leuven, Belgium

1 INTRODUCTION

To understand the etiology of Parkinsons disease (PD) we argue that it is important to understand where the initial defects in the disease originate. The pathophysiology and the progression of PD can be studied in patients by using different radionuclide imaging techniques[1] and suggest a reduction of nerve terminals and synapses as one of the early symptoms of the disease.

Parkinson's Disease. **http://dx.doi.org/10.1016/B978-0-12-803783-6.00007-9**

Synapses are essential for neuronal function, proper maintenance of these synapses is critical for proper brain function, and defects in synaptic maintenance have been connected to neurodegenerative diseases. Synapses exert their function remotely from the biosynthetic processes that the cell body provides. Despite the connection with the cell body, synapses are largely isolated structures and must operate in a semiautonomous way. Therefore, a number of processes operate independently at synapses, including local membrane and protein turnover,[2–4] compartmentalized metabolic capability,[5] and local modulatory signal cascades that are supported by synaptic protein scaffolds.[6] This implies that synaptic proteins and organelles are used and reused multiple times, potentially resulting in damage to these proteins and organelles. These defective components, over time, could decrease the efficiency of neuronal communication[7,8] and potentially lead to neurodegeneration. In PD a role for failing protein and organelle quality control systems is emerging,[7] and in combination with synaptic defects, the data are starting to point to defects in synaptic homeostasis as one of the contributing factors to the disease.[9] These connections between PD and synaptic homeostasis have emerged from studying the genes causative to the disease. Indeed, while the majority of PD cases are sporadic, around 10% of the PD cases are inherited and caused by mutations in genes. The identification and characterization of the function of these genes allows us to better understand the etiology of PD.

In this chapter we focus on the function of the synapse. We describe what is currently known about the molecular mechanisms that exert synaptic function in relevance to PD. Then we present an overview of the data that are obtained by studying the function of the genetic forms of PD. Table 7.1 summarizes the PD genes in connection to synaptic function and dysfunction (see also Chapter 1).

2 THE SYNAPTIC VESICLE CYCLE AND PD

The brain consists of millions of interconnected neurons that are highly polarized cells. They contain dendritic arbors and axons that carry synapses. These synapses exercise their function by releasing neurotransmitters into the synaptic cleft (Fig. 7.1C). Neurotransmitters are packed in presynaptic vesicles. Neurotransmitter release is controlled by Ca^{2+}-mediated fusion of a presynaptic pool of vesicles (the readily releasable pool). The release of neurotransmitters will induce a response in the postsynaptic cell.

Since during a burst of activity many vesicles are used, the resupply of vesicles to synapses is critical to maintain a normal healthy vesicle pool (Fig. 7.1).[10–13] Axonal transport (Fig. 7.1A) will deliver new material to synapses, including proteins and organelles. Newly arrived vesicles are filled with neurotransmitter (Fig. 7.1B). However, during intense activity, axonal

TABLE 7.1   Mutations in Proteins With Synaptic Function That Cause Familial Parkinsonism

Symbol	Gene/protein	Mutation	Inheritance	Function at the synapse	Protein interaction
PARK1/4	SNCA/SNCA (α-synuclein)	A30P, E46K, H50Q, G51N, A53T	AD	Clathrin-mediated endocytosis Neurotransmitter release Exosome release	CSPα Synaptobrevin-2/VAMP2 Endophilin A1 Synapsin III Synphilin-1
PARK2	PARK2/Parkin	Numerous duplications and missense mutations	AR	Neurotransmitter release Clathrin-mediated endocytosis	Endophilin A1 Synaptotagmin XI Miro PINK1 Synphilin-1
PARK7	PARK7/DJ-1	dup 168–185, A39S, E64D, D149A, Q163L, L166P, M261I	AR	Energy homeostasis	Drp1 Synaptophysin
PARK8	DARDARIN/ LRRK2	N1437H, R1441H/G/C, Y1699C, G2019S, I2020T	AD	Neurotransmitter release Clathrin-mediated endocytosis Endolysosomal trafficking Exosome release	NSF α-Synuclein Endophilin A1 Clathrin Dynamin Rab5 Rab7 Vps35
PARK9	ATP13A2/ ATP13A2	M810R, G877R, missense, small insertions, and deletions	AR	Exosome release	ESCRT complex

(Continued)

TABLE 7.1 Mutations in Proteins With Synaptic Function That Cause Familial Parkinsonism (cont.)

Symbol	Gene/protein	Mutation	Inheritance	Function at the synapse	Protein interaction
PARK17	VPS35/VPS35	D620N	AD	Endolysosomal trafficking	Retromer complex (VPS 29/26)
					LRRK2
					AMPA receptor
PARK19	DNAJC6/DNAJC6 (Auxilin)	Q734X	AR	Clathrin-mediated endocytosis	HSC70
					Clathrin
PARK20	SYNJ1/SYNJ1 (Synaptojanin 1)	R258Q	AR	Clathrin-mediated endocytosis	Endophilin A1
					Dynamin
					Clathrin
PARK21	DNAJC13/RME-8	N855S	AD	Clathrin-mediated endocytosis	HSC70
				Endosomal sorting/trafficking	Snx-1

AD, autosomal dominant; AR, autosomal recessive.

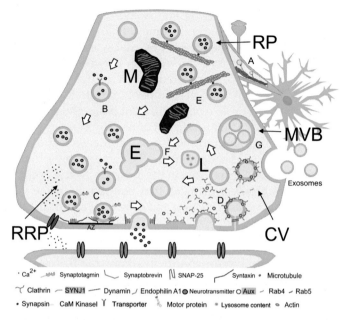

FIGURE 7.1 **The synaptic vesicle cycle and vesicle secretion in the synapse.** (A) Transport vesicles are brought from the cell body to the synapse via Kinesin-mediated transport on microtubules. (B) Newly endocytosed vesicles at the synapse are loaded with neurotransmitters. (C) The readily releasable pool of vesicles *(RRP)* can fuse with the plasma membrane via the SNARE complex and release their content in the synaptic cleft. (D) After vesicle fusion with the plasma membrane to release the neurotransmitter, Clathrin-mediated endocytosis can recycle vesicles through Clathrin-coated vesicles *(CV)*. (E) Some of these vesicles are stored in the reserve pool *(RP)*, anchored via Synapsin to the Actin cytoskeleton. These vesicles can be retrieved via an ATP-dependent process. The ATP can be locally produced by the mitochondria (M). (F) Local turnover occurs via endolysosomal pathway E (endosomes) L (lysosome). (G) Multivesicular bodies *(MVB)* fuse with membranes and secrete exosomes. Parkinsons disease *(PD)* genes are highlighted in yellow.

transport is not able to sustain the replenishment of needed vesicles. After vesicle fusion (Fig. 7.1C), new vesicles are mostly replenished by endocytic mechanisms (Fig. 7.1D, see also later) and they are then refilled locally with neurotransmitters (Fig. 7.1B). This active pool of vesicles is the recycling pool. The readily releasable pool and the recycling pool of vesicles are needed for neuronal communication under moderate stimulation conditions. However, under high-frequency stimulation, vesicles from a third pool, the reserve pool (RP), are recruited (Fig. 7.1E). Vesicles from the different pools need to be fully functional and they need to be able to efficiently fuse with the membrane in a regulated calcium-dependent manner. To ensure that proteins on synaptic vesicles are functional and the vesicles are fusion competent, dysfunctional proteins need to be recognized and sorted for degradation. Recent work has suggested that vesicles can occasionally fuse with sorting endosomes for replenishment. New vesicles that then form from the endosome contain functional proteins[14,15] (Fig. 7.1F). There is also evidence for the existence of a "super-pool" of vesicles, which are vesicles that can move in between different synaptic boutons/varicosities. Calculations predict that a single synapse can contribute around 4% of their total vesicle pool/minute to the superpool.[16] Hence, synaptic vesicles cycle trough different pools and proper synaptic communication requires a balanced pool of healthy vesicles.

There are some direct and indirect lines of evidence that defects in the synaptic vesicle cycle could contribute to PD. Some proteins that cause familial forms of PD play a direct role in the synaptic vesicle cycle, whereas others merely seem to affect proteins that play a role in the vesicle cycle.[17–24] However, their exact function and how they contribute to the disease are not well understood.

2.1 Synaptic Vesicle Fusion/Neurotransmitter Release and the Links With PD

When a neuron is excited, an action potential across the membranes is propagated and invades the synapse. After reaching the presynaptic terminal, voltage-gated ion channels allow calcium ion entry (Fig. 7.2A). This Ca^{2+} binds to Synaptotagmin, a protein present on synaptic vesicles, and within milliseconds, synaptic vesicles fuse with the plasma membrane to trigger neurotransmitter release[25] (Figs. 7.1C, 7.2B, and 7.2C). Vesicles release neurotransmitter at specialized regions that appear electron dense under the electron microscope, called active zones (Figs. 7.1C, 7.2B,C). Structural proteins like Piccolo, Bassoon, or Bruchpilot in *Drosophila* form the backbone of the active-zone cytomatrix where the vesicles dock.[26–28] Vesicle fusion with the plasma membrane is facilitated by soluble N-ethylmaleimide-sensitive fusion attachment protein receptors (SNAREs) through high-affinity interactions (Figs. 7.1C and 7.2C). SNAREs are classified depending on the their properties. In this classification the SNAREs are subdivided into two categories: the R-SNAREs, which contribute an arginine (R), and the Q-SNAREs, which contribute a glutamine (Q) to the SNARE four-helix complex. In synaptic vesicle membrane fusion, the R-SNARE Synaptobrevin/ vesicle-associated membrane protein (VAMP) is located on the vesicles and the Q-SNARE Syntaxin and synaptosomal-associated protein-25 (SNAP-25) are located on the plasma membrane. Different actions are required for efficient vesicle–plasma membrane interactions in which the SNAREs are involved. This will result in full vesicle fusion and neurotransmitter release. Vesicle fusion requires four different steps, which are called tethering (Fig. 7.2B), docking (Fig. 7.2C), priming, and fusion (Fig. 7.2D,E).

Tethering facilitates the contact of R-SNARE and Q-SNARE proteins by moving synaptic vesicles close to the active zone.[29] Second, during

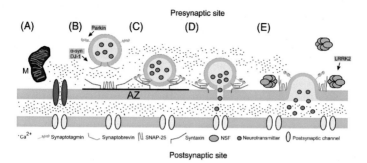

FIGURE 7.2 **Neurotransmitter release and interactions with PD-causing genes.** (A–D) Upon Ca^{2+} influx and binding of Ca^{2+} to Synaptotagmin, vesicles fuse with the plasma membrane at the active zones *(AZ)*. Ca^{2+} influx is buffered by the mitochondria *(M)*. Vesicles undergo tethering (B), docking (C), and priming and fusion (D and E) to release their neurotransmitters. (E) Afterward, NSF hexamers are recruited to disassemble the SNARE complex. Proteins of PD-related genes, highlighted in yellow, are interacting with different components of the vesicle fusion machinery.

docking, Syntaxin allows the formation of a transcomplex via zippering of a four-helix bundle between Q- and R-SNAREs.[30] This four-helix bundle is derived from three proteins. The Q-SNARE SNAP-25 is a cytoplasmic protein attached to the plasma membrane via palmitoyl side chains.[31] SNAP-25 contributes to the trans-SNARE complex with two helices,[32] whereas the R-SNARE Synaptobrevin/VAMP and Q-SNARE Syntaxin contribute each with one helix (Figs. 7.1C and 7.2C). Then during the priming phase the trans-SNARE complex is stabilized and zippering of the trans-SNARE complex enhances fusion of membranes in close proximity. Finally, during fusion, Synaptotagmin, via calcium sensing and interaction with the lipid membrane, will dimerize and lead to membrane fusion. Fusion of the synaptic vesicle membrane and the plasma membrane creates a pore for neurotransmitter release[33–35] (Fig. 7.2D). This mechanism relies on Q-SNARE clustering in close proximity to the voltage-gated calcium channel. Finally, pore expansion transforms the SNARE complex from a trans conformation to a cis conformation where the three SNARE proteins are now in the plasma membrane. SNARE complexes are very stable; however, they need to disassemble in order for Synaptobrevin/VAMP to recycle. Synaptobrevin/VAMP will recycle on newly formed synaptic vesicles, whereas (most of the) Syntaxin and SNAP-25 will stay on the plasma membrane.

The disassembly of the SNARE complex requires a hexamer of *N*-ethylmaleimide sensitive factor (NSF), which is an AAA-ATPase.[36] Binding of the SNARE complex to the NSF hexamer requires the NSF cofactor α-SNAP and energy from ATP hydrolysis (Fig. 7.2E). NSF can unwind itself from the SNARE complexes and the individual SNAREs can be reused in a novel round of exocytosis.[37,38]

Each round of vesicle exocytosis requires the assembly and disassembly of SNARE complexes. This generates highly reactive unfolded SNARE complexes that require the help of chaperone proteins to refold. Cysteine-string protein (CSP) is necessary for proper exocytosis as it binds and activates chaperones like Hsp70 via its J domain. CSP's other domain contains a cysteine string domain that is highly palmitoylated to target the protein to membranes, where it finds one of its targets, Synaptobrevin/VAMP. The importance of proper vesicle fusion in PD is reflected by the fact that some genes involved in the familial form of PD seem to play a role in vesicle fusion.

α-Synuclein (encoded by *SNCA*) is involved in inherited forms of PD. It is a protein of 140 residues that is expressed in the brain. Its N-terminal domain (residues 1–95) contains an amphipathic α-helix and allows the association with lipids; it also contains the NAC domain (residue 60–95), which is believed to be responsible for aggregation. Its C-terminal domain (residues 96–140) is acidic and unstructured and a target of posttranslational modifications. It is highly enriched in boutons, where it associates with synaptic vesicles. α-synuclein belongs to a larger protein family called the Synucleins/SNCA, which include β-synuclein and γ-synuclein. Synucleins are monomeric and unfolded when they are in solution; however, addition of phospholipids induces them to adapt an amphipathic α-helical structure that enables it to bind to membranes. α-synuclein prefers to bind to smaller vesicles like synaptic vesicles. Synucleins may be involved in neurotransmitter release,[18,39] and triple knockouts (KOs) showed reduced dopamine secretion.[40] EM micrographs of triple-*SNC* mouse KOs showed a decrease of presynaptic area that could potentially originate from defects in vesicle trafficking.[41] Proper exocytosis requires the cycling of correctly folded SNARE complex proteins, which is facilitated by CSPα. Interestingly, expression of extra α-synuclein can rescue the neurodegenerative defects of CSPα mutant mice. Furthermore, the *SNC* family is necessary for the maintenance of a continuous presynaptic SNARE-complex assembly.[42,43] α-Synuclein binds directly to the SNARE protein Synaptobrevin-2/VAMP2 and promotes assembly of SNARE complexes (Fig. 7.2). The multimeric form of α-synuclein acts as the SNARE chaperone.[44] In vivo experiments show an age-dependent neurological defect and early lethality in triple-mutant mice deficient for all *SNC* genes. There is also a reduction in assembled SNARE complexes in these mice.[42] Further evidence of α-synuclein interactions with the SNAREs came from in vitro experiments. A direct interaction between the SNAREs and α-synuclein was shown by using proteoliposomes (liposomes mixed with protein) of vesicles.[45] Indeed different vesicles with different protein content (Q-SNARE and R-SNARE proteins) can be mixed and, via FRET assays, visualized when they fuse. Using this in vitro technique a clustering of vesicles was observed when α-synuclein, but not when the A30P mutant of α-synuclein, which cannot bind lipids, was added to the mix.[46] This clustering was dependent on Synaptobrevin-2/VAMP2 and anionic lipids. However, α-synuclein did not have any effect on Ca^{2+}-triggered fusion in this assay, and therefore α-synuclein may play a role in clustering of vesicles at the active zones independent of Ca^{2+}.

Additional evidence for exocytic defects connected to PD causative genes came from the PD-associated kinase LRRK2. The LRRK2 protein is a large protein and has several independent domains, including three protein–protein-interacting domains (ankyrin, LRR, and WD40) and two enzymatic domains (kinase and GTPase). Despite the fact that LRRK2 mutant mice do not show a defect in dopamine release, maybe because of redundancy with LRRK1 in the brain, neurotransmitter release was impaired in LRRK2^{G2019S} knock-in mice.[47] Indeed LRRK2^{G2019S} knock-in mice were generated via homologous recombination,[47] LRRK2^{G2019S} mutants harbor a gain of kinase function, and in line with this notion the transgenic mice showed elevated kinase activity in their brains and synaptic exocytic defects. Similarly, neurotransmitter release was impaired at the neuromuscular junctions (NMJs) of *Drosophila Lrrk* null mutant flies and in flies that neuronally overexpress LRRK2^{G2019S}.[48] Evidence of LRRK2 involvement with the SNARE complex assembly–disassembly cycle arose from coimmunoprecipitation experiments from adult mouse brain lysates that showed that NSF coprecipitates with LRRK2.[49] These data suggest that LRRK2 may regulate NSF activity but further work is needed.

PARK2 (Parkin), another PD-related gene that causes autosomal recessive juvenile parkinsonism (AR-JP), has also been implicated in regulation of vesicle fusion. Parkin is an E3 ubiquitin ligase; it contains an N-terminal ubiquitin-like domain (UBL), a central RING0 domain, and a C-terminal RING1-IBR-RING2 domain. Parkin is able to catalyze K48-linked poly-ubiquitination for proteasome-mediated degradation. Parkin is also able to catalyze K63-linked poly-ubiquitination and mono-ubiquitination, which allows for protein–protein interactions.[50] Parkin KO mice revealed defects that are affecting synaptic plasticity via long-term depression (LTD) and long-term potentiation (LTP) in the striatal medium spiny neurons (the major dopaminergic target).[51] Also at *Drosophila* NMJs, *parkin* mutations resulted in defects in neurotransmitter release, particularly during strong stimulation paradigms.[52] Parkin's role in the regulation of vesicle fusion proteins came from a yeast-two-hybrid screen, which identified Synaptotagmin XI as an interactor of Parkin.[53] SytXI localizes to cell bodies and neurites in substantia nigra neurons. Overexpression of wild-type Parkin but not of PD-causing mutants of Parkin, Parkin^{C289G}, and Parkin^{C418R} in HEK293 cells induced more ubiquitinated SytXI, which was cleared from the cell. Hence, SytXI is ubiquitinated by Parkin for proteasomal degradation (Fig. 7.2).

Mutations in *DJ-1* cause PD and there is evidence that it is involved in exocytosis. *DJ-1* encodes a protein related to a family of molecular chaperones.[54,55] Medium spiny neurons of *DJ-1* null mutant mice have normal corticostriatal LTP, but LTD was absent. DJ-1 is found in synaptosomes and it partially colocalizes with Rab3a, which is involved in exocytosis. DJ-1 localizes with the Synaptophysin and Synaptobrevin-2/VAMP2–associated vesicles but did not bind to them. FRET assays, however, indicated that a small portion of DJ-1 is in close vicinity of Synaptophysin and Synaptobrevin-2/VAMP2.[56]

Taken together, the data suggest that the function and availability of the SNARE proteins appears to be influenced by proteins mutated in PD. It will be interesting to assess in the future how these exocytic defects contribute to PD pathology.

2.2 Synaptic Vesicle Endocytosis/Vesicle Recycling and the Links With PD

Controlled neurotransmitter release requires the availability of synaptic vesicles, and local creation and recycling of new vesicles is necessary to maintain neurotransmitter release. Different pathways involved in synaptic vesicle recovery have been proposed to exist: ultrafast endocytosis (UFE), "kiss and run" endocytosis, Clathrin-mediated endocytosis (CME), and activity-dependent bulk endocytosis (ADBE) (Fig. 7.3). Different stimulation paradigms result in the activation of one or another pathway with

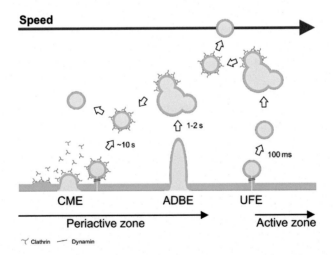

FIGURE 7.3 **Different types of synaptic vesicle reformation at synapses.** Clathrin-mediated endocytosis *(CME)*: Clathrin-coated vesicles are pinched off from the plasma membrane by Dynamin. After endocytosis the Clathrin coat is released. Activity-dependent bulk endocytosis *(ADBE)*: large amounts of presynaptic membranes are endocytosed into large endosomal/cisternal-like structures from which new synaptic vesicles can form. Ultrafast endocytosis *(UFE)* is thought to occur close to the active zone via a dynamin-related process.

different kinetics and requiring overlapping but also specific molecular components. While endocytosis is critical for synaptic vesicle reformation, it is also critically involved in regulating the abundance and availability of surface proteins, and sometimes the same molecular components are involved. For example, the N-BAR (BIN/amphiphysin/Rvs) protein Endophilin A (EndoA) is able by itself to mediate the uptake of various G-protein-coupled receptors such as α2a- and β1-adrenergic, dopaminergic D3 and D4 receptors; various tyrosine kinase, interleukin-2-receptors; and Shiga and Cholera toxins in cultured human and monkey cells.[57,58] Hence, modulation of endocytic factors that affect synaptic vesicle reformation may have broader effects than only affecting the vesicle pool.

CME is the best characterized endocytic pathway.[13,59] At the presynapse, CME is mainly dedicated to form synaptic vesicles, whereas at the postsynaptic site it is implicated in synaptic plasticity by regulating active surface receptors.[60] Clathrin coats are formed by the assembly of Clathrin triskelia made of three 190-kDa Clathrin heavy chains and three 25-kDa Clathrin light chains as cages around forming vesicles.[61]

According to ultrastructural and cell biological observations, the formation of Clathrin-coated vesicle can be divided into five stages: initiation, cargo selection, coat assembly, scission, and uncoating (Figs. 7.4 and 7.5A).[25] The protein coat that surrounds newly forming vesicles in CME is built up by the recruitment of adaptor proteins and protein complexes (like adaptor protein 2, AP-2) and accessory proteins (such as AP180 and Epsin) to the plasma membrane, which leads to the accumulation of a Clathrin lattice-like coat. After vesicles are formed, the different accessory proteins that have been recruited to the vesicle budding sites are recycled for reuse in another cycle of endocytosis.[61,62]

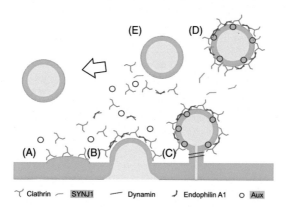

FIGURE 7.4   **Clathrin-mediated endocytosis (CME).** (A) Start of endocytosis by initiation and nucleation. (B) Cargo selection and Clathrin coat assembly. (C) Dynamin is pinching off the nascent Clathrin-coated vesicle. (D) Endophilin A1 recruits the phosphatase SYNJ1 that, together with Aux, mediates uncoating (E) Proteins of PD-related genes are highlighted in yellow.

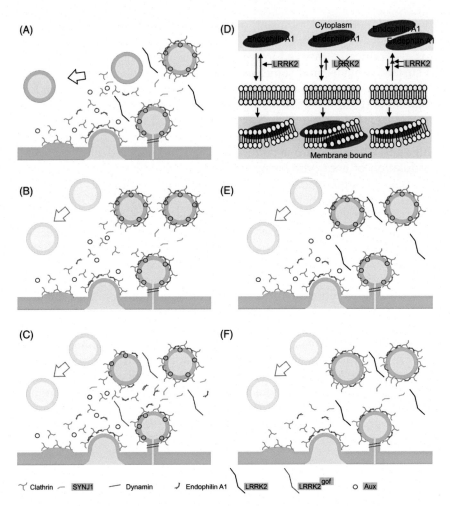

FIGURE 7.5   **Changes in vesicle pools during disease.** (A) Wild-type CME. (B) In LRRK2 loss-of-function mutants, endophilin A1 is not phosphorylated changing its properties such that it more tightly binds membranes. (C) LRRK2 gain of function results in endophilin A1 phosphorylation loosening the interaction of Endophilin A1 with membranes. (D) Both gain and loss of function in LRRK2 result in mild defects in vesicle endocytosis. (E and F) Defects in SYNJ1, Aux putatively disrupt the vesicle cycle. Proteins of PD-related genes are highlighted in yellow.

The first step is the initiation of the core complex assembly or nucleation (Fig. 7.4A). Nucleation is characterized morphologically by membrane invaginations called pits. During this step the three-layered Clathrin-coated pit is formed (Figs. 7.1D and 7.4A). The outer layer is made of a Clathrin triskelion cage, followed by a layer of adaptor proteins AP-2 and AP-180 and an inner layer made of transmembrane protein cargoes that are recognized by the adaptor proteins. EGFR pathway substrate 15 (EPS15) or Epsin and intersection may act as an initiation modules because of their preference for phosphatidylinositol 4,5-bisphosphate (PtdIns(4,5)P2) that serves as a lipid tag for endocytosis on the plasma membrane.[25,62]

During the second step (Fig. 7.4B, Cargo selection and Coat assembly), the initiation module recruits subsequently additional AP-2, the second most abundant protein of the Clathrin-coated vesicle. Since AP-2 binds Clathrin and most of the accessory proteins, it has as a major impact on the maturation of the Clathrin-coated pit. AP-2 binds to PtdIns(4,5)P2,[63] which assures its connection to the plasma membrane. In addition, AP-2 also binds to the synaptic vesicle protein Synaptotagmin together with the adaptor protein, Stoned B/Stonin.[47,64] The R-SNARE Synaptobrevin/VAMP is sorted into synaptic vesicles through another adaptor: AP-180.[65]

Next, Clathrin triskelia are recruited directly from the cytosol to sites with high concentrations of the initiation module (Figs. 7.1D, 7.4A, B). The polymerization of Clathrin stabilizes the curvature of the forming vesicle and helps organize the coated vesicle. Polymerization of Clathrin displaces EPS15 to the edge of the forming vesicle, where it is believed to induce membrane curvature at the nascent vesicle neck.

The third step is synaptic vesicle budding. BAR domain-containing protein Amphiphysin and Endophilin A1/EndoA preferentially bind to curved membranes and form a so-called neck at the Clathrin-coated vesicle.[66,67] Both proteins have an SRC homology 3 (SH3) domain that binds the proline-rich domain (PRD) of Dynamin. Clathrin-coated vesicle scission is mediated by the mechanochemical enzyme dynamin[68] (Figs. 7.1D and 7.4C). Fission of new vesicles requires directly the function of dynamin (Figs. 7.1D and 7.4C). In mammals there are three dynamin genes, dynamin 1, 2, and 3,[69] with 80% sequence identity. Dynamin 1 is expressed at high levels in neurons and it is most abundant at the synapse.[70,71] Dynamin is about 96 kDa and consists of a GTPase, a pleckstrin homology, a GTPase effector or guanidine exchange domain (GED), and a PRD. The pleckstrin homology domain is crucial for its function because it allows binding to PtdIns(4,5)P2-containing membranes and might have a curvature-sensing property.[72] At the neck, Dynamin polymerizes and undergoes a GTP hydrolysis-dependent conformational change that helps to mediate membrane fission.[73,74]

The last step of CME is to uncoat the protein coat from the new vesicle (Figs. 7.1D and 7.4D). Synaptojanin-1 (SYNJ1) contains a suppressor of actin1 (SAC1) and a 5'-phosphatase domain, which are both required for its phosphoinositol phosphatase function.[75,76] SYNJ1 can dephosphorylate different types of phosphoinositides, including PtdIns(4,5)P2, which is critical during endocytosis. The 145-kDa synaptic-enriched isoform of SYNJ1 binds to Endophilin A1 through its PRD domain at the neck of the Clathrin-coated vesicle.[77] By dephosphorylating PtdIns(4,5)P2 SYNJ1 aids in the release of AP-2, Epsin, and AP-180, all of which bind to membranes in a PtdIns(4,5)P2-dependent manner.[78,79]

Before detached synaptic vesicles can be refilled with neurotransmitter or sorted through the endolysosomal system, they have to lose their Clathrin coat. Disassembly of the Clathrin lattice is arranged by the ATPase heat shock cognate 70 (HSC70) and its cofactor, Auxilin (Aux).[80,81] While Aux is recruited to the terminal domain of the Clathrin triskelia,[82] HSC70 is recruited by Aux to the Clathrin tripod[83] (Figs. 7.1D and 7.4D). The initiation of the uncoating starts from the place where the neck was originally formed and where the Clathrin cage was not completely closed, creating an access for Aux. Aux binds Clathrin at the ankle and terminal part of the Clathrin heavy chain. Thereby, Aux places itself between the hub of the triskelia and the ankle part of the Clathrin heavy chain exposing a binding motif for HSC70.[82] With a weaker interaction within the Clathrin triskelia, HSC70 releases the C terminal of the triskelia after binding to Aux.[84] ATP hydrolysis provides the energy to disassemble the Clathrin cage and release of the synaptic vesicle.[80,83,85] Uncoated vesicles (Fig. 7.4E) can then be refilled with neurotransmitter.

During intense stimulation many presynaptic vesicles fuse with the plasma membrane. Given that CME is relatively slow, this would result in an expansion of the synapse. However, such an expansion is not obviously observed during tetanic stimulation paradigms, and instead large membrane invaginations in the surrounding region of the active zone, the periactive zone, are observed.[86–88] Synapses use ADBE that allows the retrieval of large amounts of presynaptic membranes into large endosomal/cisternal-like structures from which new synaptic vesicles can then form[89,90] (Fig. 7.3). Experiments in amphibian NMJ central nerve terminals using different fluorescent dyes showed that ADBE-derived synaptic vesicles are used to replenish the RP of synaptic vesicles.[91] Experiments in dynamin KO mice and acute inactivation of Dynamin in *Drosophila* showed that the fission of synaptic vesicles from this endosome-like structure requires Dynamin and Clathrin.[92,93]

Recently, a new endocytic mechanism, UFE, was described. Elegant experiments combining optogenetics with high-pressure freezing EM showed in *Caenorhabditis elegans* and mouse hippocampal cells invaginations of around 80-nm diameter already 50–100 ms after stimulation, between the active and the periactive zone.[94,95] Applying Latrunculin-A to disrupt Actin polymerization or Dynasore to inhibit GTPase activity of Dynamin disrupts the formation of these fast endocytic structures, indicating a crucial role of Actin and Dynamin in this pathway. Similar to the bulk endocytosis the formation of these fast endocytic endosomes itself does not require Clathrin. However, the budding of synaptic vesicles from synaptic endosomes relies on Clathrin function.[96]

Many proteins are involved in CME and a strong link between some of these proteins and PD-associated genes has been observed. Indeed, it was suggested that the levels of Synucleins and Endophilin A1 transcripts are reciprocally regulated and functionally related. An unbiased proteomic screen using *αβγ-SNC* triple KO mouse brain lysates suggested that levels of membrane-sensing proteins were upregulated with the strongest change in Endophilin A1 and the peripherally associated synaptic vesicle protein Synapsin IIb.[97] This suggests a function (possibly indirect) of αβγ -synucleins in endocytosis. Further liposome experiments suggested that monomeric but not tetrameric α-synuclein can bend membranes similar to Endophilins. Moreover, this function was abolished in the A30P α-synuclein mutant. The PD mutation possibly causes a disruption in protein folding, and this suggests that the membrane curvature is generated through the N-terminal amphipathic helix of the Synucleins.[97] Further study of the function of α-synuclein in the vesicle cycle was performed using synaptopHluorins (spH). SpH constitutes a superecliptic GFP fused to the luminal side of Synaptobrevin/VAMP and is therefore synaptic vesicle associated. Superecliptic GFP is a fluorescent protein that is quenched in the acid environment of the synaptic vesicle lumen. However, upon fusion of synaptic vesicles with the plasma membrane the lumen is exposed to the extracellular environment, and in the less acid environment the spH is unquenched. When new vesicles are endocytosed, the vesicular H+/ATPase will reacidify the vesicles and the spH will be quenched again. During repetitive stimulation, endocytosis and exocytosis occur simultaneously. Hence, to distinguish endocytosis from exocytosis, bafilomycin was used. Bafilomycin is a nonmembrane permeable blocker of the H/ATPase. Therefore, vesicles that endocytose and take up bafilomycin cannot reacidify and only the exocytic events are visible.[98] SpH measurements showed a reduction of endocytosis in *αβγ-synuclein* triple KO hippocampal neurons but not of exocytosis. They further showed that the defects in endocytosis are in CME and not in bulk endocytosis and that it acts before Dynamin. Moreover, immunohistochemical staining of α-synuclein and Clathrin in Dynamin double-KO neurons suggested that α-synuclein and Clathrin punctae colocalize. In contrast, in wild-type neurons, α-synuclein punctae localize to synapses and Clathrin localization is more dispersed and less punctated. This suggests that α-synuclein is an endocytic protein and probably has a function at a step before Dynamin.

Besides α-synuclein, also the PD-related gene *LRRK2* has been implicated in synaptic vesicle endocytosis. GST pull-down experiments using LRRK2 and functional domains of LRRK2 as a bait identified several proteins involved in presynaptic vesicle cycling, including Clathrin heavy chain and members of the AP-2 complex and Dynamin.[49] It was further shown that LRRK2 silencing resulted in defects in synaptic vesicle recycling. The various interactions of LRRK2 with presynaptic proteins suggest that LRRK2 might be part of a presynaptic protein complex.

More direct evidence of the function of LRRK2 in endocytosis came from observations that *Drosophila* and human Lrrk/LRRK2 can directly phosphorylate EndoA/Endophilin A1, an essential component of the endocytic machinery as outlined earlier.[19,99] EndoA/Endophilin A1 forms a dimer, and upon membrane binding, several helices insert in the membrane. The Lrrk/LRRK2 phosphorylation site in EndoA/Endophilin A1 (S75 in flies and S75 and T73 in humans) is located specifically in such a helix that can dive into the membrane. Adding a negative charge as a result of phosphorylation would push EndoA/Endophilin A1 away from the membrane, thus affecting its functional properties.[100] Consistently, phosphorylated EndoA/Endophilin A1 (or phosphomimetic EndoA/Endophilin A1) is incapable of creating long thin tubules from artificial giant vesicles in vitro, whereas wild-type EndoA/Endophilin A1 and phosphodead EndoA/Endophilin A1 behave normal.[99] Also, in vivo phosphomimetic EndoA/Endophilin A1 does not bind as tightly to membranes as does phosphodead EndoA/Endophilin A1, suggesting that Lrrk/LRRK2 regulates aspects of the function of EndoA in vivo as well (Fig. 7.5B–D). Indeed, EndoA/Endophilin A1 phospho-mutants display (mild) defects in endocytosis. However, what is interesting is that both phosphodead EndoA and phosphomimetic EndoA are showing very similar (mild) defects in synaptic vesicle endocytosis, suggesting that both LRRK2 mutants (Fig. 7.5C) that harbor increased kinase activity (LRRK2[G2019S]) as well as mutants (Fig. 7.5B) that harbor less kinase activity can result in similar synaptic defects as a result of EndoA phosphorylation defects (Fig. 7.5D). Both types of LRRK2 mutants have been detected in PD patients.

Further evidence that defects in vesicle endocytosis may be connected to PD came from Parkin, another PD-causing protein. Interestingly, Parkin and Endophilin A1 show functional interactions. At the transcriptional level an upregulation of Parkin transcripts was observed in SH3GL2, SH3GL1, and SH3GL3 (*endophilin A*) triple-KO mice.[101] Furthermore, HEK293T that coexpress HA-Ubi, Parkin, and either Endophilin A1, SYNJ1, or Dynamin indicated that Parkin can ubiquitinate all three proteins.[101] This suggests that binding of Parkin and Endophilin A1 possibly results in mono-ubiquitination of Endophilin A1, Dynamin, and SYNJ1. Therefore, Parkin may have a regulatory function on Endophilin A1-containing endocytic protein complexes.[101]

Recently, *SYNJ1* was found mutated in an autosomal recessive early-onset form of parkinsonism. The mutation is located in the SAC1 domain of SYNJ1. These findings provide another connection to synaptic vesicle endocytosis and particularly to Endophilin A1. Since SYNJ1 very tightly binds EndoA/Endophilin A1, its phosphoinositide phosphatase activity, essential for synaptic vesicle recycling, is stimulated when binding to lipids and to EndoA/Endophilin A1[17] (Fig. 7.5A). The mutations in the SYNJ1 SAC1 domain block the ability to dephosphorylate specific phosphoinositides[23,102,103] (Fig. 7.5E). Given that SYNJ1 is synapse enriched, the data implicate defective phosphoinositide dephosphorylation in PD. However, further research is needed to understand the functional consequences of this mutation.

Further connections between CME and PD exist. Mutations in *DNAJC6*, the gene that encodes for Aux,[104] and in receptor-mediated endocytosis (RME-8), another J-domain containing protein,[105] have been identified. Aux is selectively expressed in neurons and is enriched in nerve terminals where it participates to remove Clathrin from newly formed vesicles. As such, Aux acts at a step in endocytosis close to SYNJ1 in the uncoating of the protein coat from newly formed vesicles (Fig. 7.5F).[106] Studies in *C. elegans* indicate that RME-8 is involved in endocytosis, and data obtained with fruit flies suggest that, like Aux, RME-8 works as a cochaperone with Hsc70 in endocytosis.[107,108] Further studies are now needed to connect these proteins to mechanisms in PD.

The current data are exciting as several proteins mutated in PD are known to play an active role in endocytosis. α-synuclein may act before Dynamin during initiation, cargo selection, or coat assembly. Interestingly, connections between PD-causing genes and the endocytotic protein Endophilin A1 exist as well. While Endophilin A1 itself has not been found mutated in PD patients, the protein is phosphorylated by LRRK2 and binds to Parkin and SYNJ1. Finally, a complex of proteins encompassing Aux and RME-8 may act via Hsc70 in vesicle uncoating. Interestingly, Hsc70-4 has also been connected to endosomal microautophagy[109] (see Chapter 6), thus potentially providing further links between synaptic vesicle endocytosis and protein clearance mechanisms as well. While a role for endocytosis in PD is emerging, an important outstanding question is if and how defective endocytosis contributes to pathology.

2.3 PD, the Reserve Pool of Vesicles, and the Link With the Mitochondria

Synaptic vesicles can be subdivided into three functional pools as outlined earlier.[110] Ten to twenty percent of the SVs belong to the readily releasable pool and the recycling pool, also called the exo-endocycling pool (ECP), whereas the majority of vesicles reside in the RP (Fig. 7.1E). ECP vesicles fuse mostly during mild stimulation to release neurotransmitters, while during high-frequency stimulation, also RP vesicles are recruited.[110,111]

When in the RP, vesicles are immobilized by synapsins that crosslink the vesicles to the actin cytoskeleton (Fig. 7.6A).[112] When synapsin is phosphorylated by PKA (cAMP-dependent protein kinase), the vesicles are released and can participate in release. Recent evidence[113] connects α-synuclein to this process. Both the absence and aggregation of

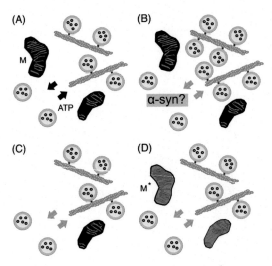

• Synapsin CaM Kinasel ∘ Actin ⦿ Neurotransmitter M Mitochondria M* Defect mitochondria

FIGURE 7.6   **Recruitment of a RP of vesicles defects in PD.** (A) Under wild-type conditions the RP and cycling vesicle pool are balanced. (B) α-Synuclein overexpression and α-synuclein loss-of-function mutants cause an upregulation and redistribution of synapsin, which potentially affects vesicle recruitment of vesicles from the RP. (C) Reduction of the number of mitochondria and (D) mitochondrial dysfunction potentially cause a reduction of vesicle recruitment from the RP. Proteins of PD-related genes are highlighted in yellow.

α-synuclein induces increased levels and redistribution of Synapsin III in dopaminergic synapses. Lack of α-synuclein in *SNCA* null mice resulted in depolarization-dependent dopamine release together with an increase and redistribution of Synapsin III. It was proposed that the absence of α-synuclein results in increased density of Synapsin III in presynaptic terminals. This potentially alters the proper clustering of synaptic vesicles. Aggregation or reduced levels of α-synuclein could induce an accumulation of Synapsin III and excessive clustering of synaptic vesicles in the RP as a result (Fig. 7.6B). In dopaminergic neurons this defect may cause a reduction of dopamine release.

The recruitment of RP vesicles is also strongly dependent on the availability of ATP at synapses. ATP fuels several steps in the vesicle cycling (eg, NSF-dependent SNARE unwinding or Aux/Hsc70-dependent clathrin coat uncaging), but it appears that transport of the vesicles is most sensitive to lower ATP levels (see later).[114] Given the heavy reliance of neurons on mitochondria for ATP production, these organelles are connected to vesicle mobilization, particularly from the RP.[114] Interestingly, several genes mutated in PD affect mitochondrial function and motility (including *PINK1*, *Parkin*, and *DJ-1*; see also other chapters in this book), suggesting that either when mitochondria are not properly delivered to synapses upon neuronal activity, or the synaptic mitochondria are dysfunctional, vesicle recruitment from the RP would be hampered.

Oxidative phosphorylation and the ATP-synthase are responsible for ATP production in mitochondria. Inhibition of these processes leads to reduced cytosolic ATP in neurons.[115–117] In mice this reduced ATP leads to a reduced evoked synaptic release in isolated synaptic terminals from neurons (synaptosomes). In addition, three-dimensional electron microscopy reconstitution of stimulated mice hippocampal synaptosomes shows that the number of Clathrin-coated vesicles correlates with the number of mitochondria.[118] Hence, synaptic vesicle cycling (exocytosis and endocytosis) might be coupled by the availability and/or the ability of synapses to produce ATP.[118] Further work using genetic mutants that block the transport of mitochondria to synapses has also permitted the study of the role of these organelles at synapses. *Miro, milton,* and *drp1* mutants have been used to that extend in fruit flies.[119–121] Miro and Milton are involved directly in mitochondrial transport (see also Chapter 4), whereas Drp1 (dynamin-related protein) is needed for mitochondrial fission. The data suggest that when fission is inhibited, mitochondria fail to be properly transported into synaptic terminals. This reduction of synaptic mitochondria correlates well with the impaired recruitment of synaptic vesicles to the RP. The functional consequence of this is a defect in neurotransmission during intense stimulation (Fig. 7.6C). For example, in *drp1* mutants, the defects in delivering mitochondria to synapses appear to hamper local energy supply and inhibit RP vesicle mobilization, a process dependent on the myosin ATPase complex that moves vesicles over actin filaments.[121] Applying exogenous ATP to the synapse restores neurotransmission,[119,121] further indicating that mitochondrial ATP production was necessary during tetanic stimulation. To what extent a single synapse relies on mitochondria for maintenance of synaptic function could be addressed in future work.[122] This idea is particularly relevant in the context of dopaminergic neurons as these cells are extremely branched (the total length of the neurites of a single dopaminergic neuron in human brain can reach 4.5 m), suggesting a reliance on mitochondrial transport (see also Chapter 4).

As outlined earlier, and in Chapter 1, mutations in at least three genes causative to PD affect mitochondrial function (*Parkin, PINK1,* and *DJ-1*[54,123–126]) and the activities of these three PD proteins appear to at least in part to overlap. Since the overexpression of either DJ-1 or of Parkin rescues aspects of the phenotypes of *pink1* mutant flies.[127–129] Mutations in at least *pink1* and *parkin* also cause defects in synaptic transmission.[51,114] *Parkin* encodes an E3 ubiquitin ligase and localizes in the cytosol until it gets recruited to damaged mitochondria. The mitochondrial Ser/Thr kinase PINK1 is imported into mitochondria, where proteolitic cleavage occurs. When mitochondria are damaged, PINK1 stabilizes on the outer mitochondrial membrane and recruits Parkin. Under endogenous conditions, PINK1 also maintains the activity of Complex I of the electron transfer chain (ETC). *Pink1* and *parkin* mutant animals display mitochondrial defects and these are extensively described in Chapters 2–5. At *Drosophila* NMJs, *pink1* as well as *parkin* mutations induce a defect in the recruitment of vesicles from the RP, a defect rescued by adding exogenous ATP to the motor neurons.[52,114] *DJ-1* encodes a protein related to a family of molecular chaperones.[54,55,130] *Drosophila* mutants in *DJ-1* show age-dependent reduction of mitochondrial DNA and respiration, which leads to reduced ATP levels.[127] The reduction of ATP levels in DJ-1 mutant animals suggests RP defects at the synapse; however, electrophysiological studies to assess RP vesicle mobilization have not yet been performed.[131,132] Nonetheless, the data indicate that the mitochondrial defects induced by mutations in PD causative genes have a functional effect on the efficiency of synaptic transmission at least in part because of lower levels of ATP production.

More direct connections between proteins mutated in PD and ATP energy metabolism at the synapse are emerging as well. α-Synuclein and parkin were shown to interact with Synphilin-1 based on a yeast-two-hybrid screen.[133] Recent work suggests that Synphilin-1 can bind to ATP and that overexpression of *synphilin-1* in HEK293T cells significantly increases cellular ATP levels, whereas knockdown of Synphilin-1 reduced cellular ATP concentrations.[134] Therefore, Synphilin-1 seems to play a role in cellular energy homeostasis by binding and regulating the cellular energy molecule ATP. Synphilin-1 localizes to presynaptic terminals,[135] but whether the protein exercises the same function in regulating ATP levels at the synapse as seen in HEK293T cells and how the PD proteins are specifically related to its function require further testing. What is known is that (1) both α-synuclein and Parkin coimmunoprecipitates with Synphillin-1; (2) Synphilin-1 colocalizes with α-synuclein; (3) cotransfection of the two proteins in HEK293 cells leads to the formation of excessive inclusion bodies (Lewy bodies) containing both proteins[136,137]; and (4) Parkin ubiquitinates Synphilin-1, whereas pathogenic Parkin mutations fail to do so. These observations suggest intricate interactions between Synphilin-1-dependent ATP buffering and PD pathways.

The findings described earlier are mainly focused on the role of mitochondria in the production of cellular energy. While ATP production is one of the major tasks of mitochondria, they are also intimately involved in the buffering of calcium (Fig. 7.2A),[138] a process that is critical at synapses given the steep dependence of neurotransmitter release on calcium levels.[139,140] Mitochondria are also integrating numerous signaling pathways, and they are a major source of ROS that serve as signaling molecules, but can also cause cellular damage.[141,142] While beyond the scope of this chapter, it will be interesting to see how these different processes collectively impinge on neurotransmitter release. It will also be interesting to assess synaptic heterogeneity: Are all synapses relying similarly on all these mitochondrial functions or are there synapse- and neuron-type-specific differences that may also explain selective vulnerability of some cells in PD (eg, dopaminergic neurons) but not others?

2.4 Endolysosomal Trafficking and PD

As outlined earlier several of the genes causative to PD play a role in exo- and endocytosis, but additional connections between PD and trafficking of cargo in the endolysosomal trafficking system exist as well (see also Chapter 6).

The trafficking of cargoes through early and late endosomes and lysosomes process is essential for many cellular and physiological processes like nutrition provision, antigen presentation, pathogen clearance, pigmentation, substrate degradation, as well as receptor-mediated cell signaling. In addition, endocytosed cargo is also sorted in the endolysosomal system and cargoes are recognized and sent to specific cellular locations, including their degradation in the lysosome. The endolysosomal system can be divided into four different compartments: early endosomes, recycling endosomes, and late endosomes/multivesicular bodies (MVB) and lysosomes. The trafficking between these different compartments is regulated by a group of small GTPases termed Rabs. These Rabs are present in many cells and define specific trafficking steps; for example, in most cells, trafficking to early endosomes is mediated by Rab5, to recycling endosomes by Rab4 and Rab11 and to late endosomes by Rab7 and Rab9.[143,144]

Endocytosed proteins in the early endosome can either follow the degradative route through the late endosome and the lysosome or recycle back to the plasma membrane either directly from endosomes or through the recycling endosomes (Fig. 7.1F). Live-cell imaging of early endosomes indicated that they can mature into late endosomes by shedding the early factor Rab5 and EEA1 (early endosome antigen 1) and recruiting the later regulator Rab7.[145] Rab5 stimulates the formation of endosomes in nonneuronal cells by recruiting several effector molecules.[146] Recruitment of phosphatidylinositol-3-kinases, PI(3)-kinases p85/p110, and Vps34/p150 by active Rab5 triggers enrichment of phosphatidylinositol-3-phosphate (PI(3)P) on the early endosomal membrane.[147] Tethering and fusion of endocytic vesicles with early endosomes is mediated by EEA1 and rabenosyn-5, which binds via its FYVE zinc-finger domain to PI(3)P.[148–150] Fusion of the late endosome with the lysosome is mediated by the Syntaxin-7, Syntaxin-8, Vps10p-tail-interactor-1b (Vti1b) and Endobrevin/VAMP8 complex. Substitution experiments and sequence and structural comparisons suggest that this function might be mediated by Syntaxin-1 and SNAP25 in neuronal cells.[151]

The trafficking of cargo is very much dependent on the cell type. In particular, in neurons that are extremely polarized, the importance of correct and efficient trafficking is apparent. Hence, in neurons, the endo-somal system is specialized and likely further adaptations depending on the neuronal subtype exist as well and merit further investigation.[152] For example, EEA1 is localized only to the somatodendritic domain of neu-rons but seems to be absent from axonal terminals[153] despite the fact that an endosomal compartment exists at presynaptic terminals. Both EM and immunolabeling studies using Rab5 antibodies and tools to detect PI(3)P (GFP-2xFYVE) demonstrated the presence of endosomes at synapses.[154] At the *Drosophila* NMJ, genetic analyses and functional studies indicated the strong dependence of the synapse on the presence of an endosom-al compartment.[14,154] Flies expressing a dominant negative Rab5 mutant (Rab5^{S43N}) or a dominant negative Rab35 (Rab35^{S22N}) where vesicles cannot fuse into an endosomal compartment (see later) showed decreased neu-rotransmitter release.[14,154]

Synaptic endosomes have been implicated in a number of specific functions that are essential for the proper working of the synaptic termi-nal. During synaptic vesicle reformation, plasma membrane components may accidently be endocytosed as well. Endosomal trafficking of synaptic vesicles is thought to be required to clear such plasma membrane proteins like SNAP25 and Syntaxin-1 from synaptic vesicles.[155] These findings are strengthened by manipulating Syntaxin-13, a member of the early endo-somal fusion machinery. The block in endosomal function again resulted in an inhibition of synaptic vesicle recycling[156] apparently by reducing the size of the readily releasable vesicle pool needed for neurotransmitter release.[155]

Endosomal sorting in HeLa cells and of synaptic vesicles at *Drosophila* NMJs also requires the function of Rab35.[14,157] At synapses the GTPase ac-tivity of Rab35 is regulated by the GTPase-activating protein (GAP) Sky-walker (Sky). Loss of function of *sky* results in increased Rab35 activity and a significant increase in the trafficking of synaptic vesicles to endosomes.

On the basis of a number of experiments, the model suggests that syn-aptic endosomes serve as sorting stations for synaptic vesicle proteins that are destined for degradation. For example, a Synaptobrevin/VAMP-ubiq-uitin fusion protein is less effectively degraded in *sky* null mutants, sug-gesting that endosomal trafficking of synaptic vesicles is capable of sorting ubiquitinated synaptic vesicle proteins away from the vesicle cycle toward degradation at the lysosome.[14] As a consequence, "older ubiquitinated" proteins are degraded and therefore the composition of the synaptic pro-tein pool gets restored. This rejuvenation was seen using a Synaptobrevin/VAMP that was fused to a fluorescent timer that slowly converts from blue to red fluorescence. The ratio of blue (young) to red (old) fluorescence yields a measure of the age of the Synaptobrevin/VAMP protein pool. In *sky* mutants the blue-to-red ratio is significantly increased, indicating a younger synaptic vesicle-associated protein pool and this effect is blocked when the trafficking to the lysosome is inhibited.[15] These observations also help to explain why *sky* mutants harbor increased neurotransmitter release and a larger readily releasable pool as was also seen in different contexts.[155] Together, these studies suggest that synaptic recycling pathways are involved in the regulation of synaptic vesicle-associated protein turnover. It is conceivable that such mechanisms of endosomal sorting, but also different forms of autophagy or proteasomal degradation, are critical to maintain synaptic function. In addition, the efficiency of these quality control mechanisms may decline during ageing, thus potentially contributing to synaptic and neuronal degeneration.

Endosomal sorting of membrane proteins is also under the control of the retromer complex, a trimer of Vps35p, Vps29p, and Vps26p, that mediates cargo selection of membrane proteins. Interestingly, mutations that cause PD have been found in VPS35, a member of the retromer complex. Vps35 is expressed in the neurons, and its localization to dendritic spines has been linked to trafficking of excitatory AMPA-type glutamate receptors (AMPARs). VPS35 overexpression alters synaptic recycling, excitatory synaptic transmission, and AMPAR trafficking. In contrast, dopamine neuron-like cells produced from induced pluripotent stem cells of human PD patients who carry the dominant PD mutation D620N in VPS35 show defects in synaptic transmission and AMPAR recycling, indicating that D620N acts as a VPS35 loss-of-function mutation.[158] Knockdown of VPS35 in *Drosophila* induces abnormal synaptogenesis, and the expression of VPS35 (D620N) causes locomotor defects and loss of dopaminergic neurons.[159,160]

Interestingly, typical endolysosomal sorting defects have been found in PD-patient derived cells and in PD animal models. LRRK2 contains several functions, and besides a role in synaptic vesicle reformation it may also act on endosomes. LRRK2 binds via its GTPase domain directly to Clathrin-light chains (CLCs). Clathrin itself is recruited by the FYVE domain containing hepatocyte growth factor-regulated tyrosine kinase substrate (Hrs) to the endosome.[161] Furthermore, genomic-encoded HA-LRRK2 colocalizes with CLC on endosomes, where they may participate in protein sorting. Disruption of LRRK2-CLC by knocking down one of the components leads to, first, cell morphological alterations and, second, in dendritic spine morphology alterations by hyperactivation of the small GTPase Rac1.[162] It is, however, unclear if this same interaction also takes place on endosomes located at synapses.

Several other studies implicate LRRK2 function in multiple steps of the endolysosomal trafficking. Examinations of postmortem brains from patients with neurodegenerative disorders detected LRRK2 in α-synuclein-positive brainstem lewy bodies. The development of these abnormal protein aggregates is strongly related with alteration in the lysosomal pathway. Double immunostaining demonstrated that LRRK2-positive granular structures often colocalize with the late endosomal marker Rab7B, further suggesting a function of LRRK2 in the endolysosomal pathway.[163] LRRK2 (Lrrk) studies in *Drosophila* animal and mammalian cell culture models further link LRRK2 function to late endosome maturation and fusion with the lysosomes, caused by reduced Rab7 activity.[164,165] Interestingly, LRRK2 defects in endolysosomal trafficking can be rescued by expression of wild-type *VPS35*, but not by a familial PD-associated mutant, suggesting a common pathway between LRRK2 and the retromer in endolysosomal trafficking.[166] In rat primary neurons and *Drosophila*, the overexpression of the autosomal dominant D620N VPS35 mutation causes degeneration of dopaminergic neurons.[159,167] Disruption of Cathepsin D trafficking, a lysosomal enzyme required for α-synuclein degradation, leads to α-synuclein accumulation likely in late endosomes/lysosomes in patient-derived fibroblasts, in *Drosophila* and in HEK cells.[168–170] DNAJC13/RME-8, which has recently been linked to PD, further tightens the connection between the retromer complex, endosomal trafficking, and parkinsonism.[171] Sorting nexin dimer 1 (SNX-1), a component of the retromer complex, interacts via its BAR domain with DNAJC13/RME-8 and downregulation of RME-8 causes an increase in highly branched endosomal tubules and impaired SNX-1 membrane association.[172] Immunofluorescent and Immunoelectron microscopic analysis revealed that SNX-1 in the hippocampus localizes to postsynaptic densities and endosomes in dendritic spines.[173] However, whether RME-8 is specifically required at the synapse needs further analysis.

3 EXOSOME RELEASE

Healthy synapses require exchange of material for intercellular signaling and they need to dispose their waste content. This is thought to occur in part through the release of exosomes. Exosomes are also important transport vehicles for mRNA, miRNA, receptors, signaling molecules, and lipids between cells to establish cell-to-cell communication during development, immunological response, tissue repair, and neuronal communication.[174–177] Although this is an emerging field, it appears that some PD-causing genes are involved in this process. We will therefore describe briefly the molecular mechanisms of exosome formation and release and explain the role of the PD-causing genes that may be affecting this process.

Exosomes are extracellular vesicles secreted by cells with a size ranging from 40 to 100 nm. The biogenesis of the exosomes occurs at the early endosome. It starts with the formation of intralumenal vesicles in MVB, a type of late endosome. These intraluminal vesicles are formed by invagination of the late endosomal limiting membrane into the lumen (Figs. 7.1G and 7.7). Formation of these intraluminal vesicles as well as cargo sorting requires the sequential action of endosomal-sorting complexes required for transport (ESCRT)-0, -I, and -II and the assembly of ESCRT-III.[178,179]

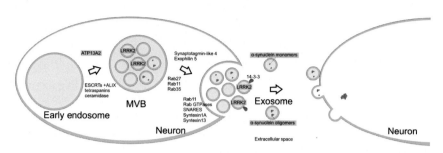

FIGURE 7.7 **Exosome formation and secretion.** Exosomes are derived from early endosomes. These endosomes become MVB by generating intraluminal vesicles. Biogenesis of intraluminal vesicles requires both ESCRT/ALIX and ceramidase/tetraspanin. Transport of the MVB to and docking of the MVB on the plasma membrane requires Rab11, Rab27, and Rab35. Docking also requires the Rab27 effectors synaptotagmin-like 4 and Exophilin-5. MVB fusion to the plasma membrane requires, among others, Rab11 and Syntaxin-1A, and Syntaxin-13. After fusion the intraluminal vesicles are released in the extracellular space, and these secreted vesicles can fuse with the membrane of other cells. LRRK2 and α-synuclein are packed in exosomes. Proteins of PD-related genes are highlighted in yellow.

Finally, the formed neck of the intraluminal vesicle is stabilized by binding of the AAA–adenosine triphosphatase Vps4 to ESCRT-III and the interaction of Vps2/Chmp2 and Snf7/Chmp4.[180] Upon interaction with Syntenin, the ESCRT-III-associated protein ALIX (apoptosis-linked gene 2-containing protein x) promotes intraluminal budding of vesicles in endosomes[181] (Fig. 7.7). Apart from the ESCRT machinery, also ceramidase and tetraspanins are involved in the biogenesis of intraluminal vesicles.[182] However, it is still unknown whether these mechanisms act simultaneously or sequentially.

Exosomes in the CNS have been shown to function in disposal of unneeded cell components as well as in the signaling to neighboring cells.[183,184] Neuronal homeostasis is important for synaptic communication since it allows anterograde and retrograde transfer of information among synapses. Exosomes are well placed to be involved in performing this function and it is conceivable that exosomes contribute to induce local changes in synaptic plasticity. In response to glutamatergic synaptic activity, cultures of mature cortical neurons and hippocampal neurons stimulate the release of exosomes from the somatodendritic postsynaptic compartment. These exosomes are neuronal exosomes since they carry the cell adhesion molecule L1, the GPI-anchored prion protein, as well as the GluR2/3 subunits of the AMPA receptor.[185,186]

Fibroblasts from patients with loss-of-function mutations in ATP13A2/PARK9 show a decreased amount of intraluminal vesicles in MVBs. ATP13A2 loss-of-function causes Kufor-Rakeb syndrome and is characterized by a juvenile-onset parkinsonism. Overexpressed and endogenous ATP13A2 localize to MVBs, indicating a potential function in exosome biogenesis. In line with this view, lentiviral overexpression of ATP13A2 in mouse primary cortical neurons, H4 and SHSY5Y cells, results in increased secretion of exosomes,[187,188] and further studies suggest that ATP13A2 cooperates with the ESCRT complex to control exosome secretion.[188] However, more work is needed to disentangle the role of ATP13A2 in exosome formation and lysosomal trafficking.

Protein sorting occurs at the late endosome and MVB. Proteins can be sorted by the ESCRT machinery based on whether they are ubiquitinated but also based on whether they are SUMOylated, phosphorylated, or glycosylated. Three ESCRT subcomplexes, ESCRT-0, -I, and -II, recognize and sequester ubiquitinated membrane proteins at the endosomal delimiting membrane. The proteins are subsequently deubiquitinated, and together with ESCRT-III they are sequestered in the intralumenal vesicles of the MVB.

Transport of the MVB to and docking of the MVB on the plasma membrane requires several proteins. Rab11, Rab27, and Rab35 have been implicated in the transport of MVB to the plasma membrane,[189] whereas Rab11 and Rab31, the R-SNARE protein YKT6,[190] the R-SNARE protein VAMP7/TIVAMP,[191] Rab27b and Rab27 effector molecules, Synaptotagmin-like 4, and Exophilin-5 are involved in MVB docking at the plasma membrane[189] (Fig. 7.7). The fusion of the MVBs with the plasma membrane likely requires Rab GTPases and SNAREs. Rab27a, Rab27b, and Rab35 have all been implicated in this mechanism in different cell types (Fig. 7.7). At the *Drosophila* NMJ the Q-SNARE Syntaxin-1A has been implicated in exosome release, but the precise mechanism remains enigmatic.[192] Live imaging showed that dendritic exosomes require Rab11 and Syntaxin-13 for the fusion with the plasma membrane.[193]

A proteomic analysis of exosome preparations from patients with genetic and sporadic forms of PD showed that 23 exosome-associated proteins were differentially present in the PD samples. The regulator of exosome biogenesis Syntenin 1 and Integrin b1 were among these proteins. This study suggests that exosomes are differently regulated or enriched for specific proteins in response to the disease.[194]

Analysis of cultured mouse neurons suggests that LRRK2-positive exosomes are released into the cerebrospinal fluid (CSF). Furthermore, release of LRRK2-containing exosomes from HEK293 cells requires the interaction with the 14-3-3 protein. Indeed, difopein (*di*meric *fo*urteen-three-three *pe*ptide *in*hibitor), a 14-3-3 inhibitor, results in the formation of exosomes without LRRK2. Since difopein does not have an effect on total exosome secretion,[195] LRRK2 probably requires 14-3-3 protein to be secreted in exosomes from HEK293 cells.

While the functional relevance of the secretion of LRRK2 via exosomes has not been elucidated, the release and spreading of α-synuclein via this mechanism may underlie the propagation of α-synuclein pathology in the brain. Indeed, α-synuclein aggregates and it is believed that the protein produced in seeding cells can spread in the brain to cause noncell autonomous pathology. The secretion of α-synuclein in exosomes could be one way how α-synuclein pathology can propagate.[196] Protein fragment complementation assays in human H4 neuroglioma cells showed that α-synuclein oligomers are present in and on the outside of exosomes. Interestingly, these exosomes are more prone to internalization and more toxic than exosome-free α-synuclein oligomers.[197] Moreover, ganglioside lipids GM1- or GM3-containing exosomes accelerated α-synuclein aggregation.[198] These exosomes are released when the cytosolic Ca^{2+} concentration rises. Cytosolic Ca^{2+} concentration can be raised by thapsigargin or ionomycin treatment. SH-SY5Y cells containing inducible α-synuclein or β-galactosidase (as a control) treated with thapsigargin or ionomycin will secrete increased α-synuclein exosomes but not β-galactosidase exosomes.[196] Therefore, α-synuclein is secreted together with exosomes. Interestingly, the transient coexpression of α-synuclein and LRRK2 in HEK293 cells did not reveal a colocalization of these proteins in exosomes, suggesting that they are derived from exosomes with different origins.[195]

4 SUMMARY

Numerous connections between PD-causing genes and a role for the gene products in the regulation of synaptic function and vesicle trafficking exist (Table 7.1). In this chapter we have reviewed how proteins mutated in PD affect aspects of exocytosis, endocytosis, vesicle trafficking, and vesicle mobilization. Some of the proteins are directly involved in these processes: for example, Aux, SYNJ1, and RME-8 mutants affect uncoating of synaptic vesicles. Other PD-related proteins directly interact with the presynaptic machinery: for example, α-synuclein and DJ-1 interact with Synaptobrevin/VAMP, Parkin ubiquitinates synaptotagmin, and LRRK2 interacts with NSF and phosphorylates Endophilin A1. Others affect vesicle trafficking in an indirect fashion: for example, ATP is needed for vesicle mobilization and PINK1 and Parkin maintain mitochondrial function, while α-synuclein and Parkin interact with Synphilin-1, an ATP binding protein. The challenge that lay ahead of us is to now understand how defects in the vesicle cycle contribute to the pathogenesis of PD.

A slowing of the vesicle cycle may harbor acute effects on the release of transmitters, but it may also propagate over time, thereby affecting protein turnover mechanisms like endolysosomal trafficking and autophagy. Indeed synaptic vesicle proteins are sorted at endosomes, and synaptic vesicles have also been shown to be engulfed by autophagosomes.[199] These turnover mechanisms are needed to ensure that damaged proteins and organelles are removed as they may hamper the normal workings of the synapse. Slowed vesicle trafficking and slowed turnover mechanisms

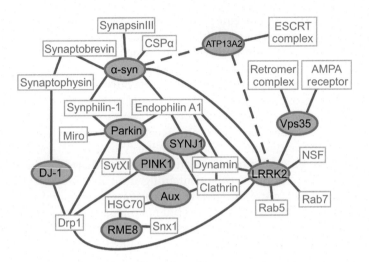

FIGURE 7.8 **Presynaptic PD interactome.** Different interactions between PD-causing proteins *(yellow background)* and different proteins present at synapse *(blue)*. Direct and indirect interactions between proteins are based on literature search.

could result in the buildup of damaged components causing proteopathic stress, synaptic dysfunction, and demise.

Vesicle trafficking defects may also have an effect on the progression of PD. As we discussed, α-synuclein can spread from cell to cell, and one of the mechanisms thought to be involved in this process is the production and release of exosomes. Membrane and vesicle trafficking and the production of intraluminal vesicles are critical aspects of exosome formation, and proteins implicated in PD have been connected to these processes as well.

In this chapter we explained the physiological processes at the synapse required for synaptic communication, and we discussed the involvement of various PD-associated proteins in maintaining synaptic function (Table 7.1). PD-causing proteins function in different processes; however, ultimately they are interconnected. We have built an evidence-based PD–protein interaction map illustrated in Fig. 7.8, providing a framework for future study. While our knowledge of PD largely stems from the familial forms of the disease, and the mutated genes responsible, we surmise that similar defects are also at play in sporadic cases of the disease.

References

1. Arena JE, Stoessl AJ. Optimizing diagnosis in Parkinsons disease: Radionuclide imaging. *Parkinsonism Relat Disord.* 2016;22 suppl 1:S47–51.
2. Hanus C, Ehlers MD. Secretory outposts for the local processing of membrane cargo in neuronal dendrites. *Traffic.* 2008;9(9):1437–1445.
3. Wojcik SM, Brose N. Regulation of membrane fusion in synaptic excitation-secretion coupling: speed and accuracy matter. *Neuron.* 2007;55(1):11–24.

4. Sutton MA, Schuman EM. Dendritic protein synthesis, synaptic plasticity, and memory. *Cell.* 2015;127(1):49–58.
5. Ly CV, Verstreken P. Mitochondria at the synapse. *Neuroscientist.* 2006;12(4):291–299.
6. Collins MO, Grant SGN. Supramolecular signalling complexes in the nervous system. *Subcell Biochem.* 2007;43:185–207.
7. Caberlotto L, Nguyen T-P. A systems biology investigation of neurodegenerative dementia reveals a pivotal role of autophagy. *BMC Syst Biol.* 2014;8(1):65.
8. Fernández-Chacón R, Wölfel M, Nishimune H, et al. The synaptic vesicle protein CSPα prevents presynaptic degeneration. *Neuron.* 2004;42(2):237–251.
9. Anglade P, Vyas S, Herrero MT, et al. Apoptosis and autophagy in nigral neurons of patients with Parkinson' s disease. *Histol. Histopathol.* 1997;1:25–31.
10. Bittner GD, Kennedy D. Quantitative aspects of transmitter release. *J Cell Biol.* 1970;47(3):585–592.
11. Clark AW, Hurlbut WP, Mauro A. Changes in the fine structure of the neuromuscular junction of the frog caused by black widow spider venom. *J Cell Biol.* 1972;52(1):1–14.
12. Ceccarelli B, Hurlbut WP, Mauro A. Depletion of vesicles from frog neuromuscular junctions by prolonged tetanic stimulation. *J Cell Biol.* 1972;54(1):30–38.
13. Heuser JE, Reese TS. Evidence for recycling of synaptic vesicle membrane during transmitter release at the frog neuromuscular junction. *J Cell Biol.* 1973;57(2):315–344.
14. Uytterhoeven V, Kuenen S, Kasprowicz J, Miskiewicz K, Verstreken P. Loss of Skywalker reveals synaptic endosomes as sorting stations for synaptic vesicle proteins. *Cell.* 2011;145(1):117–132.
15. Fernandes AC, Uytterhoeven V, Kuenen S, et al. Reduced synaptic vesicle protein degradation at lysosomes curbs TBC1D24/sky-induced neurodegeneration. *J Cell Biol.* 2014;207(4):453–462.
16. Staras K, Branco T, Burden JJ, et al. A vesicle superpool spans multiple presynaptic terminals in hippocampal neurons. *Neuron.* 2010;66(1):37–44.
17. Verstreken P, Koh T-W, Schulze KL, et al. Synaptojanin is recruited by endophilin to promote synaptic vesicle uncoating. *Neuron.* 2003;40(4):733–748.
18. Nemani VM, Lu W, Berge V, et al. Increased expression of alpha-synuclein reduces neurotransmitter release by inhibiting synaptic vesicle reclustering after endocytosis. *Neuron.* 2010;65(1):66–79.
19. Arranz AM, Delbroek L, Van Kolen K, et al. LRRK2 functions in synaptic vesicle endocytosis through a kinase-dependent mechanism. *J Cell Sci.* 2015;128(3):541–552.
20. Harris TW, Hartwieg E, Horvitz HR, Jorgensen EM. Mutations in synaptojanin disrupt synaptic vesicle recycling. *J Cell Biol.* 2000;150(3):589–600.
21. Ringstad N, Gad H, Löw P, et al. Endophilin/SH3p4 is required for the transition from early to late stages in clathrin-mediated synaptic vesicle endocytosis. *Neuron.* 1999;24(1):143–154.
22. Cremona O, Di Paolo G, Wenk MR, et al. Essential role of phosphoinositide metabolism in synaptic vesicle recycling. *Cell.* 1999;99(6):1041–1052.
23. Krebs CE, Karkheiran S, Powell JC, et al. The Sac1 domain of SYNJ1 identified mutated in a family with early-onset progressive parkinsonism with generalized seizures. *Hum Mutat.* 2013;34(9):1200–1207.
24. Quadri M, Fang M, Picillo M, et al. Mutation in the SYNJ1 gene associated with autosomal recessive, early-onset Parkinsonism. *Hum Mutat.* 2013;34(9):1208–1215.
25. Südhof TC. Neurotransmitter release: The last millisecond in the life of a synaptic vesicle. *Neuron.* 2013;80(2009):675–690.
26. Kantardzhieva A, Peppi M, Lane WS, Sewell WF. Protein composition of immunoprecipitated synaptic ribbons. *J Proteome Res.* 2012;11(2):1163–1174.
27. Siksou L, Rostaing P, Lechaire J-P, et al. Three-dimensional architecture of presynaptic terminal cytomatrix. *J. Neurosci.* 2007;27(26):6868–6877.
28. Kittel RJ, Wichmann C, Rasse TM, et al. Bruchpilot promotes active zone assembly, Ca2+ channel clustering, and vesicle release. *Science.* 2006;312(5776):1051–1054.

29. Whyte JRC, Munro S. Vesicle tethering complexes in membrane traffic. *J Cell Sci.* 2002;115(13):2627–2637.
30. Hanson PI, Heuser JE, Jahn R. Neurotransmitter release — four years of SNARE complexes. *Curr Opin Neurobiol.* 1997;7(3):310–315.
31. Gonzalo S, Greentree WK, Linder ME. SNAP-25 is targeted to the plasma membrane through a novel membrane- binding domain. *J Biol Chem.* 1999;274(30):21313–21318.
32. Sørensen JB, Matti U, Wei S-H, et al. The SNARE protein SNAP-25 is linked to fast calcium triggering of exocytosis. *Proc Natl Acad Sci USA.* 2002;99(3):1627–1632.
33. Dai H, Shen N, Araç D, Rizo J. A quaternary SNARE-synaptotagmin-Ca(2 +)-phospholipid complex in neurotransmitter release. *J Mol Biol.* 2007;367(3):848–863.
34. Bai J, Wang CT, Richards DA, Jackson MB, Chapman ER. Fusion pore dynamics are regulated by synaptotagmin*t-SNARE interactions. *Neuron.* 2004;41(6):929–942.
35. Giraudo CG. A clamping mechanism involved in SNARE-dependent exocytosis. *Science.* 2006;313(5787):676–680.
36. Clary DO, Griff IC, Rothman JE. SNAPs, a family of NSF attachment proteins involved in intracellular membrane fusion in animals and yeast. *Cell.* 2015;61(4):709–721.
37. Hayashi T, Yamasaki S, Nauenburg S, Binz T, Niemann H. Disassembly of the reconstituted synaptic vesicle membrane fusion complex in vitro. *EMBO J.* 1995;14(10):2317–2325.
38. Sollner T, Whiteheart SW, Brunner M, et al. SNAP receptors implicated in vesicle targeting and fusion. *Nature.* 1993;362(6418):318–324.
39. Liu S, Ninan I, Antonova I, et al. α-Synuclein produces a long-lasting increase in neurotransmitter release. *EMBO J.* 2004;23(22):4506–4516.
40. Anwar S, Peters O, Millership S, et al. Functional alterations to the nigrostriatal system in mice lacking all three members of the synuclein family. *J Neurosci.* 2011;31(20):7264–7274.
41. Greten-Harrison B, Polydoro M, Morimoto-Tomita M, et al. -Synuclein triple knockout mice reveal age-dependent neuronal dysfunction. *Proc Natl Acad Sci.* 2010;107(45):19573–19578.
42. Burré J, Sharma M, Tsetsenis T, Buchman V, Etherton MR, Südhof TC. Alpha-synuclein promotes SNARE-complex assembly in vivo and in vitro. *Science.* 2010;329(5999):1663–1667.
43. Chandra S, Gallardo G, Fernández-Chacón R, Schlüter OM, Südhof TC. α-Synuclein cooperates with CSPα in preventing neurodegeneration. *Cell.* 2005;123(3):383–396.
44. Burré J, Sharma M, Südhof TC. α-Synuclein assembles into higher-order multimers upon membrane binding to promote SNARE complex formation. *Proc Natl Acad Sci USA.* 2014;111(40):E4274–E4283.
45. Diao J, Ishitsuka Y, Lee H, et al. A single vesicle-vesicle fusion assay for in vitro studies of SNAREs and accessory proteins. *Nat Protoc.* 2012;7(6):921–934.
46. Diao J, Burré J, Vivona S, et al. Native α-synuclein induces clustering of synaptic-vesicle mimics via binding to phospholipids and synaptobrevin-2/VAMP2. *Elife.* 2013;2:e00592.
47. Yue M, Hinkle K, Davies P, et al. Progressive dopaminergic alterations and mitochondrial abnormalities in LRRK2 G2019S knock in mice. *Neurobiol Dis.* 2015;78:172–195.
48. Lee S, Liu H-P, Lin W-Y, Guo H, Lu B. LRRK2 kinase regulates synaptic morphology through distinct substrates at the presynaptic and postsynaptic compartments of the Drosophila neuromuscular junction. *J Neurosci.* 2010;30(50):16959–16969.
49. Piccoli G, Condliffe SB, Bauer M, et al. LRRK2 controls synaptic vesicle storage and mobilization within the recycling pool. *J Neurosci.* 2011;31(6):2225–2237.
50. Dawson TM, Ko HS, Dawson VL. Genetic animal models of Parkinsons disease. *Neuron.* 2010;66(5):646–661.
51. Kitada T, Pisani A, Porter DR, et al. Impaired dopamine release and synaptic plasticity in the striatum of Parkin-/- mice. *J Neurochem.* 2009;110:613–621.
52. Haddad DM, Vilain S, Vos M, et al. Mutations in the intellectual disability gene Ube2a cause neuronal dysfunction and impair parkin-dependent mitophagy. *Mol Cell.* 2013;50(6):831–843.

53. Huynh DP, Scoles DR, Nguyen D, Pulst SM. The autosomal recessive juvenile Parkinson disease gene product, parkin, interacts with and ubiquitinates synaptotagmin XI. *Hum Mol Genet*. 2003;12(20):2587–2597.

54. Bonifati V, Rizzu P, Squitieri F, et al. DJ-1(PARK7), a novel gene for autosomal recessive, early onset parkinsonism. *Neurol Sci*. 2003;24(3):159–160.

55. Klein C, Lohmann-Hedrich K. Impact of recent genetic findings in Parkinsons disease. *Curr Opin Neurol*. 2007;20(4):453–464.

56. Usami Y, Hatano T, Imai S, et al. DJ-1 associates with synaptic membranes. *Neurobiol Dis*. 2011;43(3):651–662.

57. Renard H-F, Simunovic M, Lemiere J, et al. Endophilin-A2 functions in membrane scission in clathrin-independent endocytosis. *Nature*. 2015;517(7535):493–496.

58. Boucrot E, Ferreira APA, Almeida-Souza L, et al. Endophilin marks and controls a clathrin-independent endocytic pathway. *Nature*. 2015;517(7535):460–465.

59. McMahon HT, Boucrot E. Molecular mechanism and physiological functions of clathrin-mediated endocytosis. *Nat Rev Mol Cell Biol*. 2011;12(8):517–533.

60. Kittler JT, Chen G, Honing S, et al. Phospho-dependent binding of the clathrin AP2 adaptor complex to GABAA receptors regulates the efficacy of inhibitory synaptic transmission. *Proc Natl Acad Sci USA*. 2005;102(41):14871–14876.

61. Brodsky FM, Chen CY, Knuehl C, Towler MC, Wakeham DE. Formation and function of clathrin-coated vesicles. *Annu Rev Cell Dev Biol*. 2001;17(1):517–568.

62. Saheki Y, De Camilli P. Synaptic vesicle endocytosis. *Cold Spring Harb Perspect Biol*. 2012;4(9):a005645.

63. Beck KA, Keen JH. Self-association of the plasma-membrane-associated clathrin assembly protein AP-2. *J Biol Chem*. 1991;266:4437–4441.

64. Diril MK, Wienisch M, Jung N, Klingauf J, Haucke V. Stonin 2 is an AP-2-dependent endocytic sorting adaptor for synaptotagmin internalization and recycling. *Dev Cell*. 2015;10(2):233–244.

65. Koo SJ, Puchkov D, Haucke V. AP180 and CALM: Dedicated endocytic adaptors for the retrieval of synaptobrevin 2 at synapses. *Cell Logist*. 2011;1(4):168–172.

66. Sundborger A, Soderblom C, Vorontsova O, Evergren E, Hinshaw JE, Shupliakov O. An endophilin–dynamin complex promotes budding of clathrin-coated vesicles during synaptic vesicle recycling. *J Cell Sci*. 2011;124(1):133–143.

67. Wigge P, Kîhler K, Vallis Y, et al. Amphiphysin heterodimers: Potential role in clathrin-mediated endocytosis. *Mol Biol Cell*. 1997;8(10):2003–2015.

68. Kosaka T, Ikeda K. Reversible blockage of membrane retrieval and endocytosis in the garland cell of the temperature-sensitive mutant of Drosophila melanogaster, shibirets1. *J Cell Biol*. 1983;97(2):499–507.

69. Cao H, Garcia F, McNiven MA. Differential distribution of dynamin isoforms in mammalian cells. In: Bonifacino JS, ed. *Mol Biol Cell*. 1998;9(9):2595–2609.

70. Nakatax T, Iwamoto A, Noda Y, Takemura R, Yoshikura H, Hirokawa N. Predominant and developmentally regulated expression of dynamin in neurons. *Neuron*. 1991;7(3):461–469.

71. Ferguson SM, Brasnjo G, Hayashi M, et al. A selective activity-dependent requirement for dynamin 1 in synaptic vesicle endocytosis. *Science*. 2007;316(5824):570–574.

72. Liu J, Fukuda K, Xu Z, et al. Structural basis of phosphoinositide binding to kindlin-2 protein pleckstrin homology domain in regulating integrin activation. *J Biol Chem*. 2011;286(50):43334–43342.

73. Hinshaw JE, Schmid SL. Dynamin self-assembles into rings suggesting a mechanism for coated vesicle budding. *Nature*. 1995;374(6518):190–192.

74. Roux A, Uyhazi K, Frost A, De Camilli P. GTP-dependent twisting of dynamin implicates constriction and tension in membrane fission. *Nature*. 2006;441(7092):528–531.

75. Perera RM, Zoncu R, Lucast L, De Camilli P, Toomre D. Two synaptojanin 1 isoforms are recruited to clathrin-coated pits at different stages. *Proc Natl Acad Sci USA*. 2006;103(51):19332–19337.

76. Dittgen T, Nimmerjahn A, Komai S, et al. Lentivirus-based genetic manipulations of cortical neurons and their optical and electrophysiological monitoring in vivo. *Proc Natl Acad Sci USA.* 2004;101:18206–18211.

77. Ringstad N, Nemoto Y, De Camilli P. Differential expression of endophilin 1 and 2 dimers at central nervous system synapses. *J Biol Chem.* 2001;276(44):40424–40430.

78. Milosevic I, Giovedi S, Lou X, et al. Recruitment of endophilin to clathrin-coated pit necks is required for efficient vesicle uncoating after fission. *Neuron.* 2011;72(4):587–601.

79. Cremona O, De Camilli P. Phosphoinositides in membrane traffic at the synapse. *J Cell Sci.* 2001;114(6):1041–1052.

80. Schlossman D, Schmid S, Braell W, Rothman J. An enzyme that removes clathrin coats: purification of an uncoating ATPase. *J Cell Biol.* 1984;99(2):723–733.

81. Ungewickell E, Ungewickell H, Holstein SEH, et al. Role of auxilin in uncoating clathrin-coated vesicles. *Nature.* 1995;378(6557):632–635.

82. Fotin A, Cheng Y, Grigorieff N, Walz T, Harrison SC, Kirchhausen T. Structure of an auxilin-bound clathrin coat and its implications for the mechanism of uncoating. *Nature.* 2004;432(7017):649–653.

83. Xing Y, Böcking T, Wolf M, Grigorieff N, Kirchhausen T, Harrison SC. Structure of clathrin coat with bound Hsc70 and auxilin: mechanism of Hsc70-facilitated disassembly. *EMBO J.* 2010;29(3):655–665.

84. Rapoport I, Boll W, Yu A, Böcking T, Kirchhausen T. A motif in the clathrin heavy chain required for the Hsc70/auxilin uncoating reaction. In: Schmid S, ed. *Mol Biol Cell.* 2008;19(1):405–413.

85. Böcking T, Aguet F, Harrison SC, Kirchhausen T. Single-molecule analysis of a molecular disassemblase reveals the mechanism of Hsc70-driven clathrin uncoating. *Nat Struct Mol Biol.* 2011;18(3):295–301.

86. Koenig JH, Ikeda K. Synaptic vesicles have two distinct recycling pathways. *J Cell Biol.* 1996;135(3):797–808.

87. Richards DA, Guatimosim C, Betz WJ. Two endocytic recycling routes selectively fill two vesicle pools in frog motor nerve terminals. *Neuron.* 2000;27(3):551–559.

88. Clayton EL, Cousin MA. Quantitative monitoring of activity-dependent bulk endocytosis of synaptic vesicle membrane by fluorescent dextran imaging. *J Neurosci Methods.* 2009;185(1):76–81.

89. Royle SJ, Lagnado L. Endocytosis at the synaptic terminal. *J Physiol.* 2003;553(Pt 2):345–355.

90. Cheung G, Cousin MA. Adaptor protein complexes 1 and 3 are essential for generation of synaptic vesicles from activity-dependent bulk endosomes. *J Neurosci.* 2012;32(17):6014–6023.

91. Cheung G, Jupp OJ, Cousin MA. Activity-dependent bulk endocytosis and clathrin-dependent endocytosis replenish specific synaptic vesicle pools in central nerve terminals. *J Neurosci.* 2010;30(24):8151–8161.

92. Kasprowicz J, Kuenen S, Swerts J, Miskiewicz K, Verstreken P. Dynamin photoinactivation blocks Clathrin and α-adaptin recruitment and induces bulk membrane retrieval. *J Cell Biol.* 2014;204(7):1141–1156.

93. Wu Y, O'Toole ET, Girard M, et al. A dynamin 1-, dynamin 3- and clathrin-independent pathway of synaptic vesicle recycling mediated by bulk endocytosis. In: Jahn R, ed. *Elife.* 2014;3:e01621.

94. Watanabe S, Rost BR, Camacho-Pérez M, et al. Ultrafast endocytosis at mouse hippocampal synapses. *Nature.* 2013;504(7479):242–247.

95. Watanabe S, Liu Q, Davis MW, et al. Ultrafast endocytosis at Caenorhabditis elegans neuromuscular junctions. In: Marder E, ed. *Elife.* 2013;2:e00723.

96. Watanabe S, Trimbuch T, Camacho-Pérez M, et al. Clathrin regenerates synaptic vesicles from endosomes. *Nature.* 2014;515(7526):228–233.

97. Westphal CH, Chandra SS. Monomeric synucleins generate membrane curvature. *J Biol Chem.* 2013;288(3):1829–1840.

98. Vargas KJ, Makani S, Davis T, Westphal CH, Castillo PE, Chandra SS. Synucleins regulate the kinetics of synaptic vesicle endocytosis. *J Neurosci.* 2014;34(28):9364–9376.

99. Matta S, Van Kolen K, da Cunha R, et al. LRRK2 controls an EndoA phosphorylation cycle in synaptic endocytosis. *Neuron.* 2012;75(6):1008–1021.

100. Ambroso MR, Hegde BG, Langen R. Endophilin A1 induces different membrane shapes using a conformational switch that is regulated by phosphorylation. *Proc Natl Acad Sci USA.* 2014;111:6982–6987.

101. Cao M, Milosevic I, Giovedi S, De Camilli P. Upregulation of parkin in endophilin mutant mice. *J Neurosci.* 2014;34(49):16544–16549.

102. Quadri M, Fang M, Picillo M, et al. Mutation in the SYNJ1 gene associated with autosomal recessive, early-onset Parkinsonism. *Hum Mutat.* 2013;34(9):1208–1215.

103. Olgiati S, De Rosa A, Quadri M, et al. PARK20 caused by SYNJ1 homozygous Arg-258Gln mutation in a new Italian family. *Neurogenetics.* 2014;15(3):183–188.

104. Edvardson S, Cinnamon Y, Ta-Shma A, et al. A deleterious mutation in DNAJC6 encoding the neuronal-specific clathrin-uncoating co-chaperone auxilin, is associated with juvenile parkinsonism. *PLoS One.* 2012;7(5):e36458.

105. Vilariño-Güell C, Rajput A, Milnerwood AJ, et al. DNAJC13 mutations in Parkinson disease. *Hum Mol Genet.* 2014;23(7):1794–1801.

106. Ahle S, Ungewickell E. Auxilin, a newly identified clathrin-associated protein in coated vesicles from bovine brain. *J Cell Biol.* 1990;111(1):19–29.

107. Chang HC, Hull M, Mellman I. The J-domain protein Rme-8 interacts with Hsc70 to control clathrin-dependent endocytosis in Drosophila. *J Cell Biol.* 2004;164(7):1055–1064.

108. Eisenberg E, Greene LE. Multiple roles of auxilin and hsc70 in clathrin-mediated endocytosis. *Traffic.* 2007;8(6):640–646.

109. Uytterhoeven V, Lauwers E, Maes I, et al. Hsc70-4 deforms membranes to promote synaptic protein turnover by endosomal microautophagy. *Neuron.* 2015;88(4):735–748.

110. Denker A, Rizzoli SO. Synaptic vesicle pools: an update. *Front Synaptic Neurosci.* 2010;2:135.

111. Kuromi H, Kidokoro Y. Tetanic stimulation recruits vesicles from reserve pool via a cAMP-mediated process in Drosophila synapses. *Neuron.* 2000;27(1):133–143.

112. Hilfiker S, Pieribone VA, Czernik AJ, Kao HT, Augustine GJ, Greengard P. Synapsins as regulators of neurotransmitter release. *Philos Trans R Soc B Biol Sci.* 1999;354(1381):269–279.

113. Zaltieri M, Grigoletto J, Longhena F, et al. α-synuclein and synapsin III cooperatively regulate synaptic function in dopamine neurons. *J Cell Sci.* 2015;128(13):2231–2243.

114. Morais Va, Verstreken P, Roethig A, et al. Parkinsons disease mutations in PINK1 result in decreased Complex I activity and deficient synaptic function. *EMBO Mol Med.* 2009;1(2):99–111.

115. Sakaba T, Neher E. Involvement of actin polymerization in vesicle recruitment at the calyx of Held synapse. *J Neurosci.* 2003;23(3):837–846.

116. Heidelberger R. Adenosine triphosphate and the late steps in calcium-dependent exocytosis at a ribbon synapse. *J Gen Physiol.* 1998;111(2):225–241.

117. Heidelberger R, Sterling P, Matthews G. Roles of ATP in depletion and replenishment of the releasable pool of synaptic vesicles. *J Neurophysiol.* 2002;88(1):98–106.

118. Ivannikov MV, Sugimori M, Llinás RR. Synaptic vesicle exocytosis in hippocampal synaptosomes correlates directly with total mitochondrial volume. *J Mol Neurosci.* 2013;49(1):223–230.

119. Guo X, Macleod GT, Wellington A, et al. The GTPase dMiro is required for axonal transport of mitochondria to drosophila synapses. *Neuron.* 2005;47(3):379–393.

120. Górska-Andrzejak J, Stowers RS, Borycz J, Kostyleva R, Schwarz TL, Meinertzhagen IA. Mitochondria are redistributed in Drosophila photoreceptors lacking Milton, a kinesin-associated protein. *J Comp Neurol.* 2003;463(4):372–388.

121. Verstreken P, Ly CV, Venken KJT, Koh T-W, Zhou Y, Bellen HJ. Synaptic mitochondria are critical for mobilization of reserve pool vesicles at Drosophila neuromuscular junctions. *Neuron*. 2005;47(3):365–378.

122. Ivannikov MV, Harris KM, Macleod GT, Mitochondria:. Enigmatic stewards of the synaptic vesicle reserve pool. *Front Synaptic Neurosci*. 2010;2:145.

123. Kitada T, Asakawa S, Hattori N, et al. Mutations in the parkin gene cause autosomal recessive juvenile parkinsonism. *Nature*. 1998;392(6676):605–608.

124. Rogaeva E, Johnson J, Lang AE, et al. Analysis of the pink1 gene in a large cohort of cases with parkinson disease. *Arch Neurol*. 2004;61(12):1898–1904.

125. Bonifati V, Rohé CF, Breedveld GJ, et al. Early-onset parkinsonism associated with PINK1 mutations: frequency, genotypes, and phenotypes. *Neurology*. 2005;65(1):87–95.

126. Valente EM, Abou-Sleiman PM, Caputo V, et al. Hereditary early-onset Parkinson's disease caused by mutations in PINK1. *Science*. 2004;304(5674):1158–1160.

127. Hao L-Y, Giasson BI, Bonini NM. DJ-1 is critical for mitochondrial function and rescues PINK1 loss of function. *Proc Natl Acad Sci USA*. 2010;107(21):9747–9752.

128. Clark IE, Dodson MW, Jiang C, et al. Drosophila pink1 is required for mitochondrial function and interacts genetically with parkin. *Nature*. 2006;441(7097):1162–1166.

129. Park J, Lee SB, Lee SB, et al. Mitochondrial dysfunction in Drosophila PINK1 mutants is complemented by parkin. *Nature*. 2006;441(7097):1157–1161.

130. Goldberg MS, Pisani A, Haburcak M, et al. Nigrostriatal dopaminergic deficits and hypokinesia caused by inactivation of the familial parkinsonism-linked gene DJ-1. *Neuron*. 2005;45(4):489–496.

131. Zhang L, Shimoji M, Thomas B, et al. Mitochondrial localization of the Parkinsons disease related protein DJ-1: Implications for pathogenesis. *Hum Mol Genet*. 2005;14(14):2063–2073.

132. Moore DJ, Dawson VL, Dawson TM. Lessons from Drosophila models of DJ-1 deficiency. *Sci Aging Knowl Environ*. 2006;2006(2).

133. Chung KK, Zhang Y, Lim KL, et al. Parkin ubiquitinates the alpha-synuclein-interacting protein, synphilin-1: implications for Lewy-body formation in Parkinson disease. *Nat Med*. 2001;7(10):1144–1150.

134. Li T, Liu J, Smith WW. Synphilin-1 binds ATP and regulates intracellular energy status. *PLoS One*. 2014;9(12):e115233.

135. Ribeiro CS, Carneiro K, Ross Ca, Menezes JRL, Engelender S. Synphilin-1 is developmentally localized to synaptic terminals, and its association with synaptic vesicles is modulated by α-synuclein. *J Biol Chem*. 2002;277:23927–23933.

136. Wakabayashi K, Engelender S, Yoshimoto M, Tsuji S, Ross CA, Takahashi H. Synphilin-1 is present in Lewy bodies in Parkinsons disease. *Ann Neurol*. 2000;47(4):521–523.

137. Engelender S, Kaminsky Z, Guo X, et al. Synphilin-1 associates with alpha-synuclein and promotes the formation of cytosolic inclusions. *Nat Genet*. 1999;22(1):110–114.

138. Montero M, Alonso MT, Carnicero E, et al. Chromaffin-cell stimulation triggers fast millimolar mitochondrial Ca2+ transients that modulate secretion. *Nat Cell Biol*. 2000;2(2):57–61.

139. Billups B, Forsythe ID. Presynaptic mitochondrial calcium sequestration influences transmission at mammalian central synapses. *J Neurosci*. 2002;22(14):5840–5847.

140. David G, Barrett EF. Mitochondrial Ca(2 +) uptake prevents desynchronization of quantal release and minimizes depletion during repetitive stimulation of mouse motor nerve terminals. *J Physiol*. 2003;548(Pt 2):425–438.

141. Murphy MP. How mitochondria produce reactive oxygen species. *Biochem J*. 2009;417(1):1–13.

142. Ray PD, Huang B-W, Tsuji Y. Reactive oxygen species (ROS) homeostasis and redox regulation in cellular signaling. *Cell Signal*. 2012;24(5):981–990.

143. Sönnichsen B, De Renzis S, Nielsen E, Rietdorf J, Zerial M. Distinct membrane domains on endosomes in the recycling pathway visualized by multicolor imaging of Rab4, Rab5, and Rab11. *J Cell Biol*. 2000;149(4):901–913.

144. Barbero P, Bittova L, Pfeffer SR. Visualization of Rab9-mediated vesicle transport from endosomes to the trans-Golgi in living cells. *J Cell Biol*. 2002;156(3):511–518.

145. Rink J, Ghigo E, Kalaidzidis Y, Zerial M. Rab conversion as a mechanism of progression from early to late endosomes. *Cell*. 2005;122(5):735–749.

146. de Renzis S, Sönnichsen B, Zerial M. Divalent Rab effectors regulate the sub-compartmental organization and sorting of early endosomes. *Nat Cell Biol*. 2002;4(2):124–133.

147. Christoforidis S, McBride HM, Burgoyne RD, Zerial M. The Rab5 effector EEA1 is a core component of endosome docking. *Nature*. 1999;397:621–625.

148. Lawe DC, Patki V, Heller-Harrison R, Lambright D, Corvera S, The FYVE. domain of early endosome antigen 1 is required for both phosphatidylinositol 3-phosphate and Rab5 binding. Critical role of this dual interaction for endosomal localization. *J Biol Chem*. 2000;275(5):3699–3705.

149. Nielsen E. Rabenosyn-5, a novel rab5 effector, is complexed with hVPS45 and recruited to endosomes through a FYVE finger domain. *J Cell Biol*. 2000;151:601–612.

150. Gaullier J-M, Simonsen A, D'Arrigo A, Bremnes B, Stenmark H, Aasland R. FYVE fingers bind PtdIns(3)P. *Nature*. 1998;394(6692):432–433.

151. Antonin W. A SNARE complex mediating fusion of late endosomes defines conserved properties of SNARE structure and function. *EMBO J*. 2000;19(23):6453–6464.

152. Chan C-C, Scoggin S, Wang D, et al. Systematic discovery of Rab GTPases with synaptic functions in Drosophila. *Curr Biol*. 2011;21(20):1704–1715.

153. Wilson JM, de Hoop M, Zorzi N, Toh BH, Dotti CG, Parton RG. EEA1, a tethering protein of the early sorting endosome, shows a polarized distribution in hippocampal neurons, epithelial cells, and fibroblasts. *Mol Biol Cell*. 2000;11(8):2657–2671.

154. Wucherpfennig T, Wilsch-Bräuninger M, González-Gaitán M. Role of Drosophila Rab5 during endosomal trafficking at the synapse and evoked neurotransmitter release. *J Cell Biol*. 2003;161(3):609–624.

155. Hoopmann P, Punge A, Barysch S.V., et al. Endosomal sorting of readily releasable synaptic vesicles. *Proc Natl Acad Sci USA*. 2010;107(16):19055–19060.

156. Rizzoli SO, Betz WJ. Effects of 2-(4-morpholinyl)-8-phenyl-4H-1-benzopyran-4-one on synaptic vesicle cycling at the frog neuromuscular junction. *J Neurosci*. 2002;22(24):10680–10689.

157. Kouranti I, Sachse M, Arouche N, Goud B, Echard A. Rab35 regulates an endocytic recycling pathway essential for the terminal steps of cytokinesis. *Curr Biol*. 2006;16(17):1719–1725.

158. Munsie LN, Milnerwood a J, Seibler P, et al. Retromer-dependent neurotransmitter receptor trafficking to synapses is altered by the Parkinsons Disease VPS35 mutation p.D620N. *Hum Mol Genet*. 2014;24(10):1–13.

159. Wang H, Toh J, Ho P, Tio M, Zhao Y, Tan E-K. In vivo evidence of pathogenicity of VPS35 mutations in the Drosophila. *Mol Brain*. 2014;7(1):73.

160. Korolchuk VI, Schütz MM, Gómez-Llorente C, et al. Drosophila Vps35 function is necessary for normal endocytic trafficking and actin cytoskeleton organisation. *J Cell Sci*. 2007;120(Pt 24):4367–4376.

161. Raiborg C, Grønvold Bache K, Mehlum A, Stang E, Stenmark H. Hrs recruits clathrin to early endosomes. *EMBO J*. 2001;20(17):5008–5021.

162. Schreij AMA, Chaineau M, Ruan W, et al. LRRK 2 localizes to endosomes and interacts with clathrin-light chains to limit Rac 1 activation. *EMBO Rep*. 2015;16(1):79–86.

163. Higashi S, Moore DJ, Yamamoto R, et al. Abnormal localization of leucine-rich repeat kinase 2 to the endosomal-lysosomal compartment in Lewy body disease. *Ann Neurol*. 2010;68(9):994–1005.

164. Gómez-Suaga P, Churchill GC, Patel S, Hilfiker S. A link between LRRK2, autophagy and NAADP-mediated endolysosomal calcium signalling. *Biochem Soc Trans*. 2012;40(5):1140–1146.

165. Gómez-Suaga P, Rivero-Ríos P, Fdez E, et al. LRRK2 delays degradative receptor trafficking by impeding late endosomal budding through decreasing Rab7 activity. *Hum Mol Genet*. 2014;23(25):6779–6796.

166. MacLeod Da, Rhinn H, Kuwahara T, et al. RAB7L1 interacts with LRRK2 to modify intraneuronal protein sorting and Parkinsons disease risk. *Neuron*. 2013;77(3):425–439.
167. Tsika E, Glauser L, Moser R, et al. Parkinsons disease-linked mutations in VPS35 induce dopaminergic neurodegeneration. *Hum Mol Genet*. 2014;23(17):4621–4638.
168. Sevlever D, Jiang P, Yen S-HC, Cathepsin D. is the main lysosomal enzyme involved in the degradation of alpha-synuclein and generation of its carboxy-terminally truncated species. *Biochemistry*. 2008;47(36):9678–9687.
169. Follett J, Norwood SJ, Hamilton Na, et al. The Vps35 D620N mutation linked to Parkinsons disease disrupts the cargo sorting function of retromer. *Traffic*. 2014;15(2):230–244.
170. Miura E, Hasegawa T, Konno M, et al. VPS35 dysfunction impairs lysosomal degradation of α-synuclein and exacerbates neurotoxicity in a Drosophila model of Parkinsons disease. *Neurobiol Dis*. 2014;71:1–13.
171. Vilariño-Güell C, Rajput A, Milnerwood AJ, et al. DNAJC13 mutations in Parkinson disease. *Hum Mol Genet*. 2014;23(7):1794–1801.
172. Freeman CL, Hesketh G, Seaman MNJ. RME-8 coordinates the WASH complex with the retromer SNX-BAR dimer to control endosomal tubulation. *J Cell Sci*. 2014;127(Pt 9):2053–2070.
173. Fukaya M, Fukushima D, Hara Y, Sakagami H. EFA6A, a guanine nucleotide exchange factor for Arf6, interacts with sorting nexin-1 and regulates neurite outgrowth. *J Neurochem*. 2014;129(1):21–36.
174. Mittelbrunn M, Sánchez-Madrid F. Intercellular communication: diverse structures for exchange of genetic information. *Nat Rev Mol Cell Biol*. 2012;13(5):328–335.
175. Chivet M, Javalet C, Hemming F, et al. Exosomes as a novel way of interneuronal communication. *Biochem Soc Trans*. 2013;41(1):241–244.
176. Sahoo S, Losordo DW. Exosomes and cardiac repair after myocardial infarction. *Circ Res*. 2014;114(2):333–344.
177. Gutiérrez-Vázquez C, Villarroya-Beltri C, Mittelbrunn M, Sánchez-Madrid F. Transfer of extracellular vesicles during immune cell-cell interactions. *Immunol Rev*. 2013;251(1):125–142.
178. Saksena S, Wahlman J, Teis D, Johnson AE, Emr SD. Functional reconstitution of ESCRT-III assembly and disassembly. *Cell*. 2015;136(1):97–109.
179. Wollert T, Wunder C, Lippincott-Schwartz J, Hurley JH. Membrane scission by the ESCRT-III complex. *Nature*. 2009;458(7235):172–177.
180. Adell MAY, Vogel GF, Pakdel M, et al. Coordinated binding of Vps4 to ESCRT-III drives membrane neck constriction during MVB vesicle formation. *J Cell Biol*. 2014;205(1):33–49.
181. Baietti MF, Zhang Z, Mortier E, et al. Syndecan-syntenin-ALIX regulates the biogenesis of exosomes. *Nat Cell Biol*. 2012;14(7):677–685.
182. Trajkovic K, Hsu C, Chiantia S, et al. Ceramide triggers budding of exosome vesicles into multivesicular endosomes. *Science*. 2008;319:1244–1247.
183. Lotvall J, Valadi H. Cell to cell signalling via exosomes through esRNA. *Cell Adh Migr*. 2007;1(3):156–158.
184. Simons M, Raposo G. Exosomes—vesicular carriers for intercellular communication. *Curr Opin Cell Biol*. 2009;21(4):575–581.
185. Fauré J, Lachenal G, Court M, et al. Exosomes are released by cultured cortical neurones. *Mol Cell Neurosci*. 2006;31(4):642–648.
186. Lachenal G, Pernet-Gallay K, Chivet M, et al. Release of exosomes from differentiated neurons and its regulation by synaptic glutamatergic activity. *Mol Cell Neurosci*. 2011;46(2):409–418.
187. Kong SMY, Chan BKK, Park J, et al. Parkinson's Disease linked human PARK9/ATP13A2 maintains zinc homeostasis and promotes αSynuclein externalisation via exosomes. *Hum Mol Genet*. 2014;23(11):1–47.
188. Tsunemi T, Hamada K, Krainc D. ATP13A2/PARK9 regulates secretion of exosomes and α-synuclein. *J Neurosci*. 2014;34(46):15281–15287.

189. Ostrowski M, Carmo NB, Krumeich S, et al. Rab27a and Rab27b control different steps of the exosome secretion pathway. *Nat Cell Biol*. 2010;12(1):19–30.
190. Gross JC, Chaudhary V, Bartscherer K, Boutros M. Active Wnt proteins are secreted on exosomes. *Nat Cell Biol*. 2012;14(10):1036–1045.
191. Fader CM, Sánchez DG, Mestre MB, Colombo MI. TI-VAMP/VAMP7 and VAMP3/cellubrevin: two v-SNARE proteins involved in specific steps of the autophagy/multivesicular body pathways. *Biochim Biophys Acta—Mol Cell Res*. 2009;1793(12):1901–1916.
192. Koles K, Nunnari J, Korkut C, et al. Mechanism of evenness interrupted (Evi)-exosome release at synaptic boutons. *J Biol Chem*. 2012;287(20):16820–16834.
193. Park M, Salgado JM, Ostroff L, et al. Plasticity-induced growth of dendritic spines by exocytic trafficking from recycling endosomes. *Neuron*. 2006;52(5):817–830.
194. Tomlinson PR, Zheng Y, Fischer R, et al. Identification of distinct circulating exosomes in Parkinsons disease. *Ann Clin Transl Neurol*. 2015;2(4):353–361.
195. Fraser KB, Moehle MS, Daher JPL, et al. LRRK2 secretion in exosomes is regulated by 14-3-3. *Hum Mol Genet*. 2013;22(24):4988–5000.
196. Emmanouilidou E, Melachroinou K, Roumeliotis T, et al. Cell-produced -synuclein is secreted in a calcium-dependent manner by exosomes and impacts neuronal survival. *J Neurosci*. 2010;30(20):6838–6851.
197. Danzer KM, Kranich LR, Ruf WP, et al. Exosomal cell-to-cell transmission of alpha synuclein oligomers. *Mol Neurodegener*. 2012;7(1):42.
198. Grey M, Dunning CJ, Gaspar R, et al. Acceleration of α-synuclein aggregation by exosomes. *J Biol Chem*. 2015;290(5):2969–2982.
199. Binotti B, Pavlos NJ, Riedel D, et al. The GTPase Rab26 links synaptic vesicles to the autophagy pathway. *Elife*. 2015;4:e05597.

Neuroinflammation as a Potential Mechanism Underlying Parkinsons Disease

C. Cebrián*, D. Sulzer*,**

*Department of Neurology, Columbia University Medical Center,
New York, NY, United States; **Departments of Psychiatry and
Pharmacology, Columbia University Medical Center, New York State
Psychiatric Institute, New York, NY, United States

OUTLINE

Parkinson's Disease. http://dx.doi.org/10.1016/B978-0-12-803783-6.00008-0

1 INTRODUCTION

Parkinsons disease (PD), the most common movement disorder,[1] is characterized by the progressive loss of the neuromelanin-containing dopamine (DA) neurons of the substantia nigra (SN) pars compacta (SNc). This cell death produces a variety of motor symptoms including slowness of movement, rigidity, and tremor,[2] as well as nonmotor symptoms such as anxiety, depression, sleep disturbances, constipation, and excessive salivation.[3] In addition to the loss of SN DA neurons, as well as neuromelanin-containing neorpeinphrine neurons of the locus coeruleus (LC), the classic neuropathological characteristic of PD is the presence of intraneuronal inclusions containing high levels of the protein alpha-synuclein (α-syn), known as Lewy bodies or Lewy neurites.[4–6]

Neuroinflammation has been long suggested to play an important role in the massive loss of DA cells during PD. Extensive microgliosis in the SNc of PD patients was initially documented by Charles Foix in 1925,[7] who drew outstanding illustrations of activated microglia in the PD brain, along with extracellular remnants of neuromelanin and Lewy bodies. This observation was ignored for decades, probably because the involvement of an immune response in the central nervous system (CNS) was thought to be a rare event, as the brain was considered an "immune privileged" organ. This assumption has only been recently superseded as numerous studies have described how cellular BBB permeability is regulated as a stress response.[8–11]

McGeer et al.[12,13] confirmed Foix's results by showing activated microglia in the SN of postmortem human samples. Following those reports, neuroinflammation has developed into a major focus of PD research: postmortem studies of the brain, analyses of proinflammatory cytokines in serum and cerebrospinal fluid (CSF), PD risk factor associations with cytokine and major histocompatibility complex (MHC) class II (MHC-II) polymorphisms, and epidemiological studies of nonsteroidal antiinflammatory use have each demonstrated a link between neuroinflammation and PD.[1,14–17] However, whether neuroinflammation is

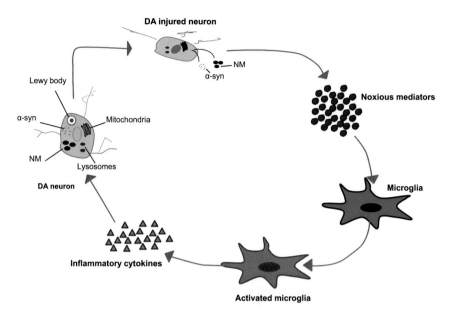

FIGURE 8.1 Stressed dopaminergic *(DA)* neurons release damaging factors (alpha-synuclein (*α-syn*) and neuromelanin *(NM)* among others) that activate microglia, which in turn produce inflammatory cytokines that further harm neurons, causing a "vicious cycle" of neuroinflammation and neurodegeneration.

a cause of PD pathogenesis or a secondary stress response remains an elusive issue.

Systemic neuroinflammation in patients has been reported to exaggerate pathogenic events associated with PD, intensify the motor symptoms,[18,19] and exacerbate neuronal dysfunction.[1] Stressed neurons activate microglia that release factors that further damage neurons[14,16,17,20] causing a "vicious cycle" of neuroinflammation and neurodegeneration (Fig. 8.1). Antiinflammatory therapies have been explored to treat PD, although to date these have shown limited success. Inflammation is primarily a stress response, and it has not been ruled out that in some instances neuroinflammation may actually provide neuroprotection.[21]

Different genetic mutations have been related to the presence of PD. Some of these also promulgate neuroinflammatory components, suggesting that neuroinflammation might play a role not only in idiopathic but also in genetic PD.[22]

In this chapter, we review current knowledge on neuroinflammatory processes in PD, with emphasis on cellular and molecular events associated with neuroinflammation involved in the degeneration of DA neurons.

2 THE ROLE OF GLIAL CELLS IN PD

2.1 Microglia

The role of immune surveillance in the CNS is primarily performed by the innate immune system via microglia, monocyte-derived resident macrophages of the brain that account for 5–20% of the total glial cell population.[23,24] Microglia are normally in a resting stage, but can be activated by brain lesions in neurodegenerative diseases including PD.[25]

Microglia in the CNS respond to environmental stress and immunological challenges by scavenging excess neurotoxins and removing dying cells and cellular debris.[9,26,27] Proinflammatory mediators released by activated glia act on their cognate receptors expressed on microglia and further exacerbate this response. Upon activation, microglia produce proinflammatory molecules to generate a neuroinflammatory response that mediates the removal of damaged cells by phagocytosis. Activated microglia produce prostaglandins, chemokines, cytokines, complement proteins, pronases, reactive oxygen species (ROS), and reactive nitrogen species including nitric oxide (NO). A sustained production of these substances can exert deleterious effects on susceptible populations by enhancing oxidative stress and activating cell-death pathways.[28] Microglia can also provide neurotrophic support for neurons by releasing neurotrophic factors, such as glial-cell-line-derived neurotrophic factor (GDNF) and brain-derived neurotrophic factor.[29,30]

Importantly for neurodegenerative diseases, microglial proliferation and activation may be triggered by protein aggregates (including α-syn) bacterial or viral infections, or traumatic brain injury.[31]

Brain imaging has revealed that microglia are active in the SN, pons, basal ganglia, striatum and frontal and temporal cortical regions of PD subjects compared to age-matched controls.[32,33] Postmortem immunohistological analysis of PD brains demonstrates an upregulation of the glycoprotein HLA-DR (human leukocyte antigen class II), a constituent of MHC-II that is involved in microglial regulation.[12,34,35]

It has been suggested that microglia acquire different phenotypes as PD progresses.[36] Microglial expression of CD68 in postmortem SN of PD patients has been correlated to disease duration[35] and in later stages, these microglia are thought to display highly phagocytic activity that may contribute to DA neuronal death.[36,37]

In contrast, microglial HLA–DR expression was positively correlated with α-syn deposition, but not clinical severity or disease progression.[35] Indeed, the number of HLA–DR immunoreactive microglia in patients over a range of disease duration did not change.[27] In PD animal models microglia upregulate MHC-II,[37] which may reflect presentation of self or foreign antigens to tissue-specific T cells.

There are several suggestions for the means by which microglia are activated in PD. Misfolded or aggregated proteins in Lewy bodies of diseased nigral DA neurons might elicit a cycle of microglial activation and increased inflammatory mediators in the SN, a region where the microglial density is the highest in the rodent brain and fairly high in the human brain.[13,29,38–41] This could provide a tertiary hit required for PD-associated features to spread to neighboring neurons.[42]

The ability of extracellular α-syn to activate microglia has been demonstrated in vitro.[43–47] Indeed, activation of microglia by α-syn is sufficient to induce DA cell death in culture via ROS[48] and tumor necrosis factor alpha (TNF-α) production,[49] although it is not yet clear how this activation occurs. There are multiple reports demonstrating that α-syn aggregation in one neuron can spread to other neurons,[50–52] which may indicate a prion-like mechanism as well as possible spread of the protein from one cell to another,[53] perhaps via vesicular exocytosis or via exosomes. Alternatively, remnants of the protein after neuronal destruction may activate microglia.

A special case of microglial activation in PD might be elicited by neuromelanin (NM), the pigment present in catecholamine neurons of SN and LC. Since Foix's 1925 publication on the basal ganglia and PD,[7] NM has been known to be present extracellularly following the death of the neurons. Microglia can be activated by human NM isolated from DA neurons, both in vitro[47,54] and in vivo.[55] The microglia rapidly phagocytose and degrade NM, while releasing massive amounts of cytokines and hydrogen peroxide.[54] This could provide a vicious cycle in which NM from dying targeted neurons activates microglia, leading to more cellular stress.

Another role for microglia that to date has been little studied is the activation of T cells by the release of cytokines, which could contribute to an immunological cascade in PD.[56]

2.2 Astrocytes

Astrocytes are the most abundant cells in the CNS.[38] They perform numerous essential functions, including trophic support for neurons, maintenance of the extracellular ion balance, production of the antioxidant glutathione, repair of scarred brain and spinal cord tissue,[57] synthesis of amino acids, control of extracellular pH,[58] and glutamate turnover, which is important to protect against excitotoxicity. Astrocytes further play roles in neuroinflammation as a bridge between the vascular and nervous systems, acting as a key regulatory element of neuronal activity and cerebral blood flow,[59] as well as limiting the spread of inflammatory cells and infectious agents from damaged areas to healthy parenchyma.[60] Activated astroglia can undergo "reactive astrogliosis," leading to cellular hypertrophy[57,60] and act as regulatory immune cells due to antigen presentation and the release pro- and antiinflammatory molecules, including cytokines, chemokines, complement neurotrophic factors.[57,61–63]

Astrocytes are heterogeneously distributed within the mesencephalon in healthy individuals: while the density of astrocytes is low in PD SNc, it is high in the ventral tegmental area and the catecholaminergic cell group A8, areas that are less affected than the SNc in PD.[16,57] It is possible that vulnerable neurons in patients with PD have a reduced number of surrounding astroglial cells, which would disrupt normal glutathione-mediated detoxification of free radicals.

Numerous studies suggest a link between α-syn and astrocytes in PD.[64] Nonfibrillized α-syn accumulates in the cytoplasm of protoplasmic but not fibrous astrocytes early in disease.[65] In addition, astrocytes with accumulated α-syn produce proinflammatory cytokines and chemokines and neuroinflammatory mediators, such as IFN-γ and TNF-α, which in turn activate more astrocytes and microglia.[66]

An additional contributory factor to astrocyte dysfunction in PD is likely to be the dysregulation of astrocyte-specific functions of recessive PD genes, such as DJ-1 and parkin. For example, DJ-1 mutations or deletions that cause a rare form of autosomal recessive parkinsonism[67] protect neurons via an astrocyte-mediated mechanism.[68] The role of DJ-1 in astrocyte-mediated neuroprotection is specific to a mechanism involving mitochondrial complex I and is independent of the oxidative stress response.[64] Several publications suggest that mutations in the parkin gene lead to astrocytic dysfunction by compromising the ability of astrocytes to cope with UPR-induced stress, and that this dysfunction in astrocytes contributes to neuronal death. Indeed, reactive astrocytes and the UPR have been shown to contribute to neuronal survival in a mouse model of PD.[69]

2.3 Oligodendrocytes

The best known function of oligodendrocytes is to produce myelin, but they also release neurotrophic factors.[70] Oligodendrocytes do not seem to play a role in promoting inflammation, but they can be damaged by inflammatory processes[71–73] as observed in multiple sclerosis.[74] Oligodendrocytes may provide a source of complement proteins in human brain and thus could contribute to pathogenesis of Alzheimers disease, multiple sclerosis, and progressive supranuclear palsy, in which complement-activated oligodendrocytes are abundant.[75]

Oligodendrocytes accumulate α-syn, possibly released from neurons, during the course of multiple system atrophy.[76] Mice that overexpress α-syn in oligodendrocytes (a model of multiple system atrophy) display mitochondrial stress and axonal swelling.[77]

Oligodendrocytes also modulate axonal transport within myelinated axons[78] and provide metabolic support to axons through the delivery of lactate,[79] the disruption of which results in axonal damage.[80] Note, however, that neurons that show Lewy pathology in PD are mostly unmyelinated, including the neurons of the autonomic system and those of the olfactory bulb.[70,81]

3 INVOLVEMENT OF PROINFLAMMATORY SUBSTANCES AND OXIDATIVE STRESS IN PD

3.1 Proinflammatory Substances

Microglial activation results in the release of proinflammatory cytokines and chemokines.[16] These substances include TNF-α, epidermal growth factor, transforming growth factor alpha and beta, and interleukins (IL) 1β, 6, and 2, each of which are elevated in the striatum of patients with PD.[82,83] TNF-α, IL-1β, and interferon gamma (IFN-γ) have been detected in the SN of patients.[84] Furthermore, DA neurons express receptors for these cytokines: for example, TNF-α receptor 1 is expressed on DA neurons[85] and the expression of this receptor increases in PD.[83] Cytokines with antiinflammatory or repair functions such as IL-10 are also reported in PD.[86–88]

Cytokines are not only elevated in the CNS of PD patients but also in their peripheral circulation.[89] IL-2,[90] TNF-α, IL-6,[91] and chemokine (C-C motif) ligand 5 (CCL-5)[92] have been found at higher levels in the blood and serum of PD patients than control individuals, and PD CSF contains high levels IL-1β.[86,93]

The release of proinflammatory cytokines in vitro induces expression of MHC proteins in glia[94,95] and neurons[47,96,97] and further activates glia and lymphocytes that in turn release cytokines, chemokines, ROS, etc.[98,99] Therefore, the presence of proinflammatory compounds could promote a vicious cycle that produces neuronal death.

3.2 Oxidative Stress

It is widely thought that oxidative stress contributes to the cascade leading to DA cell degeneration in PD and that it is linked to other components of the degenerative process, including mitochondrial dysfunction, excitotoxicity, nitric oxide toxicity, and inflammation.[100] Proinflammatory cytokines, such as TNF-α, IL-1β, and IFN-γ, induce expression of nitric oxide synthase (iNOS)[101,102] or cyclo-oxygenase 2 (COX-2),[103] enzymes that produce toxic reactive species that would interact with all cells. A general increase in oxidative stress is indicated by elevated nicotinamide adenine dinucleotide phosphate (NADPH) oxidase and myeloperoxidase in PD patients and PD animal models.[104–106] It is nonetheless possible that SN and LC neurons exhibit particularly high levels of oxidative stress. For example, SN neurons display a CD23-mediated increase in iNOS in PD.[107]

One explanation for selective oxidative stress in PD is that the SN, LC, and other targeted neurons that are lost in PD possess a high calcium conductance that drives pace-making activity in these neurons.[108] The activity of CaV1.3 channels that typically drive SN pace-making promotes oxidative stress at the mitochondria in SN neurons.[108]

A second possible source of oxidative stress in SN and LC neurons in particular is the presence of high levels of cytosolic DA. In neuronal culture, there is a linear relationship between neuronal death and cytosolic DA, and the level of cytosolic DA is dependent on high calcium load driven by the CaV1.3 channel responsible for SN tonic firing.[109] High cytosolic levels of DA in SNc murine neurons in vitro trigger neuronal MHC-I expression.[47] SN DA cell death also requires the presence of α-syn, and so may stem for a toxic interaction between DA and a-syn[110] and the ability of DA-modified α-syn to block normal protein turnover via the chaperone-mediated autophagy pathway.[111] It may be that such modifications initiate the inflammatory cascades discussed previously.

The reasons that SN and LC are subject to a high degree of oxidative stress may be related to their energy metabolism or their high content of catecholamines. DA cells are endowed with a range of protective mechanisms; however, these may be overwhelmed by excessive oxidative load. While the DA precursor, L-DOPA, appears to not accelerate disease progression, some controversy remains on whether it may in some cases cause additional oxidative stress.[100]

4 INFILTRATION OF LYMPHOCYTES IN PD

In contrast to the innate immune system, little is known about acquired inflammatory processes in PD. "Naïve" T cells are activated by antigen-presenting cells via T cell receptors and other surface molecules, as well as soluble mediators including cytokines and chemokines, secreted by innate immune system cells. T lymphocytes are activated through their T-cell receptor by recognizing cognate antigen on MHC molecules and receiving costimulation and appropriate cytokine signaling from antigen presenting cells. The ensuing response needs to be sustained through the interaction between the T cell receptor (TCR) and the corresponding MHC molecule until memory T cells are generated.[112]

Following activation, T cells produce cytokines that have an autocrine effect facilitating their own maturation according to their subtype. These include helper T cells, cytotoxic T cells, memory T cells, regulatory T cells, natural killer T cells, and γδ T cells.[113] Helper T cells, which display CD4 surface proteins and are known as CD4+, regulate immune response by their cytokine expression and promotion of antibody and cytotoxic T cell production. Cytotoxic T cells express CD8$^+$ on their surface and destroy virus-infected or tumor cells. Memory T cells persist in tissue peripheries long after initial exposure to an antigen to prevent a repeat infection. Regulatory T cells are required for the maintenance of immunological tolerance by shutting down T cell-mediated immunity and suppressing auto-reactive T cells that escaped negative selection during development. During the activation process, the regulatory T cell/effector T cell balance determines the inflammatory response.[114]

In healthy CNS, the BBB is thought to limit the entry of T cells into the parenchyma. During disease states, however, activation of endothelial cells of the blood vessels by cytokines and increased expression of adhesion molecules on endothelial cells and lymphocytes set the stage for an adaptive response that promotes the entry of T-cells across the BBB into the brain,[115–117] particularly following tropic signals from the brain to attract immune mediators to the site of injury.[118]

Surprisingly, data suggest that total numbers of T cells are *decreased* in the blood of PD patients as well as in rat models of PD[119–121] due in part to a decreased CD45RA$^+$ "naïve" CD4$^+$ subpopulation. Important subpopulations of T cells, including CD4$^+$ and CD8$^+$ T cells, are however *increased* in the blood of PD patients. These include the CD45RO$^+$ "memory" CD4$^+$ subpopulation of T cells.[122] In addition to CD45RO, T cells in PD express Fas, which are both markers for activated T cells.[123] Baba et al. reported significantly decreased CD4$^+$:CD8($^+$)T cell ratios, fewer CD4($^+$)CD25($^+$) T cells, and significantly increased ratios of IFN-γ-producing and IL-4-producing T cells in blood of PD patients, consistent with neuroinflammatory process in the brain.[124] These authors confirm a decrease in helper T cells, particularly the CD4($^+$) CD45RA($^+$) "naïve" population, as well as a decrease in B cells, and an increase in activated CD4($^+$) CD25($^+$) lymphocytes in blood of PD patients.[119] γδ T cells, which are mainly activated locally and not in lymphoid organs, were also elevated in the CSF and blood of PD patients.[125]

In the 1980s McGeer and coworkers[126] identified CD8$^+$ cytotoxic T lymphocytes in the SN of a PD patient. In a recent immunohistochemical analysis of several leukocyte markers in the SN of patients with PD, Brochard et al.[127] reported that although no B cells or natural killer cells were detectable, there were substantially higher densities of CD8$^+$ and CD4$^+$ T cells in the SNc of patients with PD than healthy individuals. These SNc T cells were in close contact with blood vessels, suggesting migration from the bloodstream, and near NM containing DA neurons, indicating a possible interaction between the lymphocytes and DA neurons. Elevated CD8$^+$ and CD4$^+$ T cells were not detected in the red nucleus, a nearby region unaffected in PD, suggesting that this infiltration is selective for diseased brain regions. The presence of CD8+ T cells has also been recently observed in the SN of PD postmortem brains and aged-matched controls, in close apposition to SN DA neurons that contained NM and expressed MHC-I.[47]

Recruitment of lymphocytes to brain parenchyma has been reproduced in several animal models of PD. T-cell infiltration was observed as an early event preceding CD68 expression when α-syn is overexpressed via rAAV in mouse SN.[128] The type of T-cell response varied depending upon cell death: in the absence of cell death, the response was primarily composed of CD4+ T-cells, whereas when cell death occurred, T-cell infiltration was delayed and the ratio of CD4+ to CD8+ T-cells was decreased.[37] In the 1-methyl-4-phenyl-1,2,3,6-tetrahydropyridine (MPTP) model, CD4+ T-cell infiltration peaked before that of CD8+ cells, but CD8+ cells were more numerous than CD4+ T-cells.[127] In the rAAV–α-syn model, the T-cell response correlated with the peak of microgliosis, while in the MPTP model, T-cells infiltrated the brain after microglia cell numbers peaked and cell death had occurred. While the differences are likely related to the more acute nigral cell death with MPTP than the rAAV–α-syn overexpression model, in both studies, minimal infiltration of T-cells into the striatum was observed, suggesting that T-cells hone to sites where cell body death occurs rather than to the DA terminals.

Further support for a role of T-cells in PD-like nigral degeneration comes from studies in mice with severe combined immunodeficiencies that lack of T-cells. These mice are relatively resistant to MPTP intoxication, and the transfer of T-cells from nitrated α-syn-immunized mice accelerated MPTP-driven neurodegeneration.[129] Brochard et al.[127] further dissected the immune response by using TCRβ-, CD4−, and CD8-deficient mice as MPTP recipients, and found that the transfer of CD4+ T-cells accelerated nigral degeneration, whereas CD8+ T-cells did not play a significant role.[127]

This reveals that T-cells are not only relevant, but are also required for neurodegeneration in PD. It has recently been reported that a deficiency of MHC-II results in attenuation of both microgliosis and loss of dopaminergic neurons in a mouse model of PD, supporting the pivotal role of CD4+ T-cells in potentiating microglial activation and favoring neurodegeneration in PD.[130] Experiments addressing the phenotype of pathogenic CD4+ T-cells involved in PD have shown that both Th1 and Th17 autoreactive cells are important for the promotion of neuronal loss.[131] Importantly, it has also been demonstrated that dopamine receptor D3 (D3R), expressed in CD4+ T-cells, is fundamental in inducing the loss of DA neurons in the SNc of a PD mouse model.[132,133]

These data indicate that peripheral cells can enter the brain parenchyma during the neurodegenerative process and that changes in the BBB function might occur in the brains of patients with PD. For example, an increased density of endothelial cells in the SN of patients with PD[134] and pathological changes in the microanatomy of capillaries in the brains of PD patients have been reported.[135] To our knowledge, T cell activation or proliferation in human brain has neither been demonstrated nor disproven to date.

5 INVOLVEMENT OF PD-RELATED GENES IN NEUROINFLAMMATION

Multiple genes have been identified to cause monogenetic forms of PD. Several of these genes are implicated in the regulation of mitochondrial activity, some of which also play a part in the inflammatory process in the CNS.

5.1 α-syn

Elevated expression levels of α-syn in the brain can lead to the development of PD.[136,137] This enhanced expression increases the deposition of soluble α-syn into insoluble aggregates.[136] Pathogenic mutations in α-syn (notably A53T) show increased rates of self-assembly and fibrillization, and are considered gain of function mutations, resulting in autosomal dominant inheritance.[138] Overexpression of the A53T mutant of α-syn results in accumulation of α-syn in the mitochondria, which has been associated with complex-I inhibition and aging.[139]

It was recently suggested that cell–cell transmission of α-syn pathology is the basis for disease spreading between interconnected areas in the brain. While several brain regions, such as the thalamus and cortical regions, exhibit formation of Lewy bodies after injection of α-syn, the neurodegeneration appears to be selective to dopaminergic neurons in the SN.[52] Watson et al. have shown that in α-syn overexpressing mice there is a selective early inflammatory response in regions of the nigrostriatal pathway despite the presence of high levels of α-syn in other brain regions.[140]

It has been suggested that specific factors that may involve an increase in the expression of Toll-like receptors (TLRs) mediate α-syn-induced inflammatory responses in the SN, which could explain the selective vulnerability of dopaminergic neurons in PD.[140]

Transgenic mice overexpressing of human α-syn exhibit increased vulnerability of dopaminergic neurons to lipopolysaccharide (LPS)-induced inflammation, and this inflammatory response leads to the accumulation of insoluble α-syn aggregates and the formation of cytoplasmatic α-syn inclusions bodies within DA neurons.[141] LPS-challenged transgenic mice expressing mutant α-syn (A53T) showed nitration and aggregation of α-syn and degeneration of DA neurons. Furthermore, those transgenic mice have prolonged microglial activation, expressing iNOS and NADPH oxidase enzymes that generate nitric oxide and superoxide, respectively.[142] It has been suggested that α-syn plays a role in neuroinflammation and microglial activation in PD.[143] Microglial cells that were exposed to extracellular α-syn show increased proinflammatory phenotype, with increased IL-1α, IL-1β, TNF-α, and IL-6 secretion. When microglia was treated with the A53T mutation of α-syn, this effect was exacerbated.[45,144]

Microglial activation following phagocytosis of aggregated α-syn enhances DA neurodegeneration through production of ROS, in a NADPH oxidase dependent manner.[44] The extent of microglial activation is correlated with the degree of dopaminergic neurotoxicity induced by both wild-type and mutant α-syn. Exposure to mutated α-syn also induced greater production of ROS than WT α-syn. These results suggest that microglia play an important role in mediating the enhanced neurotoxicity induced by mutant forms of α-syn.[48]

Mice that selectively express mutated α-syn in astrocytes developed an early onset movement disability, and demonstrated dysfunctional astrocytes. These astrocytes were able to activate microglia and induce neurodegeneration.[145]

A possible mechanism underlying the effect of α-syn on immune activation is the regulation of Nurr1, a transcription factor belonging to the orphan nuclear receptor family. The expression of Nurr1 is reduced in α-syn expressing cells, as demonstrated in mice,[146] and PD patients.[147] Nurr1 is a potential regulator of cytokine and growth factor action through regulating the inflammatory response. Moreover, Nurr1 has a neuroprotective effect because of its inhibition of the production of inflammatory mediators in microglia.[148]

Recent results suggest that passive immunization with monoclonal antibodies against α-syn may be of therapeutic relevance in patients with PD,[149] providing further evidence for the involvement of immune activity in disease progression.[22]

5.2 Leucine-Rich Repeat Kinase 2 (LRRK2)

Leucine-rich repeat kinase 2 (LRRK2) is a large gene composed of 51 exons coding a 2527 amino acid cytoplasmic protein that consists of a leucine-rich repeat toward the amino terminus of the protein and a kinase domain toward the carboxyl terminus. LRRK2 contains many domains capable of protein–protein interactions.[150–152]

To date, there are more than 50 different missense and nonsense mutations reported in LRRK2, and at least 16 are associated with PD. The relatively common G2019S mutation has been most extensively studied.[150,153] LRRK2 can localize to the outer membrane of the mitochondria[154] where it might mediate mitochondrial fragmentation and dysfunction, effects exacerbated in the G2019S mutation that may lead to neurotoxicity.[155] Fibroblasts acquired from a LRRK2 mutated patient (G2019S) show decreased mitochondrial membrane potential and intracellular ATP levels.[156] This was also recently reported with induced pluripotent stem cells (iPSC) from PD patients carrying LRRK2 mutations. Mutant LRRK2 iPSCs exhibited mitochondrial dysfunction and mitochondrial DNA damage.[157,158]

High LRRK2 expression has been noted in macrophages and monocytes,[159] as well as T-cells,[160] leading to the speculation of a functional role for LRRK2 in the immune system.[22] Moehle et al.[161] have shown that microglial activation triggers LRRK2 expression, and that LRRK2 inhibition by either small-molecule kinase inhibitors or shRNA, attenuates microglial proinflammatory response and decreases TNFα and NO levels following LPS activation.[161] This suggests that LRRK2 plays an important role in mediating proinflammatory responses in microglia cells. It was recently demonstrated that LRRK2 acts through regulation of p38 MAPK and NF-jB signaling pathways to stimulate microglial inflammatory responses. p38 MAPK plays a crucial role in regulating oxidative stress-induced cell death and survival.[162] LRRK2 knockdown attenuated the inflammatory response to LPS and reduces cytokine and NO secretion, and LRRK2 overexpression increased the response.[163] Mutated LRRK2 microglia displayed decreased IL-10 production in response to LPS stimuli.[164] Together, these results suggest that LRRK2 gain of function mutations could lead the microglia to a proinflammatory phenotype that changes the microenvironment of brain and triggers and/or enhances PD pathogenesis.

Some patients carrying LRRK2 mutations show neuronal degeneration in the SN, and some of the patients exhibited Lewy body pathology.[165] Similarly, double mutant mice carrying mutated α-syn and mutated LRRK2 showed increased striatal neurodegeneration compared to mice carrying only mutated α-syn,[166] suggesting a possible connection between LRRK2 and α-syn pathologies.

5.3 Parkin

Parkin is a 465 amino acid protein with a molecular mass of 52 kDa, that is highly expressed in the heart, testis, brain, and skeletal muscle.[167] Parkin is an E3 ubiquitin ligase,[168] and mutations in this protein result in the loss of E3 activity.[169]

Parkin plays a role in mitochondrial activity and integrity,[170] and in mammalian cells, parkin can be recruited to dysfunctional mitochondria, where it mediates the engulfment and degradation of deficient mitochondria via macroautophagy, a process known as mitophagy.[171] Parkin-null Drosophila display disordered mitochondria,[172] and parkin$^{-/-}$ mice exhibit decreased mitochondrial respiratory capacity and increased oxidative stress.[173] Mitochondrial quality control is important for vital energy production processes, and has been linked to apoptosis-related processes such as the release of cytochrome c,[174] suggesting that impairments in mitophagy could have detrimental effects on cells. Parkin$^{-/-}$ mice, however, do not show significant loss of DA neurons in the SNc.[175] Moreover, Parkin$^{-/-}$ mice bearing mutated α-syn do not display a more severe phenotype compared to mice carrying only mutated α-syn.[176] These findings suggest that parkin-mediated toxicity is not necessarily synuclein-dependent, in contrast to LRRK2, and may require additional factors to induce marked neurodegeneration.[22]

Parkin is abundantly expressed in microglia, and parkin null mice have an increased number of microglia.[177] Additionally, parkin loss of function in microglia results in enhanced toxicity to DA neurons after rotenone treatment.[177] A possible mediator for the effect of parkin on inflammation is the TNF-α receptor-associated factor (TRAF) 2/6 signaling pathway. TRAF2 and TRAF6 are essential mediators of cytokine signaling by regulating c-Jun N-terminal kinase (JNK) and nuclear factor-jB (NF-jB) signaling. Parkin expression is inversely related to TRAF2/6 expression, and it promotes the proteasomal degradation of TRAF2/6 and protects inflammatory signaling in response to cytokine activation. These results suggest that TRAF2/6 may be a physiological target of parkin, and they provide an explanation for why loss of parkin is detected in PD and its relation to neuroinflammation.[178]

Parkin levels are decreased in microglia in response to LPS of TNF-α activation in an NF-jB dependent manner. In addition, activated macrophages from Parkin-null mice expressed increased levels of TNF-α, IL-1β, and iNOS mRNA compared to WT macrophages.[179] The loss of parkin function increases the vulnerability of DA neurons in the SN to inflammation-related degeneration triggered by i.p. injection of low dose LPS.[15] These reports indicate that parkin may act as an antiinflammatory factor, and that reduced parkin levels, or mutant parkin with a loss of function phenotype, may be proinflammatory.

5.4 PTEN-Induced Putative Kinase 1 (PINK1)

PINK1 gene encodes a 581 amino acid protein with an N-terminal mitochondrial targeting motif and a highly conserved kinase domain homologous to the serine/threonine kinases of the calcium–calmodulin family.[180] Mutations in PINK1 gene account for 1–8% of the early onset autosomal recessive PD cases.[181,182] PINK1 is involved in cell respiration,[183] protein folding, and degradation,[184,185] and in several mitochondrial functions, such as fission/fusion dynamics,[186,187] trafficking[188] and calcium signaling.[189] In addition, PINK1 is required for recruitment of parkin to the mitochondria with impaired membrane potential.[190,191]

PINK1 loss of function impairs the mitochondrial respiratory chain and ATP production, and induces neuronal aggregation of α-syn.[192] Conversely, PINK1 overexpression can rescue α-syn-induced neuronal degeneration in a fly model,[193] as well as aSyn-induced mitochondrial fragmentation in a Caenorhabditis elegans model.[194] These reports emphasize the important role of PINK1 in modulation of mitochondrial activity under normal conditions, as well under disease conditions.

PINK1 deficiency results in calcium overload within mitochondria, causing ROS production in the mitochondria and cytosol.[183] Moreover, PINK1 knockdown induced mitophagy, that is, lysosomal degradation of mitochondria.[184] PINK$^{-/-}$ embryonic fibroblasts showed decreased basal and inflammatory cytokine-induced NF-jB activity. In addition, PINK$^{-/-}$ mice had increased levels of IL-1β, IL 12, and IL-10 in the striatum after peripheral stimulation with LPS.[195] Kim and coworkers have shown that in organotypic cultures, PINK1$^{-/-}$ slices expressed higher mRNA levels of TNF-α, IL1b, and IL-6 compared to those of WT slices.[196] These results indicate that PINK1 deficiency increases the production of proinflammatory cytokines, leading to a neurotoxic effect. Another possible mechanism for this effect is that PINK1 directly interacts with two members of the IL-1β-mediated downstream signaling pathway, TRAF6 and transforming growth factor b activated kinase-1 (TAK1), and positively regulates their activation leading to enhanced IL-1β-mediated cytokine production.[197] These reports demonstrate that PINK1 plays a role in the proinflammatory response in PD, possibly through interactions with the mitochondria.

5.5 DJ-1

DJ-1 encodes a small 189 amino acid protein that is ubiquitously expressed and highly conserved throughout diverse species.[198,199] DJ-1 is present in the cytoplasm, nucleus,[199,200] and mitochondria[201] of the cells, and it is widely thought that oxidation promotes the mitochondrial localization of DJ-1.[202] The loss of functional DJ-1 causes 1–2% of autosomal recessive early onset PD cases.[67,203] It has been suggested that DJ-1 plays a role in maintenance of mitochondrial complex-I activity[204] as well as in maintenance of mitochondrial morphology through fission/fusion dynamics,[205] and that loss of DJ-1 function impairs nigrostriatal dopaminergic function.[206–208]

Similarly to PINK1, DJ-1 overexpression can also rescue α-syn-induced mitochondrial fragmentation in a C. elegans model.[194] Neuronal cells that carry mutant forms of DJ-1, become more susceptible to death in parallel with the loss of oxidized forms of DJ-1,[209,210] and are more sensitive to toxins.[211] Moreover, DJ-1$^{-/-}$ mice are more vulnerable to several toxicity models of PD, such as MPTP treatment,[208] 6-hydroxydopamine (6-OHDA), and rotenone toxicity without aSyn pathology.[211,212] In each of the aforementioned mouse models, DJ-1$^{-/-}$ induced a greater loss of dopaminergic neurons.

ROS production induced by bisphenol A (BPA) compromises mitochondrial function and elevates the expression and oxidization of DJ-1. In addition, DJ-1 maintains the complex-I activity against BPA-induced oxidative stress when it is localized in mitochondria. DJ-1 also plays a role in the prevention of mitochondrial injury-induced cell death.[213]

DJ-1 is important for the proinflammatory response in astrocytes, and that DJ-1$^{-/-}$ astrocytes have neurotoxic properties, with increased NO production and enhanced induction of COX2 and IL-6.[214] The knockout of DJ attenuates astrocytes neuroprotection against 6-OHDA toxicity,[212] suggesting a protective role for astrocytes in the 6-OHDA model for PD, which depends on DJ-1 activation. Stimulation of macrophages with LPS induces an elevation in the DJ-1 protein levels.[215] DJ-1-deficient microglia have increased monoamine oxidase activity, which results in elevation of intracellular ROS, NO, and proinflammatory cytokines, leading to increased dopaminergic neurotoxicity. This phenotype is ameliorated by rasagaline, a MAO inhibitor approved for treatment in PD patients.[216] Together, these studies are consistent with an important role of DJ-1 in immune activation in PD.[22]

5.6 Others

Genetic studies have identified associations between polymorphism in immune-related genes and PD[38], particularly in the HLA genes.[217-220] Two recent meta-analyses of genome-wide association studies have confirmed an association between HLA-DRB5[221] and HLA-DRB1[220] with PD.

Other inflammation-related gene polymorphisms have been associated with PD risk. A single nucleotide polymorphism in the promoter region of the TNF-α gene (-G308A, -T1031C) increases risk of PD[222-224] by leading to overexpression of TNF-α.[225] In contrast, the frequency of the B2 haplotype of the TNF-α receptor 1 polymorphism at positions -609 and $+36$ was decreased in PD patients.[226] The CC genotype of T1030C and C857T of the TNF-α promoter has also been found to increase PD risk.[227] In addition, a significant association between some single-nucleotide polymorphisms (SNPs) in the TNF-α promoter and PD has been reported.[228]

Although the endogenous levels of IFN-γ in the healthy human brain are virtually undetectable,[229] allelic differences between early- and late-onset PD patients were reported for the IFN-γ gene I different populations,[230-232] a provocative finding, as it may influence infiltration of T-cells during progression of the disease.

Additional polymorphisms, including CD14 monocyte receptor,[233] IL-1α,[234] IL-1β,[234,235] IL-6,[236] IL-8,[237], and IL-18[238] are associated with PD risk. McGeer and coworkers[239] identified a significant increase in the IL-1β T genotype at position -511, but not of the IL-1α T genotype at position -889, in PD patients. The IL-1β polymorphism β*1/*1 and *1/*2 genotype was significantly more common in patients than healthy individuals,[235] although no effect was associated with polymorphisms in the genes that encode for IL-1α and IL-1 receptor antagonists. Other investigators found no association between polymorphisms of IL-1α position -889 and risk of PD.[234,240] Wahner and coworkers[224] found that the risk of PD doubled in carriers of the homozygous variant of IL-1β position -511 and TNF-α

position −308 and that the risk of developing PD for carriers of both was increased by threefold. Some studies found that specific IL-1β promoter polymorphisms lowered the risk of PD[223,241] although a recent meta-analysis did not confirm this association.[228] IL-6 polymorphisms might also increase risk of PD as the frequency of a GG polymorphism at position −174 was significantly increased in PD patients.[236] The IL-10 promoter polymorphism −819 has been associated with higher risk for early onset PD but not for sporadic PD[242] and the G1082A SNP was correlated with age of disease onset,[236] although other studies showed no association between polymorphisms −1082 or − 592 with PD.[222,228,243] An allele at position −260 of the promoter of the CD14 monocyte receptor is increased in women but not men.[233] Yet other alleles of genes associated with neuroinflammation, including allele 122 of IFN-γ, may affect the onset of PD.[230]

Neuroinflammatory responses are further impacted by microglial gene products.[1] Mutations resulting in reduced expression of Nurr1, a member of the receptor (NR)4 family of orphan nuclear receptors, are associated with late-onset familial PD.[244] Nurr1 (NR4A2) functions as a constitutively active transcription factor[245,246] that plays essential roles in microglia and astrocytes as a repressor of genes that encode proinflammatory neurotoxic factors.[148] Nurr1 can recruit CoREST corepressor complexes to nuclear NF-κB target genes and mediate the turnover of NF-κB and restore activated gene expression to a basal state. Loss of Nurr1 function in microglia and astrocytes of the SN results in exaggerated and prolonged inflammatory responses that accelerate the loss of DA neurons in response to LPS or over-expression of a mutant α-syn (A30P) associated with familial PD, suggesting that Nurr1 plays a protective role via modulating extraneuronal cells.[1]

6 POTENTIAL ANTIINFLAMMATORY TREATMENTS FOR PD

6.1 Antiinflammatory Drugs

Animal studies initially suggested that nonsteroidal antiinflammatory drugs (NSAIDs) are neuroprotective. NSAIDs scavenge free oxygen radicals and inhibit prostaglandin production by COX-1 and COX-2. For example, inhibition of COX-2 decreased microglial activation and prevented progressive degeneration in a retrograde lesion induced by 6-OHDA.[247]

A variety of initial studies indicated that NSAIDs such as aspirin or ibuprofen may reduce PD risk,[248,249] although subsequent studies found these compounds to provide limited or no benefit.[250,251] A systematic review and meta-analysis of studies published between 1966 and 2008 concluded that NSAIDs do not modify risk of PD.[252]

Statins, drugs commonly used to reduce cholesterol levels that also have antiinflammatory effects, have recently been reported to reduce the risk of PD,[253] although there are multiple mechanisms by which this could occur.

6.2 Regulation of Glial Associated Immunity

Strategies aimed at suppressing the activation of glial cells and their inflammatory properties by various drugs have been extensively tested in animal models.[254] Among these compounds, agonists of the receptor PPARγ seem to be promising candidates for therapeutic use in PD.[255] PPARγ agonists possess mechanisms of action beyond immunoregulatory properties, including regulation of mitochondrial bioenergetics, insulin signaling, glucose metabolism, and lactate production.[255] Preclinical studies have shown that pioglitazone, a PPARγ agonist that crosses the blood–brain barrier, can partly prevent DA cell loss induced by MPTP in mice[256] although this could be due to inhibition of monoamine oxidase B+.[257] Pioglitazone was also protective in a model of inflammation-induced DA neurodegeneration by intrastriatal LPS in rats.[258]

Additional strategies aimed at preventing microglial cell activation have been suggested. For example, microglial cell activation can be inhibited through activation of cell-surface α7 nicotinic acetylcholine receptors.[259,260] This could be related to the observation that cigarette smoking is associated with a lower incidence of PD[261,262]: cigarette smoke and nicotine itself have protective effects on MPTP-induced nigrostriatal pathway injury in mice and monkeys.[263–267] Although the neuroprotective effect of nicotine is not limited to antiinflammatory mechanisms,[267] these studies suggest that α7 receptor agonists, such as quinuclidines[268] or PHA-709829,[269] might target microglia-associated immunity in PD.

Another potentially useful molecular target to prevent microglial cell reaction is the purinergic $P2X_7$ ionotropic receptor, which mediates the mobilization of intracellular calcium ions.[270] This receptor plays a crucial role in the inflammatory response of microglial cells by modulating their LPS-mediated activation. The antagonist of this receptor was found to be neuroprotective in an animal model of Alzheimers disease.[271]

6.3 Immunotherapy

Neuroinflammation can be suppressed by immune-based therapeutic approaches that modify innate and adaptive inflammation.[114,272] As mentioned previously, T_{regs} can regulate T_{eff} activities by secreting immunosuppressive cytokines, such as IL-10 and TGF-β, and by directly killing effector T cells.[273] T_{reg}-induced attenuation of microglial inflammatory responses may protect against nigrostriatal DA neurodegeneration[272] in part by supplying neurotrophic factors. A study reports that glatiramer acetate, a synthetic 4-amino acid-length peptide used in MS patients, attenuated microglial activation via induction of CD4$^+$ T cells in the MPTP mouse model.[274] As a result of immunization with glatiramer acetate-specific T_{regs}, local GDNF expression was elevated in an MPTP-induced PD model.[275]

The implication of HLA-DR induction[12] and lymphocyte infiltration in PD,[126,127] and presence in serum of α-syn-specific antibodies[276,277] have promulgated exploration of vaccination protocols to treat PD. Active immunization of mice with α-syn produced specific high-affinity antibodies[278] that decreased extracellular α-syn accumulation and the rate of neuronal death. Antibodies produced after the introduction of α-syn were shown to degrade α-syn aggregates via lysosomal pathways.[279] In a recent study, prior to stereotactic delivery of a viral vector that overexpressed human α-syn in the rat SN, animals were vaccinated with recombinant α-syn. The vaccination protocol yielded a high-titer anti-α-syn response, an increased accumulation of CD4$^+$ and MHCII$^+$ microglia in the SN, increased infiltration of CD4$^+$, Foxp3$^+$ cells throughout the nigrostriatal system, and fewer pathologic aggregates in the striatum.[36] Administration of monoclonal α-syn antibody 9E4, which targets the C-terminal region of α-syn, leads to the recovery of deficits induced by α-syn accumulation.[149]

A vaccine developed to treat patients with synucleinopathies[280] targets PD-causing α-syn aggregates by inducing production of antibodies against α-syn and is presently in clinical trial.[279]

The antibiotic minocycline is an immunoregulatory drug that has been extensively tested in preclinical studies. This tetracycline analog is well known for its antimicrobial activity by protein synthesis inhibition, but has additional properties including antiinflammatory and antiapoptotic actions[281] including an inhibition of microglia and T-cell activation.[282] Minocycline provides neuroprotection in experimental models of amyotrophic lateral sclerosis, Alzheimers disease, Huntingtons disease, and PD.[281] Although initial studies gave encouraging results in models of PD,[283–285] recent investigations have raised concerns, as depending on the dose and the route of administration, minocycline in rodent and primate PD models can exacerbate neurodegeneration.[286–288] Despite the uncertainty, a randomized, double-blind, futility, 12-month, phase II trial of minocycline and creatine in patients newly diagnosed with PD has been conducted. The trial showed that neither drug should be considered futile,[289] encouraging future trials.

Immunomodulatory effects have been described in animal models and in vitro for two anti-parkinsonian drugs commonly used in humans: the monoamine oxidase B inhibitor drug pargyline[290] and selegiline (Deprenyl).[291]

The immunosuppressants cyclosporine A and FK-506 (tacrolimus) and nonimmunosuppressant derivatives of FK-506 referred to as immunophilin ligands such as pentoxifylline[292,293] have shown neuroprotective activity in MPTP-induced nigral injury[293] and the 6-OHDA rat model of

PD.[247] Clinical trials in PD patients using short-term administration of immunophilin ligands (which lack the immunosuppressive properties of the parent compounds) had limited success.[294]

The glucocorticoid dexamethasone was protective against MPTP[295] and intranigral LPS.[296] Other antiinflammatory regimens such as steroids, which are effective at arresting DA neuron loss in rodents are not likely to be suitable for long-term use in humans.

Activation of NF-κB is induced in vivo in the SNc of MPTP-intoxicated mice and within the SNc of PD patients. A cell-permeable peptide corresponding to the NF-κB essential modifier (NEMO)-binding domain of IκB kinase α or IκB kinase β inhibited NF-κB activation and suppressed nigral microglial activation, which protected both the nigrostriatal axis and improved motor function in MPTP-intoxicated mice.[297] Thus, selective inhibition of NF-κB activation by (NEMO)-binding domain peptide may be of therapeutic benefit for PD patients.

Other compounds with antiinflammatory actions that might rescue nigral DA neurons from a variety of neurotoxic insults include vasoactive intestinal peptide,[298,299] the polyphenolic flavonoid silymarin,[300] and the NMDA receptor antagonist dextromethorphan.[301] The selective iNOS inhibitors S-methylisothiourea and 1−N(G)-nitroarginine were shown to exert neuroprotecive effects on DA neurons in rats against LPS,[302,303] suggesting that free radical scavengers or iNOS inhibitors may have potential therapeutic effects in PD. Copolymer-1 immunization, which has been used effectively in patients with chronic neuroinflammatory disease such as relapsing–remitting MS, has recently been shown to exert neuroprotective effects in the nigrostriatal pathway against MPTP by immunomodulatory mechanisms, including promotion of CD4+ T cell accumulation within the SNc, suppression of microglial activation, and increased local expression of GDNF.[274,275]

At this time, antiinflammatory regimens have only been explored in patients with late-stage disease with disappointing results. There is a need to identify PD at an earlier stage to examine the effects of immune-based therapies.[304] A second challenge is to design more specific immune-based therapies. As neuroinflammation is common to many neurodegenerative diseases, targeting neuroinflammatory mediators that drive death pathways in particular cells may provide an approach to treat multiple diseases.[148,305,306] Alternatively, vaccination may reduce the buildup of α-syn or other proteins underlying pathogenesis in the CNS.

7 SUMMARY AND CONCLUSIONS

While much evidence indicates that inflammation influences PD progression,[38,307] it remains unclear whether it triggers and regulates PD neurodegeneration or is a secondary consequence of the disease. The evidence that inhibition of the inflammatory processes are neuroprotective suggests that inhibition of inflammation may provide therapeutic intervention for PD,[308] but clinical results to date are disappointing and improved approaches must be developed if they are to help patients.

Acknowledgments

For our work on neuroinflammation, we are grateful for support from the Parkinson's Disease, JPB, National Parkinsons, and Michael J Fox Foundations. We are grateful to Luigi Zecca, John Loike, and Alessandro Sette for advice and discussion on this topic.

References

1. Lee JK, Tran T, Tansey MG. Neuroinflammation in Parkinson's disease. *J Neuroimmune Pharmacol.* 2009;4(4):419–429.
2. Jankovic J. Parkinson's disease: clinical features and diagnosis. *J Neurol Neurosurg Psychiatry.* 2008;79(4):368–376.
3. Shulman LM, Taback RL, Rabinstein AA, Weiner WJ. Non-recognition of depression and other non-motor symptoms in Parkinson's disease. *Parkinsonism Relat Disord.* 2002;8(3):193–197.
4. Braak H, Del Tredici K. Pathophysiology of sporadic Parkinson's disease. *Fortsch Neurol-Psychiatr.* 2010;78(suppl 1):S2–4.
5. Del Tredici K, Braak H. Lewy pathology and neurodegeneration in premotor Parkinson's disease. *Mov. Disord.* 2012;27(5):597–607.
6. Sekiyama K, Sugama S, Fujita M, et al. Neuroinflammation in Parkinson's disease and related disorders: a lesson from genetically manipulated mouse models of alpha-synucleinopathies. *Parkinson's Dis.* 2012;2012:271732.
7. Foix C, Nicolesco J. Anatomie cérébrale. Les noyaux gris centraux et la région Mésencéphalo-sous-optique. In: Cie Me, ed. *Suivi d'un apéndice sur l'anatomie pathologique de la maladie de Parkinson.* Paris: Masson et Cie; 1925:508–538.
8. Franzen B, Duvefelt K, Jonsson C, et al. Gene and protein expression profiling of human cerebral endothelial cells activated with tumor necrosis factor-alpha. *Brain Res Mol Brain Res.* 2003;115(2):130–146.
9. Ransohoff RM, Perry VH. Microglial physiology: unique stimuli, specialized responses. *Annu Rev Immunol.* 2009;27:119–145.
10. Rezai-Zadeh K, Gate D, Town T. CNS infiltration of peripheral immune cells: D-Day for neurodegenerative disease?. *J Neuroimmune Pharmacol.* 2009;4(4):462–475.
11. Engelhardt B, Coisne C. Fluids and barriers of the CNS establish immune privilege by confining immune surveillance to a two-walled castle moat surrounding the CNS castle. *Fluids Barriers CNS.* 2011;8(1):4.
12. McGeer PL, Itagaki S, Boyes BE, McGeer EG. Reactive microglia are positive for HLA-DR in the substantia nigra of Parkinson's and Alzheimer's disease brains. *Neurology.* 1988;38(8):1285–1291.
13. McGeer PL, McGeer EG. Glial cell reactions in neurodegenerative diseases: pathophysiology and therapeutic interventions. *Alzheimer Dis Assoc. Disord.* 1998;12(suppl 2):S1–6.
14. Frank-Cannon TC, Alto LT, McAlpine FE, Tansey MG. Does neuroinflammation fan the flame in neurodegenerative diseases?. *Mol. Neurodegener.* 2009;4:47.
15. Frank-Cannon TC, Tran T, Ruhn KA, et al. Parkin deficiency increases vulnerability to inflammation-related nigral degeneration. *J Neurosci.* 2008;28(43):10825–10834.
16. Hirsch EC, Hunot S. Neuroinflammation in Parkinson's disease: a target for neuroprotection?. *The Lancet Neurol.* 2009;8(4):382–397.

17. Tansey MG, Goldberg MS. Neuroinflammation in Parkinson's disease: its role in neuronal death and implications for therapeutic intervention. *Neurobiol Dis*. 2010;37(3):510–518.
18. Kortekaas R, Leenders KL, van Oostrom JC, et al. Blood-brain barrier dysfunction in parkinsonian midbrain in vivo. *Ann Neurol*. 2005;57(2):176–179.
19. Collins LM, Toulouse A, Connor TJ, Nolan YM. Contributions of central and systemic inflammation to the pathophysiology of Parkinson's disease. *Neuropharmacology*. 2012;62(7):2154–2168.
20. Hoban DB, Connaughton E, Connaughton C, et al. Further characterisation of the LPS model of Parkinson's disease: a comparison of intra-nigral and intra-striatal lipopolysaccharide administration on motor function, microgliosis and nigrostriatal neurodegeneration in the rat. *Brain Behav Immun*. 2013;27(1):91–100.
21. Morale MC, Serra PA, L'Episcopo F, et al. Estrogen, neuroinflammation and neuroprotection in Parkinson's disease: glia dictates resistance versus vulnerability to neurodegeneration. *Neuroscience*. 2006;138(3):869–878.
22. Trudler D, Nash Y, Frenkel D. New insights on Parkinson's disease genes: the link between mitochondria impairment and neuroinflammation. *J Neural Transm*. 2015;122(10):1409–1419.
23. Colton CA. Heterogeneity of microglial activation in the innate immune response in the brain. *J Neuroimmune Pharmacol*. 2009;4(4):399–418.
24. Ousman SS, Kubes P. Immune surveillance in the central nervous system. *Nat Neurosci*. 2012;15(8):1096–1101.
25. Banati RB, Daniel SE, Blunt SB. Glial pathology but absence of apoptotic nigral neurons in long-standing Parkinson's disease. *Mov Disord*. 1998;13(2):221–227.
26. Nakamura Y. Regulating factors for microglial activation. *Biol Pharmaceut Bull*. 2002;25(8):945–953.
27. Orr CF, Rowe DB, Halliday GM. An inflammatory review of Parkinson's disease. *Prog Neurobiol*. 2002;68(5):325–340.
28. McGeer PL, McGeer EG. Inflammation and neurodegeneration in Parkinson's disease. *Parkinsonism Relat Disord*. 2004;10(suppl 1):S3–7.
29. Tufekci KU, Genc S, Genc K. The endotoxin-induced neuroinflammation model of Parkinson's disease. *Parkinson's Dis*. 2011;2011:487450.
30. Phani S, Loike JD, Przedborski S. Neurodegeneration and inflammation in Parkinson's disease. *Parkinsonism Relat Disord*. 2012;18(suppl 1):S207–209.
31. McGeer EG, Klegeris A, McGeer PL. Inflammation, the complement system and the diseases of aging. *Neurobiol Aging*. 2005;26(suppl 1):94–97.
32. Ouchi Y, Yoshikawa E, Sekine Y, et al. Microglial activation and dopamine terminal loss in early Parkinson's disease. *Ann Neurol*. 2005;57(2):168–175.
33. Gerhard A, Pavese N, Hotton G, et al. In vivo imaging of microglial activation with [11C] (R)-PK11195 PET in idiopathic Parkinson's disease. *Neurobiol Dis*. 2006;21(2):404–412.
34. Imamura K, Hishikawa N, Sawada M, Nagatsu T, Yoshida M, Hashizume Y. Distribution of major histocompatibility complex class II-positive microglia and cytokine profile of Parkinson's disease brains. *Acta Neuropathol*. 2003;106(6):518–526.
35. Croisier E, Moran LB, Dexter DT, Pearce RK, Graeber MB. Microglial inflammation in the parkinsonian substantia nigra: relationship to alpha-synuclein deposition. *J Neuroinflamm*. 2005;2:14.
36. Sanchez-Guajardo V, Barnum CJ, Tansey MG, Romero-Ramos M. Neuroimmunological processes in Parkinson's disease and their relation to alpha-synuclein: microglia as the referee between neuronal processes and peripheral immunity. *ASN Neuro*. 2013;5(2):113–139.
37. Sanchez-Guajardo V, Febbraro F, Kirik D, Romero-Ramos M. Microglia acquire distinct activation profiles depending on the degree of alpha-synuclein neuropathology in a rAAV based model of Parkinson's disease. *PLoS One*. 2010;5(1):e8784.
38. Tufekci KU, Meuwissen R, Genc S, Genc K. Inflammation in Parkinson's disease. *Adv Prot Chem Struct Biol*. 2012;88:69–132.

39. Lawson LJ, Perry VH, Dri P, Gordon S. Heterogeneity in the distribution and morphology of microglia in the normal adult mouse brain. *Neuroscience*. 1990;39(1):151–170.

40. Mittelbronn M, Dietz K, Schluesener HJ, Meyermann R. Local distribution of microglia in the normal adult human central nervous system differs by up to one order of magnitude. *Acta Neuropathol*. 2001;101(3):249–255.

41. Tansey MG, McCoy MK, Frank-Cannon TC. Neuroinflammatory mechanisms in Parkinson's disease: potential environmental triggers, pathways, and targets for early therapeutic intervention. *Exp Neurol*. 2007;208(1):1–25.

42. Sulzer D. Multiple hit hypotheses for dopamine neuron loss in Parkinson's disease. *Trends Neurosci*. 2007;30(5):244–250.

43. Klegeris A, Pelech S, Giasson BI, et al. Alpha-synuclein activates stress signaling protein kinases in THP-1 cells and microglia. *Neurobiol Aging*. 2008;29(5):739–752.

44. Zhang W, Wang T, Pei Z, et al. Aggregated alpha-synuclein activates microglia: a process leading to disease progression in Parkinson's disease. *FASEB J*. 2005;19(6):533–542.

45. Alvarez-Erviti L, Couch Y, Richardson J, Cooper JM, Wood MJ. Alpha-synuclein release by neurons activates the inflammatory response in a microglial cell line. *Neurosci Res*. 2011;69(4):337–342.

46. Beraud D, Hathaway HA, Trecki J, et al. Microglial activation and antioxidant responses induced by the Parkinson's disease protein alpha-synuclein. *J Neuroimmune Pharmacol*. 2013;8(1):94–117.

47. Cebrian C, Zucca FA, Mauri P, et al. MHC-I expression renders catecholaminergic neurons susceptible to T-cell-mediated degeneration. *Nat Commun*. 2014;5:3633.

48. Zhang W, Dallas S, Zhang D, Microglial PHOX. et al. and Mac-1 are essential to the enhanced dopaminergic neurodegeneration elicited by A30P and A53T mutant alpha-synuclein. *Glia*. 2007;55(11):1178–1188.

49. Stefanova N, Schanda K, Klimaschewski L, Poewe W, Wenning GK, Reindl M. Tumor necrosis factor-alpha-induced cell death in U373 cells overexpressing alpha-synuclein. *J Neurosci Res*. 2003;73(3):334–340.

50. Volpicelli-Daley LA, Luk KC, Patel TP, et al. Exogenous alpha-synuclein fibrils induce Lewy body pathology leading to synaptic dysfunction and neuron death. *Neuron*. 2011; 72(1):57–71.

51. Luk KC, Kehm V, Carroll J, et al. Pathological alpha-synuclein transmission initiates Parkinson-like neurodegeneration in nontransgenic mice. *Science*. 2012;338(6109):949–953.

52. Luk KC, Kehm VM, Zhang B, O'Brien P, Trojanowski JQ, Lee VM. Intracerebral inoculation of pathological alpha-synuclein initiates a rapidly progressive neurodegenerative alpha-synucleinopathy in mice. *J Exp Med*. 2012;209(5):975–986.

53. Lema Tome CM, Tyson T, Rey NL, Grathwohl S, Britschgi M, Brundin P. Inflammation and alpha-synuclein's prion-like behavior in Parkinson's disease—is there a link?. *Mol Neurobiol*. 2013;47(2):561–574.

54. Zhang W, Phillips K, Wielgus AR, et al. Neuromelanin activates microglia and induces degeneration of dopaminergic neurons: implications for progression of Parkinson's disease. *Neurotox Res*. 2011;19(1):63–72.

55. Zecca L, Wilms H, Geick S, et al. Human neuromelanin induces neuroinflammation and neurodegeneration in the rat substantia nigra: implications for Parkinson's disease. *Acta Neuropathol*. 2008;116(1):47–55.

56. Hall GL, Girdlestone J, Compston DA, Wing MG. Recall antigen presentation by gamma-interferon-activated microglia results in T cell activation and propagation of the immune response. *J Neuroimmunol*. 1999;98(2):105–111.

57. Sofroniew MV, Vinters HV, Astrocytes:. Biology and pathology. *Acta Neuropathol*. 2010;119(1):7–35.

58. Maragakis NJ, Rothstein JD. Mechanisms of disease: astrocytes in neurodegenerative disease. *Nat Clin Pract Neurol*. 2006;2(12):679–689.

59. Chung YC, Ko HW, Bok E, et al. The role of neuroinflammation on the pathogenesis of Parkinson's disease. *BMB Rep*. 2010;43(4):225–232.

60. Hamby ME, Sofroniew MV. Reactive astrocytes as therapeutic targets for CNS disorders. *Neurotherapeutics*. 2010;7(4):494–506.
61. Allaman I, Belanger M, Magistretti PJ. Astrocyte-neuron metabolic relationships: for better and for worse. *Trends Neurosci*. 2011;34(2):76–87.
62. Ricci G, Volpi L, Pasquali L, Petrozzi L, Siciliano G. Astrocyte-neuron interactions in neurological disorders. *J Biol Phys*. 2009;35(4):317–336.
63. Sidoryk-Wegrzynowicz M, Wegrzynowicz M, Lee E, Bowman AB, Aschner M. Role of astrocytes in brain function and disease. *Toxicol Pathol*. 2011;39(1):115–123.
64. Phatnani H, Maniatis T. Astrocytes in neurodegenerative disease. *Cold Spring Harbor Perspect Biol*. 2015;7(6).
65. Song YJ, Halliday GM, Holton JL, et al. Degeneration in different parkinsonian syndromes relates to astrocyte type and astrocyte protein expression. *J Neuropathol Exp Neurol*. 2009;68(10):1073–1083.
66. Lee MC, Ting KK, Adams S, Brew BJ, Chung R, Guillemin GJ. Characterisation of the expression of NMDA receptors in human astrocytes. *PLoS One*. 2010;5(11):e14123.
67. Bonifati V, Rizzu P, van Baren MJ, et al. Mutations in the DJ-1 gene associated with autosomal recessive early-onset parkinsonism. *Science*. 2003;299(5604):256–259.
68. Mullett SJ, Hinkle DA. DJ-1 knock-down in astrocytes impairs astrocyte-mediated neuroprotection against rotenone. *Neurobiol Dis*. 2009;33(1):28–36.
69. Hashida K, Kitao Y, Sudo H, et al. ATF6alpha promotes astroglial activation and neuronal survival in a chronic mouse model of Parkinson's disease. *PLoS One*. 2012;7(10):e47950.
70. Mena MA, Garcia de Yebenes J. Glial cells as players in parkinsonism: the "good," the "bad," and the "mysterious" glia. *The Neuroscientist*. 2008;14(6):544–560.
71. Yamada M, Jung M, Tetsushi K, Ivanova A, Nave KA, Ikenaka K. Mutant Plp/DM20 cannot be processed to secrete PLP-related oligodendrocyte differentiation/survival factor. *Neurochem Res*. 2001;26(6):639–645.
72. Canals S, Casarejos MJ, Rodriguez-Martin E, de Bernardo S, Mena MA. Neurotrophic and neurotoxic effects of nitric oxide on fetal midbrain cultures. *J Neurochem*. 2001;76(1):56–68.
73. Canals S, Casarejos MJ, de Bernardo S, Solano RM, Mena MA. Selective and persistent activation of extracellular signal-regulated protein kinase by nitric oxide in glial cells induces neuronal degeneration in glutathione-depleted midbrain cultures. *Mol Cell Neurosci*. 2003;24(4):1012–1026.
74. Sospedra M, Martin R. Immunology of multiple sclerosis. *Annu Rev Immun*. 2005;23:683–747.
75. Hosokawa M, Klegeris A, Maguire J, McGeer PL. Expression of complement messenger RNAs and proteins by human oligodendroglial cells. *Glia*. 2003;42(4):417–423.
76. Kisos H, Pukass K, Ben-Hur T, Richter-Landsberg C, Sharon R. Increased neuronal alpha-synuclein pathology associates with its accumulation in oligodendrocytes in mice modeling alpha-synucleinopathies. *PLoS One*. 2012;7(10):e46817.
77. Stefanova N, Kaufmann WA, Humpel C, Poewe W, Wenning GK. Systemic proteasome inhibition triggers neurodegeneration in a transgenic mouse model expressing human alpha-synuclein under oligodendrocyte promoter: implications for multiple system atrophy. *Acta Neuropathol*. 2012;124(1):51–65.
78. Witt A, Brady ST. Unwrapping new layers of complexity in axon/glial relationships. *Glia*. 2000;29(2):112–117.
79. Lamberts JT, Hildebrandt EN, Brundin P. Spreading of alpha-synuclein in the face of axonal transport deficits in Parkinson's disease: a speculative synthesis. *Neurobiol Dis*. 2015;77:276–283.
80. Lee Y, Morrison BM, Li Y, et al. Oligodendroglia metabolically support axons and contribute to neurodegeneration. *Nature*. 2012;487(7408):443–448.
81. Braak H, Braak E. Development of Alzheimer-related neurofibrillary changes in the neocortex inversely recapitulates cortical myelogenesis. *Acta Neuropathol*. 1996;92(2):197–201.

82. Mogi M, Kondo T, Mizuno Y, Nagatsu T. p53 protein, interferon-gamma, and NF-kappaB levels are elevated in the parkinsonian brain. *Neurosci Lett.* 2007;414(1):94–97.

83. Mogi M, Togari A, Kondo T, et al. Caspase activities and tumor necrosis factor receptor R1 (p55) level are elevated in the substantia nigra from parkinsonian brain. *J Neural Transm.* 2000;107(3):335–341.

84. Hunot S, Dugas N, Faucheux B, et al. FcepsilonRII/CD23 is expressed in Parkinson's disease and induces, in vitro, production of nitric oxide and tumor necrosis factor-alpha in glial cells. *J Neurosci.* 1999;19(9):3440–3447.

85. Boka G, Anglade P, Wallach D, Javoy-Agid F, Agid Y, Hirsch EC. Immunocytochemical analysis of tumor necrosis factor and its receptors in Parkinson's disease. *Neurosci Lett.* 1994;172(1–2):151–154.

86. Mogi M, Harada M, Narabayashi H, Inagaki H, Minami M, Nagatsu T. Interleukin (IL)-1 beta, IL-2, IL-4, IL-6 and transforming growth factor-alpha levels are elevated in ventricular cerebrospinal fluid in juvenile parkinsonism and Parkinson's disease. *Neurosci Lett.* 1996;211(1):13–16.

87. Nagatsu T, Mogi M, Ichinose H, Togari A. Cytokines in Parkinson's disease. *J Neural Transm Suppl.* 2000;(58):143–151.

88. Brodacki B, Staszewski J, Toczylowska B, et al. Serum interleukin (IL-2, IL-10, IL-6, IL-4), TNFalpha, and INFgamma concentrations are elevated in patients with atypical and idiopathic parkinsonism. *Neurosci. Lett.* 2008;441(2):158–162.

89. Koziorowski D, Tomasiuk R, Szlufik S, Friedman A. Inflammatory cytokines and NT-proCNP in Parkinson's disease patients. *Cytokine.* 2012;60(3):762–766.

90. Stypula G, Kunert-Radek J, Stepien H, Zylinska K, Pawlikowski M. Evaluation of interleukins, ACTH, cortisol and prolactin concentrations in the blood of patients with parkinson's disease. *Neuroimmunomodulation.* 1996;3(2–3):131–134.

91. Dobbs RJ, Charlett A, Purkiss AG, Dobbs SM, Weller C, Peterson DW. Association of circulating TNF-alpha and IL-6 with ageing and parkinsonism. *Acta Neurol Scand.* 1999;100(1):34–41.

92. Rentzos M, Nikolaou C, Andreadou E, et al. Circulating interleukin-15 and RANTES chemokine in Parkinson's disease. *Acta Neurol Scand.* 2007;116(6):374–379.

93. Blum-Degen D, Muller T, Kuhn W, Gerlach M, Przuntek H, Riederer P. Interleukin-1 beta and interleukin-6 are elevated in the cerebrospinal fluid of Alzheimer's and de novo Parkinson's disease patients. *Neurosci Lett.* 1995;202(1–2):17–20.

94. Wong GH, Bartlett PF, Clark-Lewis I, Battye F, Schrader JW. Inducible expression of H-2 and Ia antigens on brain cells. *Nature.* 1984;310(5979):688–691.

95. Neumann H, Misgeld T, Matsumuro K, Wekerle H. Neurotrophins inhibit major histocompatibility class II inducibility of microglia: involvement of the p75 neurotrophin receptor. *Proc Natl Acad Sci USA.* 1998;95(10):5779–5784.

96. Medana IM, Gallimore A, Oxenius A, Martinic MM, Wekerle H, Neumann H. MHC class I-restricted killing of neurons by virus-specific CD8+ T lymphocytes is effected through the Fas/FasL, but not the perforin pathway. *Eur J Immunol.* 2000;30(12):3623–3633.

97. Meuth SG, Herrmann AM, Simon OJ, et al. Cytotoxic CD8+ T cell-neuron interactions: perforin-dependent electrical silencing precedes but is not causally linked to neuronal cell death. *J Neurosci.* 2009;29(49):15397–15409.

98. Watkins LR, Milligan ED, Maier SF. Glial activation: a driving force for pathological pain. *Trends Neurosci.* 2001;24(8):450–455.

99. Aarli JA. Role of cytokines in neurological disorders. *Curr Med Chem.* 2003;10(19):1931–1937.

100. Jenner P. Oxidative stress in Parkinson's disease. *Ann Neurol.* 2003;53(suppl 3):S26–36; discussion S36-28.

101. Boje KM, Arora PK. Microglial-produced nitric oxide and reactive nitrogen oxides mediate neuronal cell death. *Brain Res.* 1992;587(2):250–256.

102. Chao CC, Hu S, Molitor TW, Shaskan EG, Peterson PK. Activated microglia mediate neuronal cell injury via a nitric oxide mechanism. *J Immunol.* 1992;149(8):2736–2741.

103. Vane JR, Mitchell JA, Appleton I, et al. Inducible isoforms of cyclooxygenase and nitric-oxide synthase in inflammation. *Proc Natl Acad Sci USA.* 1994;91(6):2046–2050.

104. Knott C, Stern G, Wilkin GP. Inflammatory regulators in Parkinson's disease: iNOS, lipocortin-1, and cyclooxygenases-1 and -2. *Mol Cell Neurosci.* 2000;16(6):724–739.

105. Choi J, Rees HD, Weintraub ST, Levey AI, Chin LS, Li L. Oxidative modifications and aggregation of Cu,Zn-superoxide dismutase associated with Alzheimer and Parkinson diseases. *J Biol Chem.* 2005;280(12):11648–11655.

106. Wu DC, Teismann P, Tieu K, et al. NADPH oxidase mediates oxidative stress in the 1-methyl-4-phenyl-1,2,3,6-tetrahydropyridine model of Parkinson's disease. *Proc Natl Acad Sci USA.* 2003;100(10):6145–6150.

107. Hunot S, Boissiere F, Faucheux B, et al. Nitric oxide synthase and neuronal vulnerability in Parkinson's disease. *Neuroscience.* 1996;72(2):355–363.

108. Surmeier DJ, Schumacker PT. Calcium, bioenergetics, and neuronal vulnerability in Parkinson's disease. *J Biol Chem.* 2013;288(15):10736–10741.

109. Mosharov EV, Larsen KE, Kanter E, et al. Interplay between cytosolic dopamine, cal-cium, and alpha-synuclein causes selective death of substantia nigra neurons. *Neuron.* 2009;62(2):218–229.

110. Mazzulli JR, Armakola M, Dumoulin M, Parastatidis I, Ischiropoulos H. Cellular oligo-merization of alpha-synuclein is determined by the interaction of oxidized catechols with a C-terminal sequence. *J Biol Chem.* 2007;282(43):31621–31630.

111. Martinez-Vicente M, Talloczy Z, Kaushik S, et al. Dopamine-modified alpha-synuclein blocks chaperone-mediated autophagy. *J Clin Investig.* 2008;118(2):777–788.

112. Freitas AA, Rocha B. Peripheral T cell survival. *Curr Opin Immunol.* 1999;11(2):152–156.

113. Kaiko GE, Horvat JC, Beagley KW, Hansbro PM. Immunological decision-making: how does the immune system decide to mount a helper T-cell response?. *Immunology.* 2008;123(3):326–338.

114. Ha D, Stone DK, Mosley RL, Gendelman HE. Immunization strategies for Parkinson's disease. *Parkinsonism Relat Disord.* 2012;18(suppl 1):S218–221.

115. Monahan AJ, Warren M, Carvey PM. Neuroinflammation and peripheral immune infiltration in Parkinson's disease: an autoimmune hypothesis. *Cell Transplant.* 2008;17(4):363–372.

116. Engelhardt B, Ransohoff RM. Capture, crawl, cross: the T cell code to breach the blood-brain barriers. *Trends Immunol.* 2012;33(12):579–589.

117. Coisne C, Lyck R, Engelhardt B. Live cell imaging techniques to study T cell trafficking across the blood-brain barrier in vitro and in vivo. *Fluids Barriers CNS.* 2013;10(1):7.

118. Hickey WF. Leukocyte traffic in the central nervous system: the participants and their roles. *Semin Immunol.* 1999;11(2):125–137.

119. Bas J, Calopa M, Mestre M, et al. Lymphocyte populations in Parkinson's disease and in rat models of parkinsonism. *J Neuroimmunol.* 2001;113(1):146–152.

120. Czlonkowska A, Kurkowska-Jastrzebska I, Czlonkowski A, Peter D, Stefano GB. Im-mune processes in the pathogenesis of Parkinson's disease—a potential role for microg-lia and nitric oxide. *Med Sci Monit.* 2002;8(8):RA165–177.

121. Calopa M, Bas J, Callen A, Mestre M. Apoptosis of peripheral blood lymphocytes in Parkinson patients. *Neurobiol Dis.* 2010;38(1):1–7.

122. Fiszer U, Mix E, Fredrikson S, Kostulas V, Link H. Parkinson's disease and immunolog-ical abnormalities: increase of HLA-DR expression on monocytes in cerebrospinal fluid and of CD45RO+ T cells in peripheral blood. *Acta Neurol Scand.* 1994;90(3):160–166.

123. Hisanaga K, Asagi M, Itoyama Y, Iwasaki Y. Increase in peripheral CD4 bright+ CD8 dull+ T cells in Parkinson disease. *Arch Neurol.* 2001;58(10):1580–1583.

124. Baba Y, Kuroiwa A, Uitti RJ, Wszolek ZK, Yamada T. Alterations of T-lymphocyte popu-lations in Parkinson disease. *Parkinsonism Relat Disord.* 2005;11(8):493–498.

125. Fiszer U, Mix E, Fredrikson S, Kostulas V, Olsson T, Link H. gamma delta+ T cells are increased in patients with Parkinson's disease. *J Neurol Sci.* 1994;121(1):39–45.

126. Itagaki S, McGeer PL, Akiyama H. Presence of T-cytotoxic suppressor and leucocyte common antigen positive cells in Alzheimer's disease brain tissue. *Neurosci Lett.* 1988;91(3):259–264.

127. Brochard V, Combadiere B, Prigent A, et al. Infiltration of CD4+ lymphocytes into the brain contributes to neurodegeneration in a mouse model of Parkinson disease. *J Clin Investig.* 2009;119(1):182–192.

128. Theodore S, Cao S, McLean PJ, Standaert DG. Targeted overexpression of human alpha-synuclein triggers microglial activation and an adaptive immune response in a mouse model of Parkinson disease. *J Neuropathol Exp Neurol.* 2008;67(12):1149–1158.

129. Benner EJ, Banerjee R, Reynolds AD, et al. Nitrated alpha-synuclein immunity accelerates degeneration of nigral dopaminergic neurons. *PLoS One.* 2008;3(1):e1376.

130. Harms AS, Cao S, Rowse AL, et al. MHCII is required for alpha-synuclein-induced activation of microglia, CD4 T cell proliferation, and dopaminergic neurodegeneration. *J Neurosci.* 2013;33(23):9592–9600.

131. Reynolds AD, Stone DK, Hutter JA, Benner EJ, Mosley RL, Gendelman HE, Regulatory T. cells attenuate Th17 cell-mediated nigrostriatal dopaminergic neurodegeneration in a model of Parkinson's disease. *J Immunol.* 2010;184(5):2261–2271.

132. Gonzalez H, Pacheco R. T-cell-mediated regulation of neuroinflammation involved in neurodegenerative diseases. *J Neuroinflamm.* 2014;11:201.

133. Gonzalez H, Contreras F, Prado C, et al. Dopamine receptor D3 expressed on CD4+ T cells favors neurodegeneration of dopaminergic neurons during Parkinson's disease. *J Immunol.* 2013;190(10):5048–5056.

134. Faucheux BA, Bonnet AM, Agid Y, Hirsch EC. Blood vessels change in the mesencephalon of patients with Parkinson's disease. *Lancet.* 1999;353(9157):981–982.

135. Farkas E, De Jong GI, de Vos RA, Jansen Steur EN, Luiten PG. Pathological features of cerebral cortical capillaries are doubled in Alzheimer's disease and Parkinson's disease. *Acta Neuropathol.* 2000;100(4):395–402.

136. Miller DW, Hague SM, Clarimon J, et al. Alpha-synuclein in blood and brain from familial Parkinson disease with SNCA locus triplication. *Neurology.* 2004;62(10):1835–1838.

137. Singleton AB, Farrer M, Johnson J, et al. alpha-Synuclein locus triplication causes Parkinson's disease. *Science.* 2003;302(5646):841.

138. Conway KA, Lee SJ, Rochet JC, et al. Accelerated oligomerization by Parkinson's disease linked alpha-synuclein mutants. *Ann NY Acad Sci.* 2000;920:42–45.

139. Chinta SJ, Mallajosyula JK, Rane A, Andersen JK. Mitochondrial alpha-synuclein accumulation impairs complex I function in dopaminergic neurons and results in increased mitophagy in vivo. *Neurosci Lett.* 2010;486(3):235–239.

140. Watson MB, Richter F, Lee SK, et al. Regionally-specific microglial activation in young mice over-expressing human wildtype alpha-synuclein. *Exp Neurol.* 2012;237(2):318–334.

141. Gao HM, Kotzbauer PT, Uryu K, Leight S, Trojanowski JQ, Lee VM. Neuroinflammation and oxidation/nitration of alpha-synuclein linked to dopaminergic neurodegeneration. *J Neurosci.* 2008;28(30):7687–7698.

142. Gao HM, Zhang F, Zhou H, Kam W, Wilson B, Hong JS. Neuroinflammation and alpha-synuclein dysfunction potentiate each other, driving chronic progression of neurodegeneration in a mouse model of Parkinson's disease. *Environ Health Perspect.* 2011;119(6):807–814.

143. Meulener MC, Graves CL, Sampathu DM, Armstrong-Gold CE, Bonini NM, Giasson BI. DJ-1 is present in a large molecular complex in human brain tissue and interacts with alpha-synuclein. *J Neurochem.* 2005;93(6):1524–1532.

144. Roodveldt C, Labrador-Garrido A, Gonzalez-Rey E, et al. Glial innate immunity generated by non-aggregated alpha-synuclein in mouse: differences between wild-type and Parkinson's disease-linked mutants. *PLoS One.* 2010;5(10):e13481.

145. Gu XL, Long CX, Sun L, Xie C, Lin X, Cai H. Astrocytic expression of Parkinson's disease-related A53T alpha-synuclein causes neurodegeneration in mice. *Mol Brain*. 2010;3:12.

146. Lin X, Parisiadou L, Sgobio C, et al. Conditional expression of Parkinson's disease-related mutant alpha-synuclein in the midbrain dopaminergic neurons causes progressive neurodegeneration and degradation of transcription factor nuclear receptor related 1. *J Neurosci*. 2012;32(27):9248–9264.

147. Chu Y, Le W, Kompoliti K, Jankovic J, Mufson EJ, Kordower JH. Nurr1 in Parkinson's disease and related disorders. *J Compar Neurol*. 2006;494(3):495–514.

148. Saijo K, Winner B, Carson CT, et al. A Nurr1/CoREST pathway in microglia and astrocytes protects dopaminergic neurons from inflammation-induced death. *Cell*. 2009;137(1):47–59.

149. Masliah E, Rockenstein E, Mante M, et al. Passive immunization reduces behavioral and neuropathological deficits in an alpha-synuclein transgenic model of Lewy body disease. *PLoS One*. 2011;6(4):e19338.

150. Klein C, Westenberger A. Genetics of Parkinson's disease. *Cold Spring Harbor Perspect Med*. 2012;2(1):a008888.

151. Berger Z, Smith KA, Lavoie MJ. Membrane localization of LRRK2 is associated with increased formation of the highly active LRRK2 dimer and changes in its phosphorylation. *Biochemistry*. 2010;49(26):5511–5523.

152. Venderova K, Kabbach G, Abdel-Messih E, et al. Leucine-rich repeat kinase 2 interacts with Parkin, DJ-1 and PINK-1 in a Drosophila melanogaster model of Parkinson's disease. *Hum Mol Genet*. 2009;18(22):4390–4404.

153. West AB, Moore DJ, Biskup S, et al. Parkinson's disease-associated mutations in leucine-rich repeat kinase 2 augment kinase activity. *Proc Natl Acad Sci USA*. 2005;102(46):16842–16847.

154. Biskup S, Moore DJ, Celsi F, et al. Localization of LRRK2 to membranous and vesicular structures in mammalian brain. *Ann Neurol*. 2006;60(5):557–569.

155. Wang X, Yan MH, Fujioka H, et al. LRRK2 regulates mitochondrial dynamics and function through direct interaction with DLP1. *Hum Mol Genet*. 2012;21(9):1931–1944.

156. Mortiboys H, Johansen KK, Aasly JO, Bandmann O. Mitochondrial impairment in patients with Parkinson disease with the G2019S mutation in LRRK2. *Neurology*. 2010;75(22):2017–2020.

157. Cooper O, Seo H, Andrabi S, et al. Pharmacological rescue of mitochondrial deficits in iPSC-derived neural cells from patients with familial Parkinson's disease. *Sci Transl Med*. 2012;4(141):141ra190.

158. Sanders LH, Laganiere J, Cooper O, et al. LRRK2 mutations cause mitochondrial DNA damage in iPSC-derived neural cells from Parkinson's disease patients: reversal by gene correction. *Neurobiol Dis*. 2014;62:381–386.

159. Thevenet J, Pescini Gobert R, Hooft van Huijsduijnen R, Wiessner C, Sagot YJ. Regulation of LRRK2 expression points to a functional role in human monocyte maturation. *PLoS One*. 2011;6(6):e21519.

160. Hakimi M, Selvanantham T, Swinton E, et al. Parkinson's disease-linked LRRK2 is expressed in circulating and tissue immune cells and upregulated following recognition of microbial structures. *J Neural Transm*. 2011;118(5):795–808.

161. Moehle MS, Webber PJ, Tse T, et al. LRRK2 inhibition attenuates microglial inflammatory responses. *J Neurosci*. 2012;32(5):1602–1611.

162. Runchel C, Matsuzawa A, Ichijo H. Mitogen-activated protein kinases in mammalian oxidative stress responses. *Antioxid Redox Signal*. 2011;15(1):205–218.

163. Kim B, Yang MS, Choi D, et al. Impaired inflammatory responses in murine Lrrk2-knockdown brain microglia. *PLoS One*. 2012;7(4):e34693.

164. Gillardon F, Schmid R, Draheim H. Parkinson's disease-linked leucine-rich repeat kinase 2(R1441G) mutation increases proinflammatory cytokine release from activated primary microglial cells and resultant neurotoxicity. *Neuroscience*. 2012;208:41–48.

165. Zimprich A, Biskup S, Leitner P, et al. Mutations in LRRK2 cause autosomal-dominant parkinsonism with pleomorphic pathology. *Neuron.* 2004;44(4):601–607.
166. Lin X, Parisiadou L, Gu XL, et al. Leucine-rich repeat kinase 2 regulates the progression of neuropathology induced by Parkinson's-disease-related mutant alpha-synuclein. *Neuron.* 2009;64(6):807–827.
167. Kitada T, Asakawa S, Hattori N, et al. Mutations in the parkin gene cause autosomal recessive juvenile parkinsonism. *Nature.* 1998;392(6676):605–608.
168. Shimura H, Hattori N, Kubo S, et al. Familial Parkinson disease gene product, parkin, is a ubiquitin-protein ligase. *Nat Genet.* 2000;25(3):302–305.
169. Imai Y, Soda M, Takahashi R. Parkin suppresses unfolded protein stress-induced cell death through its E3 ubiquitin-protein ligase activity. *J Biol Chem.* 2000;275(46):35661–35664.
170. Abou-Sleiman PM, Muqit MM, Wood NW. Expanding insights of mitochondrial dysfunction in Parkinson's disease. *Nat Rev.* 2006;7(3):207–219.
171. Narendra D, Tanaka A, Suen DF, Youle RJ. Parkin is recruited selectively to impaired mitochondria and promotes their autophagy. *J Cell Biol.* 2008;183(5):795–803.
172. Greene JC, Whitworth AJ, Kuo I, Andrews LA, Feany MB, Pallanck LJ. Mitochondrial pathology and apoptotic muscle degeneration in Drosophila parkin mutants. *Proc Natl Acad Sci USA.* 2003;100(7):4078–4083.
173. Palacino JJ, Sagi D, Goldberg MS, et al. Mitochondrial dysfunction and oxidative damage in parkin-deficient mice. *J Biol Chem.* 2004;279(18):18614–18622.
174. Youle RJ, van der Bliek AM. Mitochondrial fission, fusion, and stress. *Science.* 2012;337(6098):1062–1065.
175. Goldberg MS, Fleming SM, Palacino JJ, et al. Parkin-deficient mice exhibit nigrostriatal deficits but not loss of dopaminergic neurons. *J Biol Chem.* 2003;278(44):43628–43635.
176. von Coelln R, Thomas B, Andrabi SA, et al. Inclusion body formation and neurodegeneration are parkin independent in a mouse model of alpha-synucleinopathy. *J Neurosci.* 2006;26(14):3685–3696.
177. Casarejos MJ, Menendez J, Solano RM, Rodriguez-Navarro JA, Garcia de Yebenes J, Mena MA. Susceptibility to rotenone is increased in neurons from parkin null mice and is reduced by minocycline. *J Neurochem.* 2006;97(4):934–946.
178. Chung JY, Park HR, Lee SJ, et al. Elevated TRAF2/6 expression in Parkinson's disease is caused by the loss of Parkin E3 ligase activity. *Lab Investig.* 2013;93(6):663–676.
179. Tran TA, Nguyen AD, Chang J, Goldberg MS, Lee JK, Tansey MG. Lipopolysaccharide and tumor necrosis factor regulate Parkin expression via nuclear factor-kappa B. *PLoS One.* 2011;6(8):e23660.
180. Gandhi S, Muqit MM, Stanyer L, et al. PINK1 protein in normal human brain and Parkinson's disease. *Brain.* 2006;129(Pt 7):1720–1731.
181. Klein C, Lohmann-Hedrich K. Impact of recent genetic findings in Parkinson's disease. *Curr Opin Neurol.* 2007;20(4):453–464.
182. Xiromerisiou G, Dardiotis E, Tsimourtou V, et al. Genetic basis of Parkinson disease. *Neurosurg Focus.* 2010;28(1):E7.
183. Gandhi S, Wood-Kaczmar A, Yao Z, et al. PINK1-associated Parkinson's disease is caused by neuronal vulnerability to calcium-induced cell death. *Mol Cell.* 2009;33(5): 627–638.
184. Dagda RK, Cherra SJ, 3rd, Kulich SM, Tandon A, Park D, Chu CT. Loss of PINK1 function promotes mitophagy through effects on oxidative stress and mitochondrial fission. *J Biol Chem.* 2009;284(20):13843–13855.
185. Moriwaki Y, Kim YJ, Ido Y, et al. L347P PINK1 mutant that fails to bind to Hsp90/ Cdc37 chaperones is rapidly degraded in a proteasome-dependent manner. *Neurosci Res.* 2008;61(1):43–48.
186. Deng H, Jankovic J, Guo Y, Xie W, Le W. Small interfering RNA targeting the PINK1 induces apoptosis in dopaminergic cells SH-SY5Y. *Biochem Biophys Res Commun.* 2005;337(4):1133–1138.

187. Poole AC, Thomas RE, Andrews LA, McBride HM, Whitworth AJ, Pallanck LJ. The PINK1/Parkin pathway regulates mitochondrial morphology. *Proc Natl Acad Sci USA.* 2008;105(5):1638–1643.

188. Weihofen A, Thomas KJ, Ostaszewski BL, Cookson MR, Selkoe DJ. Pink1 forms a multiprotein complex with Miro and Milton, linking Pink1 function to mitochondrial trafficking. *Biochemistry.* 2009;48(9):2045–2052.

189. Marongiu R, Spencer B, Crews L, et al. Mutant Pink1 induces mitochondrial dysfunction in a neuronal cell model of Parkinson's disease by disturbing calcium flux. *J Neurochem.* 2009;108(6):1561–1574.

190. Narendra DP, Jin SM, Tanaka A, et al. PINK1 is selectively stabilized on impaired mitochondria to activate Parkin. *PLoS Biol.* 2010;8(1):e1000298.

191. Vives-Bauza C, Zhou C, Huang Y, et al. PINK1-dependent recruitment of Parkin to mitochondria in mitophagy. *Proc Natl Acad Sci USA.* 2010;107(1):378–383.

192. Liu W, Vives-Bauza C, Acin-Perez R, et al. PINK1 defect causes mitochondrial dysfunction, proteasomal deficit and alpha-synuclein aggregation in cell culture models of Parkinson's disease. *PLoS One.* 2009;4(2):e4597.

193. Todd AM, Staveley BE. Pink1 suppresses alpha-synuclein-induced phenotypes in a Drosophila model of Parkinson's disease. *Genome.* 2008;51(12):1040–1046.

194. Kamp F, Exner N, Lutz AK, et al. Inhibition of mitochondrial fusion by alpha-synuclein is rescued by PINK1, Parkin and DJ-1. *EMBO J.* 2010;29(20):3571–3589.

195. Akundi RS, Huang Z, Eason J, et al. Increased mitochondrial calcium sensitivity and abnormal expression of innate immunity genes precede dopaminergic defects in Pink1-deficient mice. *PLoS One.* 2011;6(1):e16038.

196. Kim J, Byun JW, Choi I, et al. PINK1 deficiency enhances inflammatory cytokine release from acutely prepared brain slices. *Exp Neurobiol.* 2013;22(1):38–44.

197. Lee HJ, Jang SH, Kim H, Yoon JH, Chung KC. PINK1 stimulates interleukin-1beta-mediated inflammatory signaling via the positive regulation of TRAF6 and TAK1. *Cell Mol Life Sci.* 2012;69(19):3301–3315.

198. Bandopadhyay R, Kingsbury AE, Cookson MR, et al. The expression of DJ-1 (PARK7) in normal human CNS and idiopathic Parkinson's disease. *Brain.* 2004;127(Pt 2):420–430.

199. Nagakubo D, Taira T, Kitaura H, et al. DJ-1, a novel oncogene which transforms mouse NIH3T3 cells in cooperation with ras. *Biochem Biophys Res Commun.* 1997;231(2):509–513.

200. Bader V, Ran Zhu X, Lubbert H, Stichel CC. Expression of DJ-1 in the adult mouse CNS. *Brain Res.* 2005;1041(1):102–111.

201. Zhang L, Shimoji M, Thomas B, et al. Mitochondrial localization of the Parkinson's disease related protein DJ-1: implications for pathogenesis. *Hum Mol Genet.* 2005;14(14):2063–2073.

202. Blackinton J, Ahmad R, Miller DW, et al. Effects of DJ-1 mutations and polymorphisms on protein stability and subcellular localization. *Brain Res Mol Brain Res.* 2005;134(1):76–83.

203. Abou-Sleiman PM, Healy DG, Quinn N, Lees AJ, Wood NW. The role of pathogenic DJ-1 mutations in Parkinson's disease. *Ann Neurol.* 2003;54(3):283–286.

204. Hayashi T, Ishimori C, Takahashi-Niki K, et al. DJ-1 binds to mitochondrial complex I and maintains its activity. *Biochem Biophys Res Commun.* 2009;390(3):667–672.

205. Irrcher I, Aleyasin H, Seifert EL, et al. Loss of the Parkinson's disease-linked gene DJ-1 perturbs mitochondrial dynamics. *Hum Mol Genet.* 2010;19(19):3734–3746.

206. Chen L, Cagniard B, Mathews T, et al. Age-dependent motor deficits and dopaminergic dysfunction in DJ-1 null mice. *J Biol Chem.* 2005;280(22):21418–21426.

207. Goldberg MS, Pisani A, Haburcak M, et al. Nigrostriatal dopaminergic deficits and hypokinesia caused by inactivation of the familial Parkinsonism-linked gene DJ-1. *Neuron.* 2005;45(4):489–496.

208. Kim RH, Smith PD, Aleyasin H, et al. Hypersensitivity of DJ-1-deficient mice to 1-methyl-4-phenyl-1,2,3,6-tetrahydropyridine (MPTP) and oxidative stress. *Proc Natl Acad Sci USA.* 2005;102(14):5215–5220.

209. Taira T, Saito Y, Niki T, Iguchi-Ariga SM, Takahashi K, Ariga H. DJ-1 has a role in anti-oxidative stress to prevent cell death. *EMBO Rep.* 2004;5(2):213–218.

210. Yokota T, Sugawara K, Ito K, Takahashi R, Ariga H, Mizusawa H. Down regulation of DJ-1 enhances cell death by oxidative stress, ER stress, and proteasome inhibition. *Biochem Biophys Res Commun.* 2003;312(4):1342–1348.

211. Lev N, Ickowicz D, Barhum Y, Lev S, Melamed E, Offen D. DJ-1 protects against dopamine toxicity. *J Neural Transm.* 2009;116(2):151–160.

212. Lev N, Barhum Y, Pilosof NS, et al. DJ-1 protects against dopamine toxicity: implications for Parkinson's disease and aging. *J Gerontol Series A.* 2013;68(3):215–225.

213. Ooe H, Taira T, Iguchi-Ariga SM, Ariga H. Induction of reactive oxygen species by bisphenol A and abrogation of bisphenol A-induced cell injury by DJ-1. *Toxicol Sci.* 2005;88(1):114–126.

214. Waak J, Weber SS, Waldenmaier A, et al. Regulation of astrocyte inflammatory responses by the Parkinson's disease-associated gene DJ-1. *FASEB J.* 2009;23(8):2478–2489.

215. Mitsumoto A, Nakagawa Y. DJ-1 is an indicator for endogenous reactive oxygen species elicited by endotoxin. *Free Rad Res.* 2001;35(6):885–893.

216. Trudler D, Weinreb O, Mandel SA, Youdim MB, Frenkel D. DJ-1 deficiency triggers microglia sensitivity to dopamine toward a pro-inflammatory phenotype that is attenuated by rasagiline. *J Neurochem.* 2014;129(3):434–447.

217. Hamza TH, Zabetian CP, Tenesa A, et al. Common genetic variation in the HLA region is associated with late-onset sporadic Parkinson's disease. *Nat Genet.* 2010;42(9):781–785.

218. Puschmann A, Verbeeck C, Heckman MG, et al. Human leukocyte antigen variation and Parkinson's disease. *Parkinsonism Relat Disord.* 2011;17(5):376–378.

219. Simon-Sanchez J, van Hilten JJ, van de Warrenburg B, et al. Genome-wide association study confirms extant PD risk loci among the Dutch. *Eur J Hum Genet.* 2011;19(6):655–661.

220. Ahmed I, Tamouza R, Delord M, et al. Association between Parkinson's disease and the HLA-DRB1 locus. *Mov Disord.* 2012;27(9):1104–1110.

221. Nalls MA, Plagnol V, International Parkinson Disease Genomics C,et al. Imputation of sequence variants for identification of genetic risks for Parkinson's disease: a meta-analysis of genome-wide association studies. *Lancet.* 2011;377(9766):641–649.

222. Bialecka M, Klodowska-Duda G, Kurzawski M, et al. Interleukin-10 (IL10) and tumor necrosis factor alpha (TNF) gene polymorphisms in Parkinson's disease patients. *Parkinsonism Relat Disord.* 2008;14(8):636–640.

223. Nishimura M, Mizuta I, Mizuta E, et al. Tumor necrosis factor gene polymorphisms in patients with sporadic Parkinson's disease. *Neurosci Lett.* 2001;311(1):1–4.

224. Wahner AD, Sinsheimer JS, Bronstein JM, Ritz B. Inflammatory cytokine gene polymorphisms and increased risk of Parkinson disease. *Arch Neurol.* 2007;64(6):836–840.

225. Wilson AG, Symons JA, McDowell TL, McDevitt HO, Duff GW. Effects of a polymorphism in the human tumor necrosis factor alpha promoter on transcriptional activation. *Proc Natl Acad Sci USA.* 1997;94(7):3195–3199.

226. Kruger R, Hardt C, Tschentscher F, et al. Genetic analysis of immunomodulating factors in sporadic Parkinson's disease. *J Neural Transm (Vienna).* 2000;107(5):553–562.

227. Wu YR, Feng IH, Lyu RK, et al. Tumor necrosis factor-alpha promoter polymorphism is associated with the risk of Parkinson's disease. *Am J Med Genet. Part B.* 2007;144B(3):300–304.

228. Chu K, Zhou X, Luo BY. Cytokine gene polymorphisms and Parkinson's disease: a meta-analysis. *Can J Neurol Sci.* 2012;39(1):58–64.

229. Frugier T, Morganti-Kossmann MC, O'Reilly D, McLean CA. In situ detection of inflammatory mediators in post mortem human brain tissue after traumatic injury. *J Neurotrauma.* 2010;27(3):497–507.

230. Mizuta I, Nishimura M, Mizuta E, et al. Relation between the high production related allele of the interferon-gamma (IFN-gamma) gene and age at onset of idiopathic Parkinson's disease in Japan. *J Neurol Neurosurg Psychiatry.* 2001;71(6):818–819.

231. Abdeldayem M, Anani M, Hassan H, Taha S. Association of Serum Interferon Gamma Level with Parkinson's Disease in Egyptian patients. *Egypt J Med Microbiol.* 2014;23(1).
232. Nie K, Zhang Y, Gan R, et al. Polymorphisms in immune/inflammatory cytokine genes are related to Parkinson's disease with cognitive impairment in the Han Chinese population. *Neurosci Lett.* 2013;541:111–115.
233. Lin JJ, Chen CH, Yueh KC, Chang CY, Lin SZ. A CD14 monocyte receptor polymorphism and genetic susceptibility to Parkinson's disease for females. *Parkinsonism Relat Disord.* 2006;12(1):9–13.
234. Schulte T, Schols L, Muller T, Woitalla D, Berger K, Kruger R. Polymorphisms in the interleukin-1 alpha and beta genes and the risk for Parkinson's disease. *Neurosci Lett.* 2002;326(1):70–72.
235. Mattila KM, Rinne JO, Lehtimaki T, Roytta M, Ahonen JP, Hurme M. Association of an interleukin 1B gene polymorphism (-511) with Parkinson's disease in Finnish patients. *J Med Genet.* 2002;39(6):400–402.
236. Hakansson A, Westberg L, Nilsson S, et al. Interaction of polymorphisms in the genes encoding interleukin-6 and estrogen receptor beta on the susceptibility to Parkinson's disease. *Am J Med Genet Part B.* 2005;133B(1):88–92.
237. Ross OA, O'Neill C, Rea IM, et al. Functional promoter region polymorphism of the proinflammatory chemokine IL-8 gene associates with Parkinson's disease in the Irish. *Hum Immunol.* 2004;65(4):340–346.
238. Xu X, Li D, He Q, Gao J, Chen B, Xie A. Interleukin-18 promoter polymorphisms and risk of Parkinson's disease in a Han Chinese population. *Brain Res.* 2011;1381:90–94.
239. McGeer PL, Yasojima K, McGeer EG. Association of interleukin-1 beta polymorphisms with idiopathic Parkinson's disease. *Neurosci Lett.* 2002;326(1):67–69.
240. Moller JC, Depboylu C, Kolsch H, et al. Lack of association between the interleukin-1 alpha (-889) polymorphism and early-onset Parkinson's disease. *Neurosci Lett.* 2004;359(3):195–197.
241. Nishimura M, Kuno S, Kaji R, Yasuno K, Kawakami H. Glutathione-S-transferase-1 and interleukin-1beta gene polymorphisms in Japanese patients with Parkinson's disease. *Mov Disord.* 2005;20(7):901–902.
242. Li D, He Q, Li R, Xu X, Chen B, Xie A. Interleukin-10 promoter polymorphisms in Chinese patients with Parkinson's disease. *Neurosci Lett.* 2012;513(2):183–186.
243. Pascale E, Passarelli E, Purcaro C, et al. Lack of association between IL-1beta, TNF-alpha, and IL-10 gene polymorphisms and sporadic Parkinson's disease in an Italian cohort. *Acta Neurol Scand.* 2011;124(3):176–181.
244. Le W, Pan T, Huang M, et al. Decreased NURR1 gene expression in patients with Parkinson's disease. *J Neurol Sci.* 2008;273(1–2):29–33.
245. Aarnisalo P, Kim CH, Lee JW, Perlmann T. Defining requirements for heterodimerization between the retinoid X receptor and the orphan nuclear receptor Nurr1. *J Biol Chem.* 2002;277(38):35118–35123.
246. Wang Z, Benoit G, Liu J, et al. Structure and function of Nurr1 identifies a class of ligand-independent nuclear receptors. *Nature.* 2003;423(6939):555–560.
247. Sanchez-Pernaute R, Ferree A, Cooper O, Yu M, Brownell AL, Isacson O. Selective COX-2 inhibition prevents progressive dopamine neuron degeneration in a rat model of Parkinson's disease. *J Neuroinflamm.* 2004;1(1):6.
248. Chen H, Zhang SM, Hernan MA, et al. Nonsteroidal anti-inflammatory drugs and the risk of Parkinson disease. *Arch Neurol.* 2003;60(8):1059–1064.
249. Chen H, Jacobs E, Schwarzschild MA, et al. Nonsteroidal antiinflammatory drug use and the risk for Parkinson's disease. *Ann Neurol.* 2005;58(6):963–967.
250. Hernan MA, Logroscino G, Garcia Rodriguez LA. Nonsteroidal anti-inflammatory drugs and the incidence of Parkinson disease. *Neurology.* 2006;66(7):1097–1099.
251. Hancock DB, Martin ER, Stajich JM, et al. Smoking, caffeine, and nonsteroidal anti-inflammatory drugs in families with Parkinson disease. *Arch Neurol.* 2007;64(4):576–580.

252. Samii A, Etminan M, Wiens MO, Jafari S. NSAID use and the risk of Parkinson's disease: systematic review and meta-analysis of observational studies. *Drugs Aging.* 2009;26(9):769–779.
253. Undela K, Gudala K, Malla S, Bansal D. Statin use and risk of Parkinson's disease: a meta-analysis of observational studies. *J Neurol.* 2013;260(1):158–165.
254. Hirsch EC, Breidert T, Rousselet E, Hunot S, Hartmann A, Michel PP. The role of glial reaction and inflammation in Parkinson's disease. *Ann NY Acad Sci.* 2003;991:214–228.
255. Randy LH, Guoying B. Agonism of peroxisome proliferator receptor-gamma may have therapeutic potential for neuroinflammation and Parkinson's disease. *Curr Neuropharmacol.* 2007;5(1):35–46.
256. Dehmer T, Heneka MT, Sastre M, Dichgans J, Schulz JB. Protection by pioglitazone in the MPTP model of Parkinson's disease correlates with I kappa B alpha induction and block of NF kappa B and iNOS activation. *J Neurochem.* 2004;88(2):494–501.
257. Quinn LP, Crook B, Hows ME, et al. The PPARgamma agonist pioglitazone is effective in the MPTP mouse model of Parkinson's disease through inhibition of monoamine oxidase B. *Br J Pharmacol.* 2008;154(1):226–233.
258. Hunter RL, Dragicevic N, Seifert K, et al. Inflammation induces mitochondrial dysfunction and dopaminergic neurodegeneration in the nigrostriatal system. *J Neurochem.* 2007;100(5):1375–1386.
259. Shytle RD, Mori T, Townsend K, et al. Cholinergic modulation of microglial activation by alpha 7 nicotinic receptors. *J Neurochem.* 2004;89(2):337–343.
260. De Simone R, Ajmone-Cat MA, Carnevale D, Minghetti L. Activation of alpha7 nicotinic acetylcholine receptor by nicotine selectively up-regulates cyclooxygenase-2 and prostaglandin E2 in rat microglial cultures. *J Neuroinflamm.* 2005;2(1):4.
261. Gorell JM, Rybicki BA, Johnson CC, Peterson EL. Smoking and Parkinson's disease: a dose-response relationship. *Neurology.* 1999;52(1):115–119.
262. Hernan MA, Takkouche B, Caamano-Isorna F, Gestal-Otero JJ. A meta-analysis of coffee drinking, cigarette smoking, and the risk of Parkinson's disease. *Ann Neurol.* 2002;52(3):276–284.
263. Parain K, Marchand V, Dumery B, Hirsch E. Nicotine, but not cotinine, partially protects dopaminergic neurons against MPTP-induced degeneration in mice. *Brain Res.* 2001;890(2):347–350.
264. Parain K, Hapdey C, Rousselet E, Marchand V, Dumery B, Hirsch EC. Cigarette smoke and nicotine protect dopaminergic neurons against the 1-methyl-4-phenyl-1,2,3,6-tetrahydropyridine Parkinsonian toxin. *Brain Res.* 2003;984(1–2):224–232.
265. Quik M, Vailati S, Bordia T, et al. Subunit composition of nicotinic receptors in monkey striatum: effect of treatments with 1-methyl-4-phenyl-1,2,3,6-tetrahydropyridine or L-DOPA. *Mol Pharmacol.* 2005;67(1):32–41.
266. Quik M, Chen L, Parameswaran N, Xie X, Langston JW, McCallum SE. Chronic oral nicotine normalizes dopaminergic function and synaptic plasticity in 1-methyl-4-phenyl-1,2,3,6-tetrahydropyridine-lesioned primates. *J Neurosci.* 2006;26(17):4681–4689.
267. Quik M, Parameswaran N, McCallum SE, et al. Chronic oral nicotine treatment protects against striatal degeneration in MPTP-treated primates. *J Neurochem.* 2006;98(6):1866–1875.
268. Mazurov A, Klucik J, Miao L, et al. 2-(Arylmethyl)-3-substituted quinuclidines as selective alpha 7 nicotinic receptor ligands. *Bioorg Med Chem Lett.* 2005;15(8):2073–2077.
269. Acker BA, Jacobsen EJ, Rogers BN, et al. Discovery of N-[(3R,5R)-1-azabicyclo[3.2.1]oct-3-yl]furo[2,3-c]pyridine-5-carboxamide as an agonist of the alpha7 nicotinic acetylcholine receptor: in vitro and in vivo activity. *Bioorg Med Chem Lett.* 2008;18(12):3611–3615.
270. McLarnon JG. Purinergic mediated changes in Ca2+ mobilization and functional responses in microglia: effects of low levels of ATP. *J Neurosci Res.* 2005;81(3):349–356.
271. Rampe D, Wang L, Ringheim GE. P2X7 receptor modulation of beta-amyloid- and LPS-induced cytokine secretion from human macrophages and microglia. *J Neuroimmunol.* 2004;147(1–2):56–61.

272. Kosloski LM, Ha DM, Hutter JA, et al. Adaptive immune regulation of glial homeostasis as an immunization strategy for neurodegenerative diseases. *J Neurochem.* 2010;114(5):1261–1276.

273. Huppa JB, Davis MM. T-cell-antigen recognition and the immunological synapse. *Nat Rev Immunol.* 2003;3(12):973–983.

274. Laurie C, Reynolds A, Coskun O, Bowman E, Gendelman HE, Mosley RL. CD4+ T cells from Copolymer-1 immunized mice protect dopaminergic neurons in the 1-methyl-4-phenyl-1,2,3,6-tetrahydropyridine model of Parkinson's disease. *J Neuroimmunol.* 2007;183(1–2):60–68.

275. Benner EJ, Mosley RL, Destache CJ, et al. Therapeutic immunization protects dopaminergic neurons in a mouse model of Parkinson's disease. *Proc Natl Acad Sci USA.* 2004;101(25):9435–9440.

276. Papachroni KK, Ninkina N, Papapanagiotou A, et al. Autoantibodies to alpha-synuclein in inherited Parkinson's disease. *J Neurochem.* 2007;101(3):749–756.

277. Smith LM, Schiess MC, Coffey MP, Klaver AC, Loeffler DA. alpha-Synuclein and anti-alpha-synuclein antibodies in Parkinson's disease, atypical Parkinson syndromes, REM sleep behavior disorder, and healthy controls. *PLoS One.* 2012;7(12):e52285.

278. Masliah E, Rockenstein E, Adame A, et al. Effects of alpha-synuclein immunization in a mouse model of Parkinson's disease. *Neuron.* 2005;46(6):857–868.

279. Schneeberger A, Mandler M, Mattner F, Schmidt W. Vaccination for Parkinson's disease. *Parkinsonism Relat Disord.* 2012;18(suppl 1):S11–13.

280. Schneeberger A, Mandler M, Mattner F, Schmidt W. AFFITOME(R) technology in neurodegenerative diseases: the doubling advantage. *Hum Vaccin.* 2010;6(11):948–952.

281. Kim HS, Suh YH. Minocycline neurodegenerative diseases. *Behav Brain Res.* 2009;196(2):168–179.

282. Giuliani F, Hader W, Yong VW. Minocycline attenuates T cell and microglia activity to impair cytokine production in T cell-microglia interaction. *J Leukoc Biol.* 2005;78(1):135–143.

283. Du Y, Ma Z, Lin S, et al. Minocycline prevents nigrostriatal dopaminergic neurodegeneration in the MPTP model of Parkinson's disease. *Proc Natl Acad Sci USA.* 2001;98(25):14669–14674.

284. He Y, Appel S, Le W. Minocycline inhibits microglial activation and protects nigral cells after 6-hydroxydopamine injection into mouse striatum. *Brain Res.* 2001;909(1–2):187–193.

285. Wu DC, Jackson-Lewis V, Vila M, et al. Blockade of microglial activation is neuroprotective in the 1-methyl-4-phenyl-1,2,3,6-tetrahydropyridine mouse model of Parkinson disease. *J Neurosci.* 2002;22(5):1763–1771.

286. Yang L, Sugama S, Chirichigno JW, et al. Minocycline enhances MPTP toxicity to dopaminergic neurons. *J Neurosci Res.* 2003;74(2):278–285.

287. Diguet E, Fernagut PO, Wei X, et al. Deleterious effects of minocycline in animal models of Parkinson's disease and Huntington's disease. *Eur J Neurosci.* 2004;19(12):3266–3276.

288. Quintero EM, Willis L, Singleton R, et al. Behavioral and morphological effects of minocycline in the 6-hydroxydopamine rat model of Parkinson's disease. *Brain Res.* 2006;1093(1):198–207.

289. Investigators NN-P. A randomized, double-blind, futility clinical trial of creatine and minocycline in early Parkinson disease. *Neurology.* 2006;66(5):664–671.

290. Kohutnicka M, Lewandowska E, Kurkowska-Jastrzebska I, Czlonkowski A, Czlonkowska A. Microglial and astrocytic involvement in a murine model of Parkinson's disease induced by 1-methyl-4-phenyl-1,2,3,6-tetrahydropyridine (MPTP). *Immunopharmacology.* 1998;39(3):167–180.

291. Klegeris A, McGeer PL. R-(-)-Deprenyl inhibits monocytic THP-1 cell neurotoxicity independently of monoamine oxidase inhibition. *Exp Neurol.* 2000;166(2):458–464.

292. Wersinger C, Sidhu A. Inflammation and Parkinson's disease. *Curr Drug Targets Inflamm Allergy.* 2002;1(3):221–242.

293. Liu B, Hong JS. Role of microglia in inflammation-mediated neurodegenerative diseases: mechanisms and strategies for therapeutic intervention. *J Pharmacol Exp Ther*. 2003;304(1):1–7.

294. Gold BG, Nutt JG. Neuroimmunophilin ligands in the treatment of Parkinson's disease. *Curr Opin Pharmacol*. 2002;2(1):82–86.

295. Kurkowska-Jastrzebska I, Litwin T, Joniec I, et al. Dexamethasone protects against dopaminergic neurons damage in a mouse model of Parkinson's disease. *Int Immunopharmacol*. 2004;4(10–11):1307–1318.

296. Castano A, Herrera AJ, Cano J, Machado A. The degenerative effect of a single intranigral injection of LPS on the dopaminergic system is prevented by dexamethasone, and not mimicked by rh-TNF-alpha, IL-1beta and IFN-gamma. *J Neurochem*. 2002;81(1):150–157.

297. Ghosh A, Roy A, Liu X, et al. Selective inhibition of NF-kappaB activation prevents dopaminergic neuronal loss in a mouse model of Parkinson's disease. *Proc Natl Acad Sci USA*. 2007;104(47):18754–18759.

298. Delgado M, Ganea D. Vasoactive intestinal peptide prevents activated microglia-induced neurodegeneration under inflammatory conditions: potential therapeutic role in brain trauma. *FASEB J*. 2003;17(13):1922–1924.

299. Delgado M, Ganea D. Neuroprotective effect of vasoactive intestinal peptide (VIP) in a mouse model of Parkinson's disease by blocking microglial activation. *FASEB J*. 2003;17(8):944–946.

300. Wang MJ, Lin WW, Chen HL, et al. Silymarin protects dopaminergic neurons against lipopolysaccharide-induced neurotoxicity by inhibiting microglia activation. *Eur J Neurosci*. 2002;16(11):2103–2112.

301. Liu Y, Qin L, Li G, et al. Dextromethorphan protects dopaminergic neurons against inflammation-mediated degeneration through inhibition of microglial activation. *J Pharmacol Exp Ther*. 2003;305(1):212–218.

302. Iravani MM, Kashefi K, Mander P, Rose S, Jenner P. Involvement of inducible nitric oxide synthase in inflammation-induced dopaminergic neurodegeneration. *Neuroscience*. 2002;110(1):49–58.

303. Arimoto T, Bing G. Up-regulation of inducible nitric oxide synthase in the substantia nigra by lipopolysaccharide causes microglial activation and neurodegeneration. *Neurobiol Dis*. 2003;12(1):35–45.

304. Golde TE. The therapeutic importance of understanding mechanisms of neuronal cell death in neurodegenerative disease. *Mol Neurodegener*. 2009;4:8.

305. Wyss-Coray T. Inflammation in Alzheimer disease: driving force, bystander or beneficial response?. *Nat Med*. 2006;12(9):1005–1015.

306. McGeer EG, McGeer PL. The role of anti-inflammatory agents in Parkinson's disease. *CNS Drugs*. 2007;21(10):789–797.

307. Blandini F. Neural and immune mechanisms in the pathogenesis of Parkinson's disease. *J Neuroimmune Pharmacol*. 2013;8(1):189–201.

308. Gao HM, Liu B, Zhang W, Hong JS. Novel anti-inflammatory therapy for Parkinson's disease. *Trends Pharmacol Sci*. 2003;24(8):395–401.

CHAPTER

9

Protein Translation in Parkinson's Disease

J.W. Kim*,**, L. Abalde-Atristain*,†, H. Jia*,**,
I. Martin*,‡,§§, T.M. Dawson*,†,‡,§,¶,††,‡‡,
V.L. Dawson*,**,†,‡,§,††,‡‡

*Neurodegeneration and Stem Cell Programs, Institute for Cell Engineering, Baltimore, MD, United States; **Department of Physiology, Baltimore, MD, United States; †Graduate Program in Cellular and Molecular Medicine, Baltimore, MD, United States; ‡Department of Neurology, Baltimore, MD, United States; §Solomon H. Snyder Department of Neuroscience, Baltimore, MD, United States; ¶Department of Pharmacology and Molecular Sciences, Johns Hopkins University School of Medicine, Baltimore, MD, United States; ††Adrienne Helis Malvin Medical Research Foundation, New Orleans, LA, United States; ‡‡Diana Helis Henry Medical Research Foundation, New Orleans, LA, United States; §§Jungers Center for Neurosciences Research, Parkinson Center of Oregon, Department of Neurology, Oregon Health and Science University, Portland, OR, United States

OUTLINE

Parkinson's Disease. http://dx.doi.org/10.1016/B978-0-12-803783-6.00009-2

1 PROTEIN TRANSLATION AND ITS REGULATION IN NEURONS

Protein translation is a fundamental cellular process, and its regulation serves as a major posttranscriptional regulatory mode of gene expression. Translation is a crucial part of protein homeostasis, and it is tightly linked to vital cell physiology, such as the cell cycle, metabolism, stress, and even cell death.

Translation consists of three distinct steps: initiation, elongation, and termination. Each step of translation is subject to tight regulation to meet the physiological needs of the cell. Although translational regulation is relatively less known compared to transcriptional regulation, there is growing evidence suggesting that it comprises diverse and complex regulatory mechanisms. Initiation is a crucial step to choose "what to make" by selecting messenger RNAs (mRNAs), and also to decide "how to make" by choosing the start codon and thereby choosing the open reading frame (ORF). Therefore, it is not surprising that the majority of currently known regulatory mechanisms focus on the proper control of the initiation process. During translational initiation in eukaryotes, eukaryotic initiation factors (eIFs) are recruited to the mRNAs to facilitate ribosomal loading and start codon scanning. The ribosome is a large and complex molecular machinery specialized in protein synthesis, and it is central to the translation process. Initiation is over when the ribosome and the initiation factors locate the start codon, and then elongation proceeds.[1,2] During elongation, the actual coding information is "translated" by the ribosome through codon–anticodon interaction between mRNA and aminoacyl-transfer RNA (tRNA) and subsequent peptide bond formation. Therefore, elongation is directly related to the fidelity of protein synthesis. For instance, speed of elongation influences pairing of a codon with its cognate tRNA, which is required for proper amino acid incorporation and appropriate folding of the growing polypeptide chain, underlining the importance of elongation regulation.[3] Finally, ribosomal recognition of a stop codon triggers its release from the mRNA, thereby terminating the process. Translational termination is tightly connected to ribosome recycling and reinitiation of translation, and thus regulates the balance between free and active ribosomes.[4]

1.1 Translation in Eukaryotic Cells

In eukaryotes, mature mRNAs are exported from the nucleus to the cytosol after splicing. The cap-binding eukaryotic initiation factor 4F (eIF4F), a complex that consists of eIF4E, eIF4G, eIF4A, binds to the mRNAs and activates them. With the help of initiation factors such as eIF1, eIF1A, and eIF3, the small ribosomal subunit (40S in eukaryotes) and ternary complex, comprising initiator methionyl-tRNA (Met-tRNA$_i^{Met}$), eIF2 and GTP, form the 43S preinitiation complex (PIC). The eIF4F cap complex recruits the PIC to the mRNA and forms the 48S initiation complex. Once the 48S complex is formed, it scans the mRNA until it finds a start codon. Upon recognition of the start codon, the GTP in the ternary complex is hydrolyzed with the aid of eIF5, an eIF2-specific GTPase-activating protein (GAP). eIF5 and eIF5B facilitate the dissociation of the initiation factors and the joining of the large ribosomal subunit (60S in eukaryotes). The large subunit joins to make an 80S ribosome, which has peptidyltransferase activity and thereby synthesizes polypeptide by forming peptide bonds between amino acids. With the help of eukaryotic elongation factor 2 (eEF2), an aminoacyl-tRNA is recruited to the acceptor site (A site) of the ribosome. A new peptide bond is formed between the amino acids in the peptidyl (P)- and A-sites. eEF2 also promotes ribosomal translocation, and the whole cycle repeats until the ribosome encounters a stop codon. At the stop codon, instead of a charged tRNA, release factors approach the P-site and mediate termination of protein synthesis (Fig. 9.1).[4]

eIF2 is one of the major regulatory targets in translational initiation. eIF2 is a component of the ternary complex and it binds to GTP and Met-tRNA$_i^{Met}$. More specifically, it delivers the Met-tRNA$_i^{Met}$ to the 40S small ribosomal subunit. With the help of other initiation factors, the ternary complex forms the 48S initiation complex and enables the 40S subunit to scan, find the start codon, and initiate protein synthesis. Once the 48S complex finds the start codon, GTP is hydrolyzed to GDP, which is an essential step for initiation factor dissociation and 60S large subunit joining. eIF2 is a heterotrimeric protein consisting of subunits alpha, beta, and gamma. Serine 51 on eIF2α can be phosphorylated by various kinases in response to stress signals, and phosphorylated eIF2α is resistant to the guanine nucleotide exchange factor (GEF) activity of eIF2B, therefore it remains in an inactive eIF2-GDP-tRNAi state.[1,2] Hence, it decreases active ternary complex availability and thereby reduces global protein synthesis, for example, under conditions of stress.[5] It is also known that the phosphorylation of eIF2α increases the expression of a group of genes with stress-responsive functions, such as activating transcription factor 4 (ATF4), through upstream open reading frame (uORF)-mediated translational regulation.[5,6] uORF-mediated regulation is involved

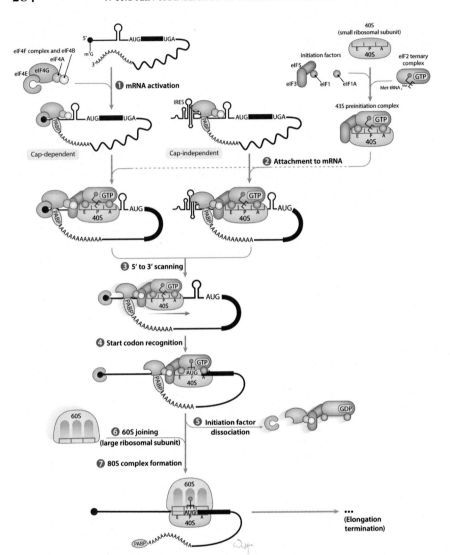

FIGURE 9.1 **Translation initiation in eukaryotic cells.** The canonical pathway of eukaryotic translation initiation is divided into seven different stages. (1) mRNA activation: mRNA is activated by eIF4F complex formation, poly-A binding protein *(PABP)* binding to the poly-A tail, and subsequent circularization of the mRNA. m⁷G: 7-methylguanosine "cap"; eIF4F cap-binding complex consists of eIF4E, eIF4G, and eIF4A. eIF4E binds directly to the m7G "cap," eIF4G works as a scaffold protein and eIF4A is an RNA helicase, whose function is assisted by eIF4B. IRES: internal ribosome entry site; there are several types of IRESs with different initiation factor requirements; the illustration depicts a certain type of IRES (eg, encephalomyocarditis virus) with eIF4G/A/B requirements. (2) Attachment to mRNA: a ternary complex consists of eIF2, Met-tRNA$_i^{Met}$ initiator tRNA, and GTP. A 43S PIC comprises a 40S small subunit, a ternary complex, eIF1, eIF1A, eIF3, and eIF5. A 43S ▶

in start codon selection; upon arrangement of the different start codons, different ORFs can be chosen, resulting in the production of different proteins. Briefly, when ternary complex availability is low, start codon recognition by the 48S complex is hampered and the balance of start codon selection is skewed toward the downstream start codons. Several stress-responsive genes have multiple uORFs encoding nonfunctional gene products. Under normal conditions, the uORFs prevent the major ORF from being translated. However, when the ternary complex is depleted, the downstream start codons are preferred, leading to the synthesis of the desired functional protein.[1,2,5] It is noteworthy that eIF2α-mediated translational regulation has been shown to play important roles in processes relevant to neurons, like synaptic plasticity or neurodegeneration.[7–9]

1.2 Translation and Neurological Disorders

Aberrant translational regulation can lead to various neurological disorders, including neurodegenerative diseases. In fact, mouse models with straight overexpression/knockout of translation initiation factors or translation regulation factors show cognitive defects.[10–12] For example, eIF4E overexpression or eukaryotic initiation factor 4E-binding protein (4E-BP) knockout animals have phenotypes reminiscent of autism-spectrum disorders.[10,11] 4E-BP acts as a negative regulator of eIF4E by directly binding to eIF4E and thereby sequesters eIF4E from the eIF4F complex formation. This suggests that neurons may be highly susceptible to translational abnormalities. Furthermore, protein aggregation-related neurodegenerative diseases—such as prion-mediated neurodegeneration or Alzheimers disease (AD)—might be linked to eIF2α-mediated protein synthesis defects.[7,13] It has been suggested that the unfolded protein response (UPR) triggered by protein aggregates increases the levels of phosphorylated eIF2α, thereby reducing global protein translation. As a result, many genes that are important for neuronal activity and synaptic function are downregulated and this eventually leads to cognitive defects and neuronal cell death. Fragile X syndrome (FXS), and fragile X-related tremor and ataxia syndrome (FXTAS) are caused by loss of function of a single gene, fragile X mental retardation 1 (FMR1).[14] FMR1 is an RNA-binding protein and it has been reported to function as translational repressor. Translational profiling identified that FMR1 target genes are important for neuronal

FIGURE 9.1 (CONT.)   PIC is recruited to the activated mRNA by interaction between initiation factors, primarily by eIF4G-eIF3 interaction. (3) 5′ to 3′ scanning: A 43S PIC joins to the activated mRNA to make a 48S initiation complex, and starts scanning. It is noteworthy that there is a type of IRES (cricket paralysis virus) that doesn't require a scanning process for initiation. (4) Start codon recognition (5) Initiation factor dissociation (6) 60S joining: after start codon recognition, eIF5 and eIF5B are known to mediate the eIF2-bound GTP hydrolysis, the initiation factor dissociation, and the 60S subunit joining. (7) 80S complex formation: after formation of an 80S complex, elongation step succeeds. *(Illustration: I-Hsun Wu)* function.[15] Recently, repeat-associated non-ATG (RAN) translation has been suggested as a key mechanism to understand triplet expansion-related homopolymeric protein disorders, such as spinocerebellar ataxia (SCA), myotonic dystrophy (DM), and amyotropic lateral sclerosis/frontotemporal dementia (ALS/FTD).[16]

Recent studies on familial forms of Parkinson's disease (PD) have revealed that there is a clear link between PD and protein translation. For example, we recently showed that LRRK2 (leucine-rich repeat kinase 2) G2019S mutation increased bulk protein synthesis, causing neurotoxicity.[17] Furthermore, PD-associated eukaryotic initiation factor 4 gamma-1 (eIF4G1) mutations were identified in genome-wide association studies (GWAS), and PTEN-induced putative kinase 1 (PINK1) was suggested to have a role in protein synthesis of mitochondrial proteins.[18,19] Although there is increasing evidence suggesting the role of disrupted protein synthesis in PD, the detailed mechanisms are still under investigation. In this chapter, we are going to introduce our current understanding on the relationship between PD and protein synthesis, with an emphasis on the effects of familial PD-linked mutations on protein translation. We will also discuss protein translation-related neurotoxic mechanisms in PD (Table 9.1).

2 PD-LINKED MUTATIONS AND THEIR IMPACT ON PROTEIN TRANSLATION

Many PD-linked genes are linked to protein translation; either their gene products have a direct involvement in protein synthesis or certain phenotypes associated with these genes are susceptible to changes in translational regulation. In this chapter, we are going to introduce known protein translation-related cellular processes associated with different PD-linked genes, and discuss different pathways involved in the translational abnormalities.

2.1 LRRK2 Mutations and Protein Translation

2.1.1 LRRK2 Mutations and PD

The PARK8 locus was identified in 2002 in a large Japanese family with autosomal dominant parkinsonism, and subsequent screening studies with familial PD groups identified mutations of LRRK2 in the PARK8 locus.[32] Up to now, LRRK2 mutations are the most common genetic cause of familial PD identified, accounting for up to 20% of specific PD cases in some populations like Ashkenazi Jews. Postmortem analysis of brains of affected individuals demonstrated that LRRK2 mutations show variable pathology regarding the presence of Lewy bodies and tau-reactive

TABLE 9.1   Summary of PD Genes Linked to Translation, Modified from Taymans et al.[20]

Gene	Protein	Mendelian inheritance	Link to protein translation
SNCA	α-synuclein	AD	Alternate transcript usage through 3′UTR, genetic interaction with EIF4G1[21,22]
PARK2	Parkin	Early-onset AR	E3 ligase substrate in translational machinery, genetic interaction with 4E-BP[23,24]
DJ-1	DJ-1	Early-onset AR	Interacts with ribosomes in *Escherichia coli*[25]
PINK1	Pten-induced kinase 1	Early-onset AR	Genetic interaction with 4E-BP, regulation of translation of respiratory chain complex mRNAs on mitochondrial outer membrane[19,26]
LRRK2	Leucine-rich repeat kinase 2	Late-onset AD	Interaction with translational regulatory proteins, kinase substrates in translation machinery, increases bulk translation, involved in miRNA pathway[27-30]
VPS35	Vacuolar protein sorting-associated protein 35	Late-onset AD	Genetic interaction with LRRK2 and EIF4G1[22,31]
EIF4G1	Eukaryotic translation initiation factor 4 gamma-1	Late-onset AD	Translation initiation factor, suggested to affect cap complex formation[18]

lesions.[33] LRRK2 is a large multidomain protein with known GTPase and kinase enzymatic activities, and multiple protein–protein interaction domains flank the enzymatic domains. There are many LRRK2 mutations reported from PD patients; six of those mutations clearly segregate with the disease, all of which reside within the enzymatic Ras of complex (ROC)—C-terminus of ROC and kinase domains.[34] This suggests that the enzymatic activities of LRRK2 are important to disease development. Among the LRRK2 mutations, G2019S in the kinase domain is the most prevalent. Globally, it constitutes 4% of familial and 1% of sporadic known cases. Studies have shown that the G2019S LRRK2 mutation augments its kinase activity, and that the increased kinase activity is toxic to neurons.[34-38] Therefore, identifying LRRK2 kinase substrates is pivotal to gain insight into LRRK2-associated pathology. Despite the importance of enzymatic activities in PD, the normal biological role of LRRK2 is still under investigation. A series of experimental evidence suggests a role for LRRK2 in vesicular trafficking, autophagy, cytoskeletal network formation, synaptogenesis, and protein translation.[27,28,31,34,38–43] However, the major pathological mechanism contributing to disease progression is still unclear. Notably, recent efforts to achieve an integrative understanding of LRRK2-mediated neurotoxicity suggest that the roles of LRRK2 converge on the regulation of protein homeostasis, including protein translation.[27–29] This allows us to hypothesize that the core disease mechanism of LRRK2 mutation is misregulation of protein homeostasis, and it also highlights the importance of understanding the effects of LRRK2 in protein translation.

2.1.2 LRRK2 in Protein Translation

A series of studies showed that LRRK2 physically and functionally interacts with the protein translation machinery. These studies also showed that the defects in protein synthesis caused by PD-linked LRRK2 mutations incur toxicity in neurons. In 2008, Imai et al. reported that LRRK2 genetically interacts with the target of rapamycin (TOR) pathway components in *Drosophila*, suggesting a role of LRRK2 in translational regulation. In the study, genetic deletion of dLRRK either aggravated the TOR suppression-mediated impaired cell growth, or suppressed TOR hyperactivation-mediated enhanced cell growth. Therefore, LRRK2 was proposed to affect global protein synthesis through the TOR pathway.[27] In addition, a subsequent study from the same group reported that G2019S/I2020T PD-linked LRRK2 variants impact translation through repressing miRNA activity. Transgenic flies expressing mutant human or fly LRRK2 showed suppression of let-7 and miR-184, thereby increasing expression of E2F1 and DP, respectively.[29] However, the molecular mechanism explaining how mutant LRRK2 disrupts the miRNA pathway remains unclear. In addition to the *Drosophila* models, Nikonova et al. reported that expression of the core components of the protein synthesis machinery is altered in LRRK2 mouse models. The authors performed microarray and subsequent gene set enrichment analysis to compare the expression profiles of LRRK2 knockout and G2019S transgenic mice. As a result, the authors reported that there was a significant increase in ribosomal gene expression in G2019S LRRK2 transgenic mice compared to LRRK2 knockout mice.[44] This also suggests that there is an increased need of protein synthesis in G2019S LRRK2 mice. In addition to that, LRRK2 has been reported to interact with the translation elongation factor eEF1A in mammalian cells. As eEF1A brings aminoacyl-tRNAs to the ribosomal A-site during elongation, it suggests the possibility that LRRK2 affects this step of translation as well.[30]

Recently, a more direct functional relationship between LRRK2 and protein translation has been suggested. Our laboratory reported that pathogenic G2019S LRRK2 increases phosphorylation of its substrate ribosomal protein s15 (RPS15), thereby increasing global protein synthesis and causing neurotoxicity.[28] In the study, we identified ribosomal proteins as targets of G2019S LRRK2, via proteomic screening and subsequent kinase assays. Among those ribosomal proteins, we showed s15 to be a "pathogenic" target, as blocking s15 phosphorylation on the identified site as being phosphorylated by LRRK2, threonine 136, substantially reduced G2019S LRRK2 neurotoxicity in both human neuron and *Drosophila* models. Furthermore, the phospho-deficient allele of s15 on threonine 136 (T136A s15) could block G2019S LRRK2-mediated neuronal injury, namely neurite shortening and cell death in human neuron models, while a phospho-mimetic version of s15 (T136D s15) could partially replicate the toxicity elicited by G2019S LRRK2. Subsequently, we confirmed phosphorylated s15-mediated neurotoxicity in *Drosophila* in vivo models; these transgenic flies showed age-dependent dopaminergic (DA) neuronal loss and resultant locomotor deficits, and blockade of s15 phosphorylation on T136 in G2019S LRRK2 transgenic flies was protective. Furthermore, the study also showed that phosphorylation of s15 enhances global protein synthesis, and that blocking this effect is neuroprotective. When we tried to dissect out the role of LRRK2 in translation more specifically, G2019S LRRK2 showed increase translation in both a cap-dependent and -independent manner in in vitro reporter assays. We also detected elevated de novo protein synthesis in ^{35}S methionine/cysteine incorporation assays. This increase in bulk translation was neutralized by coexpression of a phosphorylation-deficient form of s15. Furthermore, the elongation inhibitor anisomycin or phospho-deficient s15 rescued neurotoxicity in G2019S LRRK2 *Drosophila* models.[28] Those findings are in line with previous reports that increased protein synthesis serves as a key mechanism in pathogenic LRRK2 mutations, suggesting that protein synthesis inhibitors might have therapeutic potential for LRRK2 PD patients. However, how increased global protein synthesis selectively kills dopamine neurons is still unclear.

Ribosomal protein s15 is located on the surface of the 40S ribosomal subunit and it has been described to be essential for the export of pre-40S particles from the nucleus to the cytosol.[45,46] Its knockdown extends lifespan in *Caenorhabditis elegans*, allegedly due to the downregulation of the highly energy-demanding process of protein synthesis.[47] However, the exact function and regulatory mechanisms of the newly identified substrate s15 still remain obscure.

2.1.3 LRRK2 in Vesicular Trafficking

It is noteworthy that LRRK2 has been suggested to have a role in vesicular trafficking, and that several LRRK2 mutation models have shown trafficking-related defects[39] (see also Chapters 6 and 7). For example,

LRRK2 knockout animal models showed increased autophagy markers, suggesting a role of LRRK2 in the autophagy pathway.[48,49] Additionally, LRRK2 interactome studies revealed that Rab7L1 and GAK are LRRK2 binding partners, and that LRRK2 forms a complex with those proteins to participate in the autophagy-lysosome pathway.[31] Although the detailed pathological significance of this is still unclear, it suggests the possibility that LRRK2 works as a broad-spectrum regulator of protein homeostasis, from protein translation to trafficking and degradation.

2.2 eIF4G1 Mutations and Protein Translation

eIF4G is a scaffold component in the mRNA translation initiation complex, performing a key role in both cap-dependent and cap-independent translation initiation. Together with PABP and eIF4E, eIF4G circularizes the mRNA to stabilize it, and to also promote initiation. In conventional cap-dependent translation, eIF4G functions as a molecular bridge. Working with eIF3, eIF4G assists the 43S PIC loading to the mRNA.[1] In cap-independent translation initiation, eIF4G binds with eIF4A to form a subcomplex, thereby initiates translation by recruiting 43S complex onto internal ribosome entry site (IRES) without the cap-binding eIF4F complex.[50,51] Therefore, it has been suggested that concentration of eIF4G is the key switch regulator between cap-dependent and cap-independent translation.[52]

2.2.1 eIF4G1 Mutations in PD

There are three paralogs of eIF4G in the human genome, from eIF4G1 to eIF4G3. In 2011, Chartier-Harlin et al. reported that mutations of eIF4G1 are associated with autosomal dominant familial PD.[18] The missense mutation c.3614G > A (p.R1205H) was initially identified in a genome-wide linkage analysis of a large French family with autosomal-dominant Parkinsonism, then the same mutation was also found in another seven small families. In a subsequent screening for novel PD-associated eIF4G1 variants, additional four variants, c.1505C > T (p.A502V), c.2056G > T (p.G686C), c.3490A > C (p.S1164R), and c.3589C > T (p.R1197W), were identified in PD cases but not in control subjects. The A502V mutation was the most frequent substitution identified in the study, and it was also suggested from an ancestral founder in the haplotype analysis.[18] The subsequent neuropathological study showed that patients carrying one or double of these eIF4G1 variants manifested clinical features of dementia with Lewy bodies.[53]

After the first report of eIF4G1 mutations, other groups performed the screening of additional eIF4G1 variants. Follow-up studies validated the suggested linkage between PD and eIF4G1 mutation. However, several studies also found the initial mutants—R1205H and A502V—in their control cohorts, so the exact PD-linked mutation sites are still controversial. Tucci et al. identified c.1456C > T (p.P486S) in two PD individuals out of 150 African familial PD cases, but also detected A502V variant in two controls out of 3500 European–American samples.[54] Also, Schulte et al. identified seven novel nonsynonymous variants in six patients with classic PD phenotype in their 376 Central European PD cases. Then they assessed the frequency of novel and previously reported variants in 975 familial and sporadic PD cases from Austria, Germany, and Hungary and 1014 general controls. Four were validated but with very rare frequency. Unexpectedly, the original R1205H mutation was found in three controls only.[55] Lesage et al. identified one possible new variant (p.E462delInsGK) in two affected siblings with segregation out of 251 autosomal dominant PD cases, mostly of French origin. Two previously reported variants were found in their study, but one was in two isolated patients and the other in a control case.[56] Using whole exome sequencing of 213 Caucasian PD patients and 272 control individuals, Nuytemans et al. confirmed that the initially identified p.R1205H mutation segregated in all affected members of a 3-generation family, except for an 86-year-old member. In addition, they also identified eight novel nonsynonymous variants and two small deletions in isolated PD patients.[57] Among other screening studies for PD-linked eIF4G1 mutations from different ethnic groups including Greek, Italian, African, Chinese, Japanese, and Indian, none identified the initial R1205H and A502V mutants in either patients or control groups, nor found any novel variants with strong disease association.[58–65] Hence, studies to date indicate that eIF4G1 mutations are very rare, representing less than 1% of worldwide PD patients, and their penetrance is incomplete.[66]

2.2.2 eIF4G1 Mutations and Translation Initiation

While standard genetic approaches have not yet demonstrated that the identified eIF4G1 mutations are causal to PD, biochemical and cellular studies have suggested a potential role of eIF4G1 mutations in neurodegeneration. In coimmunoprecipitation studies of eIF4G1 mutants from transfected cell lysates, R1205H allele showed reduced interaction between eIF4G1 and eIF3E, which may hamper eIF4G1's function as a molecular bridge for 43S complex loading.[18,67] Furthermore, A502V eIF4G1 has been shown to perturb its binding to eIF4E, suggesting that cap-dependent translation may be affected. In addition, mitochondrial membrane (MOM) potential of cells overexpressing mutant eIF4G1 was substantially reduced with hydrogen peroxide treatment compared to cells overexpressing wildtype eIF4G1, suggesting that eIF4G1 mutants may have defective stress response pathways.[18]

There is also evidence showing that eIF4G1 interacts with α-synuclein, another PD gene and also a major component of Lewy bodies. In a yeast α-synuclein model, overexpression of yeast eIF4G1 homolog TIF4631 or human eIF4G1 suppressed α-synuclein toxicity, while upregulation of R1205H eIF4G1 was not able to suppress the toxicity.[22] In addition, the cellular level of wildtype eIF4G1 was reported to be crucial in its neuroprotective role in ischemia-induced neuronal death. Vosler et al. showed that calpain, a protease activated by ischemia, degrades eIF4G1 and leads to persistent protein synthesis inhibition. The authors showed that the protein synthesis rate was correlated with the viability after oxygen–glucose deprivation (OGD) in primary neurons, and eIF4G1 overexpression increased neuronal viability after OGD treatment.[68] These studies suggest that mutations of eIF4G1 may alter its function under stress conditions, resulting in neurodegeneration. Further studies on eIF4G1 mutations including genetic association to PD, as well as molecular pathologic mechanism are warranted, including examining the effects of eIF4G mutations on translation and neuronal viability in animal models.

2.3 Other PD Genes and Protein Translation

2.3.1 α-Synuclein

α-synuclein, an autosomal dominant PD gene, is best studied in terms of its cellular characteristics. Aggregated α-synuclein is a major component of Lewy bodies, and misfolding and/or excess of α-synuclein is regarded as a main trigger for α-synuclein-related neurotoxicity.[33] So far, more than 18 pathogenic missense mutations have been identified in the α-synuclein gene (*SNCA*). Along with misfolding of α-synuclein, duplications and triplications in the gene have also been observed in patients with an earlier onset and faster progression of the disease.[33,69] This pinpoints to the relevance of misregulated α-synuclein expression levels in the pathogenesis of PD. In addition to its previously discussed genetic relevance with eIF4G1, alternative α-synuclein transcript usage caused by translational misregulation has been described in the literature. For example, unbiased differential coexpression analysis of PD brain tissue detected preferential usage of an SNCA transcript with a longer 3'UTR (aSynL). Common variants associated with PD risk in this longer 3'UTR promoted aggregation and translation of aSynL, which relocalized from synapses to mitochondria.[21] In addition, presence of a target site for miR-34b, whose level was reported to be altered in PD patients, only in the longer but not shorter 3'UTR, offers a potential mechanism for differential usage of transcripts in the context of disease.[70] Accordingly, inhibition of miR-34b in SH-SY5Y cells leads to higher levels of α-synuclein and its aggregates.[71]

2.3.2 DJ-1

DJ-1 has several reported functions including protection against oxidative stress and molecular chaperone activity, and its deletions and point mutations have been found in autosomal recessive PD cases.[33] However, the exact role of DJ-1 in neurons is still unclear. A role in translation has been suggested for YajL, the bacterial homolog of DJ-1, which was reported to interact with several ribosomal proteins and to associate with translating ribosomes. A mutant form, yajL, dissociated from polysomes under stress conditions and increased the number of frameshifts during translation, suggesting its role in elongation.[25] Furthermore, DJ-1 was also identified in a *Drosophila* screen as a suppressor of the PTEN pathway and these results were verified in mammalian cells. Consistent with this, an additional study found that manipulation of the phosphoinositide 3-kinase (PI3K)/Akt pathway could modulate the phenotype associated with DJ-1 downregulation.[72] PI3K/Akt feeds into the mechanistic target of rapamycin (mTOR) pathway, which is one of the major regulators of protein synthesis.[73] However, it remains to be elucidated whether DJ-1 has any impact on the translation regulation branch of the pathway or is rather associated with the prosurvival effects instead.[74]

2.3.3 PINK1 and Parkin

Recent studies of PINK1 and Parkin, both related to autosomal recessive familial PD, have suggested that there is a converging role of those two proteins in mitochondrial quality control, including mitophagy and mitochondrial biogenesis.[75] Although the molecular mechanisms are still under investigation, there are several clues suggesting that translation-related pathways are affected in PINK1 and/or Parkin mutants. In *Drosophila*, PINK1 and Parkin were found to genetically interact with a major component of the translation machinery, 4E-BP. Upregulation of 4E-BP by genetic or pharmacological means (overexpression and treatment with rapamycin, respectively) prevented DA neuron loss and rescued both flight and climbing activities in PINK1 and Parkin mutant flies.[76] Similarly, both ribosomal protein S6 kinase (S6K) and ATG1 of the mTOR pathway were genetic modifiers of a *Drosophila* PINK1 model of PD. Overexpression of S6K or its constitutively active forms disrupted jump/flight activity, depleted ATP levels, and caused DA neuron and muscle degeneration in PINK1 siRNA flies. Either downregulation of translation or upregulation of autophagy by genetic manipulation of other components of the mTOR pathway alleviated the aforementioned phenotypes.[77] In addition, loss of function of eIF4E was able to rescue the fertility defects of a *Drosophila* Parkin mutant, as well as to suppress the decrease in the pupal viability and adult body size observed in mutant flies. Parkin and eIF4E colocalized in vivo in the egg chamber during oogenesis.[23] However, these Parkin mutant flies did not show any DA neuron degeneration, and therefore it remains to be tested whether this genetic interaction also occurs in the fly brain.

A more direct potential role of these proteins in the mRNA translation process has also been reported. For instance, AIMP2/p38, a subunit of the aminoacyl-tRNA synthetase complex, was identified as a Parkin substrate, although AIMP2 overexpression in the PC12 rat cell line did not show any noticeable effects on global translation rates.[24,78] Therefore, further investigation on the potential effects on translation is warranted.

PINK1 was also suggested to be involved in the translation-mediated response to hypoxia. Loss of PINK1 hindered the induction of hypoxia-inducible factor alpha (HIF-1α) and consequent activity in response to hypoxia, due to a decrease in synthesis of new HIF-1α protein. Concomitantly, loss of PINK1 and subsequent impaired HIF-1α production resulted in an impaired switch of cap-dependent to cap-independent translation under hypoxic stress, which correlated with the accumulation of hyperphosphorylated 4E-BP1. Overexpression of 4E-BP1 could rescue the translational defects in PINK1 mutants, suggesting that the normal role of PINK1 under hypoxic conditions is to mediate the switch from cap-dependent to cap-independent translation in a 4E-BP1-dependent manner.[26]

Finally, in line with their role in mitochondrial quality control, a more recent study found that PINK1 and Parkin directly regulate local translation of nuclearly encoded respiratory chain-related mRNAs (nRCCs) on the outer MOM. Gehrke et al. reported that particular nRCCs were enriched in mitochondrial fractions of fly neuromuscular tissue and mammalian cells, supposedly by direct association with PINK1, as suggested by cross-link immunoprecipitation assays.[19] Participation of the protein import machinery, including translocase of the outer membrane (TOM) and translocase of the inner membrane (TIM) complexes, was necessary for the process. PINK1 enabled translational activation of nRCCs on the MOM by displacing and/or directly competing with translational repressor Pumilio, as well as PINK1- and Parkin-regulated ubiquitination of translational repressor Glorund/hnRNP-F. Notably, PINK1 and Parkin mutations attenuated this process in flies and mammalian cells, pointing toward impaired translation of nRCCs as a possible underlying cause in the pathology of this form of PD.[19]

2.3.4 VPS35

VPS35 is a component of the retromer complex, which is involved in retrograde transport from endosomes to the trans-Golgi network. VPS35 mutations have been recently identified as autosomal dominant PD mutations, suggesting that they may disrupt protein homeostasis by defective protein trafficking.[79,80] Although the role of VPS35 in PD pathogenesis is still unclear, genetic interactions with other PD genes have been identified, suggesting potential common pathways underlying PD pathology. For example, overexpression of wildtype VPS35 rescued G2019S LRRK2 phenotypes in *Drosophila* and cell culture models.[31] In addition, it has been reported that it genetically interacts with eIF4G1 in yeast and worms. In a genome-wide screen for synthetic lethality, Dhungel et al. observed that deletion of VPS35 was toxic in the background of TIF4631 overexpression. Upregulation of other retromer complex components was protective against TIF4631, indicating that intact trafficking between the endosomal compartment and the Golgi network is pivotal to neutralize exacerbated protein translation. It remains to be ascertained whether such genetic interaction also exists in mammals. The potential interaction between trafficking and protein synthesis phenotypes in the context of PD will be discussed further.[22]

In summary, the aforementioned studies highlight the importance of protein synthesis in the role of diverse PD genes in health and disease (Fig. 9.2). Since current knowledge on the molecular and cellular mechanisms of PD-linked mutations is limited, further research should be conducted to determine whether there is a unified and converging mechanism underlying seemingly distinctive molecular processes. This may represent a significant therapeutic leap, as modulating those processes may be an amenable strategy for the different forms of the disease.

3 PROTEIN TRANSLATION AND NEUROTOXICITY IN PARKINSON'S DISEASE

In this section, we will discuss possible pathogenic mechanisms underlying aberrant translation associated with PD genes. Knowledge on the molecular mechanisms can suggest novel therapeutic targets and, hence, we will discuss potential opportunities for therapeutic intervention involved in translation-related defects as well.

3.1 Protein Homeostasis and PD

Although there are various pathways affected by different PD-linked mutations, protein homeostasis seems to be a common pathway involved in autosomal dominant PD (Fig. 9.3). As previously discussed, α-synuclein aggregation has been linked to protein quality control, UPR, and downstream stress signaling. LRRK2 and EIF4G1 are directly involved in translational regulation, and there is also growing evidence suggesting that PINK1 and Parkin mutations can cause defective translational regulation.

3.1.1 Translational Regulation, Metabolism and Mitochondrial Stress

The relationship between mitochondrial stress and neurodegenerative diseases has been repeatedly demonstrated by separate studies, from morphological abnormalities to functional defects. Defects in mitochondrial function can lead to higher cellular oxidative stress through increased reactive oxygen species (ROS) generation, thereby contributing to neuronal cell death. Since mitochondria act as a cellular power plant to provide

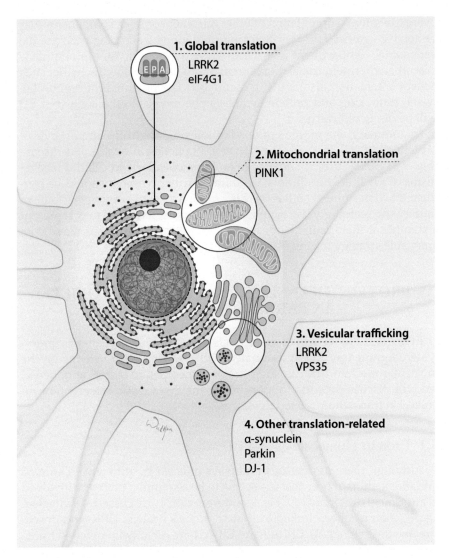

FIGURE 9.2 **Familial PD mutations and the associated cellular pathways.** Known familial PD mutations affecting protein homeostasis are depicted. (1) Global translation: LRRK2 is known to increase global translation and also affect miRNA pathway. eIF4G1 is a translation initiation factor and suggested to affect cap complex formation. (2) Mitochondrial translation: PINK1 was suggested to regulate translation of mitochondrial proteins. Of note, they are encoded in the nucleus and not in the mitochondrial genome, so mitochondrial translation here does not imply translation by mitochondrial ribosomes. (3) Vesicular trafficking: LRRK2 is known to affect vesicular trafficking and autophagy. VPS35 is a retromer complex component. (4) Other translation-related: α-synuclein has a genetic link to eIF4G1, parkin is involved in the regulation of mitochondrial protein translation with PINK1, DJ-1 ortholog was suggested to interact with ribosomes in prokaryotes. *(Illustration: I-Hsun Wu)*

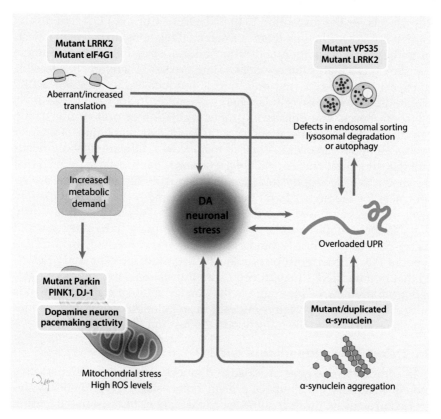

FIGURE 9.3   **Protein synthesis, protein homeostasis, and neuronal toxicity in DA neurons.** Increased protein synthesis can lead to increased metabolism and also trigger the unfolded protein response (UPR) by chaperone overload. Aberrant translation may decrease the fidelity of protein synthesis, thereby causing a higher burden to the protein folding machineries and also increasing metabolic needs by the cost of futile translation. Protein aggregation including α-synuclein aggregation can also absorb protein-folding capacity of the neuron, and thereby triggering the UPR. Vesicular trafficking/lysosomal pathways/ autophagy defects can activate the UPR and increase metabolic demands by the failure in protein quality control. Increased metabolic demands may elevate mitochondrial stress. Combined with the pacemaking activities of DA neurons, this increase may especially be detrimental to DA neurons. *(Illustration: I-Hsun Wu)*

ATP through oxidative phosphorylation (see also Chapter 2), chronically increased metabolic demands augment mitochondrial activity, causing long-term mitochondrial stress and age-dependent cell death.[33,81] Furthermore, in PD, due to the distinctive pacemaking and channel activities of *substantia nigra pars compacta* (SNc) DA neurons, it has been reported that SNc DA neurons suffer from higher levels of endogenous mitochondrial oxidative stress, especially compared to the neighboring ventral tegmental area (VTA) DA neurons. This difference has been suggested as one of the key aspects for the susceptibility of SNc DA neurons in PD.[82,83] In this regard, it is noteworthy that there is a potential relationship between defects in translational regulation and mitochondrial stress. First of all, protein translation is a highly energy-demanding process. Therefore, it is feasible to expect that the reported bulk translation increase by G2019S LRRK2 would directly increase cellular energy demands. In addition to the direct metabolic surge, any defects in proper regulation of protein translation can incur additional metabolic cost by inefficient regulation of cellular processes, and it can eventually lead to increased cellular metabolic needs and mitochondrial burden. Notably, it has previously been suggested that the distinctive electrophysiological characteristics of DA neurons are responsible for the higher basal metabolic rate and mitochondrial oxidant level in SNc, compared to VTA DA neurons.[82] This may be relevant to the SNc susceptibility in PD, and thus it is tempting to hypothesize that increased metabolic needs stand as a central toxicity mechanism in certain types of PD. Furthermore, as previously discussed, a recent study reported that mutant PINK1 may directly alter translational regulation of mitochondrial proteins.[19] This suggests that, on top of the metabolic demands caused by increased protein synthesis, misregulation of a group of genes can also contribute to the mitochondrial stress in PD.

3.1.2 Protein Quality Control and PD

In addition to energetic overload and mitochondrial stress, hyperactive protein synthesis may impact protein homeostasis. Hypothetically, increased protein synthesis requires more chaperones to assist in the folding of the increased amount of protein generated, and failure to meet this need can incur inappropriate protein folding and even aggregation, triggering UPR.[84] The importance of UPR in protein aggregation and neurodegenerative diseases has been repetitively suggested. For instance, in prion-mediated neurodegeneration and certain cases of AD, UPR activation and subsequent phospho-eIF2α-mediated downregulation of critical genes in neurons has been proposed as a toxicity mechanism.[7,13,85] Although there is no clear link between increased protein synthesis and α-synuclein aggregation yet, protein quality control defects in a hyperactive protein translation condition and its relationship to protein aggregation in PD requires further investigation. Furthermore, besides global downregulation of protein translation, UPR triggers eIF2α- and uORF-mediated upregulation of certain genes required for UPR resolution, such as ATF4.[5,6] Precise regulation of translation initiation is key to achieve uORF-mediated induction of expression. However, translational defects can also impair the uORF-mediated stress-responsive gene induction; it deters the expression of survival genes and thereby further exacerbates imbalance in protein homeostasis. In theory, hyperactive protein translation can trigger protein aggregation and concomitantly hinder protective UPR downstream gene expression, causing a detrimental effect to the cell. Furthermore, recent evidence suggests that vesicular dynamics and autophagy are defective in PD. First of all, α-synuclein aggregation has been reported to block vesicular trafficking, while overexpression of Rab3a and Rab8a showed protective effects from α-synuclein-induced toxicity.[86] As previously discussed, Rab7la is a PD risk factor (PARK16) and it has been shown to interact with LRRK2, both genetically and physically.[31,87] Rab proteins are Ras superfamily of small GTPases and also peripheral membrane proteins linked to vesicular dynamics. Retromer complex component VPS35 mutations have also been identified as a cause of autosomal dominant PD.[79,80] Moreover, it has been reported that α-synuclein inhibits autophagy by reducing autophagosome formation, and several LRRK2 genetic models showed autophagy defects in neurons as well.[48,88,89] These findings from various PD risk variants suggest a common cellular mechanism in PD: defective vesicular trafficking and consequent perturbed protein quality control. Vesicular dynamics and autophagy mechanisms are also involved in protein modification, localization, degradation, and recycling of proteins, making it a crucial part of protein homeostasis. Direct relationship between vesicular trafficking, autophagy, and translation has not been revealed yet. However, these mechanisms also suggest a potential link between protein homeostasis and Lewy body formation, a hallmark of PD.

3.1.3 Gene-Specific Effects and Protein Homeostasis

Among the PD-linked mutations, LRRK2 has been most thoroughly studied in terms of its effects on the translation machinery. The recent study on its substrate s15 provides a paradigm-shifting idea that a post-translational modification of a single ribosomal protein can have a huge impact on translational regulation.[28] In the past, the ribosome was regarded as a molecular machine as a whole, but now it is generally accepted that there are specialized ribosomes that enable gene expression in a time- and cell-specific manner. These specialized ribosomes have been hypothesized to differ in their relative abundance of ribosomal proteins, the posttranslational modifications on them or the factors that associate with them, in order to drive the differential translation of particular mRNA subsets.[90] A relevant example is Rpl28, which is located at the P site of the ribosome. Rpl28 is heavily ubiquitinylated during the S phase of the cell cycle in *Saccharomyces cerevisiae*, and this has been shown to have a stimulatory effect in translation. Similarly, the LRRK2 phosphorylation site T136 of s15 resides in its C-terminal tail, which extends to the ribosomal A site during elongation and termination of translation.[91] Therefore, it is possible that a group of genes with higher sensitivity to the translational effects caused by LRRK2 mutations are differentially regulated in PD through s15 phosphorylation. This potential expression change may distort the protein expression profile, thereby disrupting protein homeostasis. Previously, expression profiling from LRRK2 *Drosophila* models suggested a couple of potential key targets from disrupted translational regulation.[29] Additional translational target analysis from mammalian LRRK2 models should be conducted to reveal additional pathogenic candidates as well as to elucidate detailed molecular mechanisms. In addition, future structural analyses in order to elucidate the conformation of s15 in its phosphorylated versus unphosphorylated form accessing the A site may provide more detailed molecular mechanisms of LRRK2- and s15-mediated translational abnormality in PD.

3.2 mTOR Pathway and PD

Organisms require a mechanism of adaptation to the dynamic changes in their surroundings. In mammals, the mTOR pathway is one of the major signaling cascades that enable such transition at the metabolic level, facilitating the switch between anabolic and catabolic states in response to environmental cues.[92] As its name indicates, this pathway was originally identified as a TOR, a macrolide antifungal compound widely used in the clinic to avoid transplant rejections due to its antiproliferative and immunosuppressive properties.[93] More recently, some derivatives of rapamycin (known collectively as rapalogs) are being tested in clinical trials for cancer therapy.[92,94]

3.2.1 mTORC1 and mTORC2

mTOR is an atypical serine/threonine kinase and member of the PI3K-related family that comes in two flavors: mTOR complex 1 (mTORC1) and mTOR complex 2 (mTORC2). Both complexes share some subunits such as the catalytic component mTOR, but they are also accompanied by mutually exclusive binding partners. This ultimately results in two different branches of the signaling cascade with distinct inputs and outputs. mTORC1 is better characterized than mTORC2 and plays a role in several processes critical to the cell, including protein and lipid synthesis, autophagy, lysosome biogenesis, and energy metabolism.[93] This is dependent on the integration of a diverse set of cues, namely stress, energy status, oxygen, growth factors, and amino acids.[93,95] A major hub integrating most of these input is tuberous sclerosis complex (TSC), a heterodimer composed of Hamartin/TSC1 and Tuberin/TSC2.[96,97] TSC has a GAP function that controls the GTP-/GDP-binding status of Ras homolog enriched in brain (Rheb), a positive regulator of the mTOR kinase. Signals that downregulate the mTORC1 pathway stimulate the GAP activity of TSC over Rheb and, as a result, GDP-bound Rheb can no longer stimulate kinase activity. The opposite is true in the presence of cues upregulating the pathway.[93] Unlike mTORC1, the more poorly characterized mTORC2 regulates metabolism, apoptosis, survival, growth, and cytoskeletal organization.[98] It does not respond to nutrient stimulation, yet it is affected by growth factors such as insulin, and association with the ribosomes has also been reported to activate mTORC2. Effectors downstream to mTORC2 include the members of the AGC kinase family Akt, PKC-alpha, and SGK1.[99,100] Of note, despite the initial belief that mTORC2 was insensitive to rapamycin, long-term exposure to this compound ultimately inhibits this branch of the pathway too.[101]

3.2.2 Impact of mTORC1 on Protein Homeostasis

mTORC1 activity is tightly connected to the regulation of protein homeostasis. mTORC1 affects protein homeostasis in diverse ways; first, it regulates protein synthesis via the phosphorylation of a number of proteins, including two of its best-characterized downstream effectors, 4E-BP and S6K. Active mTORC1 phosphorylates 4E-BP, and phosphorylated 4E-BP is no longer able to prevent eIF4E from associating to the cap complex to stimulate translation. In addition, S6K phosphorylation by mTOR promotes translation at the level of initiation and elongation via additional effectors like ribosomal protein S6 (RPS6) and eukaryotic elongation factor 2 kinase (eEF2K). Other direct mTOR protein synthesis intermediaries include tripartite-motif containing protein-24 (TIF-1A) and MAF1.[102]

Another critical process impinging on the protein levels within a cell is autophagy. The complex required to initiate autophagy, which is composed of unc-51-like kinase 1 (ULK-1), mammalian autophagy-related gene 13 (ATG13), and focal adhesion kinase family-interacting protein of 200 kDa (FIP200), is inhibited by active mTORC1.[103] Another mTOR substrate involved in this process is death-associated protein 1 (DAP1).[104] In addition, it has been reported that active mTORC1 inhibits lysosomal biogenesis through the transcription factor EB (TFEB).[105]

3.2.3 mTOR in Disease

There are major pathological consequences resulting from the deregulation of mTOR. So far, the majority of studies have reported its dysregulation in cancer, but there are increasing evidence showing the roles of mTOR in metabolic and neurological disorders.[85,106] Notably, all the aforementioned pathologies are age-related, and this correlates with the role of mTOR in the modulation of lifespan in several organisms. It has been reviewed elsewhere that dysfunction of the mTOR pathway is related to neurodegenerative diseases, including AD, Huntington's disease, and PD.[107] In the particular case of the latter, several studies have shown some sort of interaction of PD genes with the mTOR pathway.[108] Although the exact role of mTOR in PD pathogenesis is still under debate, there is some evidence of a genetic and biochemical crosstalk between mTOR and PD proteins, suggestive of a direct involvement of this pathway in the disease.[23,26,27,76] Results from other studies, however, fail to show a direct relationship between PD genes and mTOR, and rather seem to indicate that mTOR may rather exacerbate the pathological phenotype, by aggravating defects in particular processes already gone awry due to underlying PD mutations. In addition, the modulation of mTOR with rapamycin seems to effectively alleviate disease phenotypes in several PD models.[106] These benefits have been ascribed to the action of rapamycin over different mTOR-related processes. Some studies claim that mitigation of the PD phenotypes is due to a rapamycin-induced improvement in autophagic flux, while others attribute it to rapamycin acting upon the translation branch of mTOR pathway.[92,93,106–108] Further research should be conducted to elucidate what parts of this complex pathway are specifically affected in a given PD mutant background, that is, to dissect out whether observed phenotypes can be attributed to aberrant protein synthesis, autophagy, or other mTOR-related processes, and which specific effectors downstream of mTOR are responsible for it. Given the roles of mTOR pathway in protein homeostasis, more in-depth investigation on the roles of mTOR in PD is warranted.

4 FUTURE DIRECTIONS

4.1 Metabolism and SNc Neuronal Specificity

A central question in PD has been to understand the relatively selective degeneration of SNc DA neurons. As previously discussed, higher basal metabolism and reduced mitochondrial reserve capacity have been identified as distinctive characteristics of SNc DA neurons.[83] Thus, high-cost translational defects may have more detrimental effects in them. However, a more direct and causal link between translational abnormalities and metabolic demands needs to be established to test this hypothesis. For example, in the case of increased bulk translation with G2019S LRRK2, measuring basal metabolism rate as well as mitochondrial oxidative stress with LRRK2 kinase inhibitors and/or protein synthesis inhibitors may be informative. Also, monitoring the effects of s15 phosphorylation on metabolism and mitochondrial stress will support the suggested role of s15 phosphorylation in PD pathology.

4.2 Target Identification for Translational Abnormality

Aside from bulk translational effects including increased global translation, the exact translational effects, including finding specific targets, have not been thoroughly addressed. It may be possible that there are additional susceptible genes for PD-linked mutations that can contribute to cell-type specific stress in PD. To test this hypothesis, it would be particularly interesting to survey genome-wide translation profiling from PD-linked mutations, such as G2019S LRRK2. If the target genes identified are important to DA neuronal survival, it may provide a clue to explain their sensitivity. Recently, more advanced tools for "omics" study have been developed, including ribosome profiling—a genome-wide profiling method for translational regulation—and stable isotope labeling by amino acids in cell culture (SILAC)— a quantitative proteomic tool by incorporation of nonradioactive isotopes.[109,110] These techniques will allow us to test those hypotheses through genome-wide and proteome-wide expression profiles with LRRK2 mutants. Needless to say, it will be interesting to apply the translational profiling technique to other PD-linked mutations as well. For example, translational profiling on PINK1 mutants may be informative, with a focus on mitochondrial gene expression changes. Moreover, it is still unclear how eIF4G1 mutations affect translational initiation. Thus, biochemical approaches as well as gene expression analysis can help investigate the molecular mechanisms of eIF4G1 mutations on translation. It is also of particular interest to survey the effects of α-synuclein mutations, from its suggested genetic relationship with eIF4G1 and translation. Furthermore, it might be important to test potential effects of aforementioned mutants on the expression of α-synuclein. Excessive α-synuclein expression has a clear link to PD pathology.[33] Thus, any translational defects increasing α-synuclein expression can be potentially harmful as they may exacerbate α-synuclein aggregation.

4.3 DA Neuron-Specific Expression Profiling

Although it is critical to understand DA neuron specificity in PD, SNc DA neurons are small in number. This is the reason why obtaining DA neuron-specific expression data has been challenging. Recent advancement on genetic tools to study cell-type specific expression may be particularly informative to understand PD. For example, translating ribosome affinity purification (TRAP) has been shown to provide clean and reproducible data from a specific neuronal population. Briefly, TRAP models are expressing enhanced green fluorescent protein (EGFP)-tagged ribosomal protein L10a (Rpl10a) under cell-type specific promoters.[111] For DA neurons, dopamine transporter (DAT) promoter-driven TRAP (DAT-TRAP) mice have been generated, and the expression profile was successfully generated.[112] Combined with PD mouse models, DAT-TRAP mice can provide specific gene expression changes from DA neurons in mouse brain, thereby expanding our knowledge on the effects of PD genes on gene expression including translational defects.

4.4 Translation as a New Druggable Target in PD

There are many translational inhibitors found in nature and also synthesized in the laboratory, targeting different stages in protein translation.

Therefore, it may be promising to test translational inhibitors as a PD treatment, especially in the cases of increased translation and metabolic demands. Furthermore, in regard to the well-characterized roles of mTOR in translation and metabolism, since rapamycin attenuates neurodegenerative phenotypes in diverse PD models and it is a compound already approved by the FDA for the treatment of other conditions, modulating the mTOR pathway with rapamycin seems a promising therapeutic approach for PD.[106] At the same time, new mTOR inhibitors discovered recently also have a potential as a PD treatment, with the advantage that, unlike rapamycin, some of them can completely block all pathways downstream to mTOR.[113] Collectively, these efforts will ultimately result in more specific and safer design and assignment of disease-modifying agents for patients suffering from the devastating PD.

Acknowledgments

This work was supported by grants from the NIH NS3837, the JPB Foundation and the MSCRF grant 2013-MSCRFII-0105-00 (VLD), a New York Stem Cell Foundation-Druckenmiller Fellowship and 2014-MSCRFF-0610 (IM). L.A-A is supported by a La Caixa Foundation grant.

T.M.D. is the Leonard and Madlyn Abramson Professor in Neurodegenerative Diseases. The authors acknowledge the joint participation by the Adrienne Helis Malvin and Diana Henry Helis Medical Research Foundations and their direct engagement in the continuous active conduct of medical research in conjunction with The Johns Hopkins Hospital and the Johns Hopkins University School of Medicine and the Foundation's Parkinson's Disease Programs, H-2013, H-2014, M-1, M-2, M-2013, M-2014. I-Hsun Wu created Figures 9.1, 9.2, 9.3.

References

1. Sonenberg N, Hinnebusch AG. Regulation of translation initiation in eukaryotes: mechanisms and biological targets. *Cell*. 2009;136:731–745.
2. Jackson RJ, Hellen CUT, Pestova TV. The mechanism of eukaryotic translation initiation and principles of its regulation. *Nat Rev Mol Cell Biol*. 2010;11:113–127.
3. Crombie T, Swaffield JC, Brown AJ. Protein folding within the cell is influenced by controlled rates of polypeptide elongation. *J Mol Biol*. 1992;228(1):7–12.
4. Dever TE, Green R. The elongation, termination, and recycling phases of translation in eukaryotes. *Cold Spring Harb Perspect Biol*. 2012;4(7):a013706.
5. Harding HP, Novoa I, Zhang Y, et al. Regulated translation initiation controls stress-induced gene expression in mammalian cells. *Mol Cell*. 2000;6(5):1099–1108.
6. Vattem KM, Wek RC. Reinitiation involving upstream ORFs regulates ATF4 mRNA translation in mammalian cells. *Proc Natl Acad Sci USA*. 2004;101(31):11269–11274.
7. Moreno JA, Radford H, Peretti D, et al. Sustained translational repression by eIF2α-P mediates prion neurodegeneration. *Nature*. 2012;485:507–511.
8. Kim H-J, Raphael AR, LaDow ES, et al. Therapeutic modulation of eIF2α phosphorylation rescues TDP-43 toxicity in amyotrophic lateral sclerosis disease models. *Nat Genet*. 2014;46(2):152–160.
9. Scheper GC, van der Knaap MS, Proud CG. Translation matters: protein synthesis defects in inherited disease. *Nat Rev Genet*. 2007;8(9):711–723.
10. Gkogkas CG, Khoutorsky A, Ran I, et al. Autism-related deficits via dysregulated eIF4E-dependent translational control. *Nature*. 2013;493:371–377.

11. Santini E, Huynh TN, MacAskill AF, et al. Exaggerated translation causes synaptic and behavioural aberrations associated with autism. *Nature*. 2013;493:411–415.

12. Bhattacharya A, Kaphzan H, Alvarez-Dieppa AC, Murphy JP, Pierre P, Klann E. Genetic removal of p70 S6 kinase 1 corrects molecular, synaptic, and behavioral phenotypes in fragile X syndrome mice. *Neuron*. 2012;76(2):325–337.

13. Ma T, Trinh MA, Wexler AJ, et al. Suppression of eIF2α kinases alleviates Alzheimer's disease-related plasticity and memory deficits. *Nat Neurosci*. 2013;16(9):1299–1305.

14. Sharma A, Hoeffer CA, Takayasu Y, et al. Dysregulation of mTOR signaling in fragile X syndrome. *J Neurosci*. 2010;30(2):694–702.

15. Darnell JC, Klann E. The translation of translational control by FMRP: therapeutic targets for FXS. *Nat Neurosci*. 2013;16(11):1530–1536.

16. Cleary JD, Ranum LPW. Repeat-associated non-ATG (RAN) translation in neurological disease. *Hum Mol Genet*. 2013;22:R45–51.

17. Martin I, Abalde-Atristain L, Kim JW, Dawson TM, Dawson VL. Abberant protein synthesis in G2019S LRRK2 Drosophila Parkinson disease-related phenotypes. *Fly (Austin)*. 2014;8(3):165–169.

18. Chartier-Harlin MC, Dachsel JC, Vilarino-Guell C, et al. Translation initiator EIF4G1 mutations in familial Parkinson disease. *Am J Hum Genet*. 2011;89(3):398–406.

19. Gehrke S, Wu Z, Klinkenberg M, et al. PINK1 and Parkin control localized translation of respiratory chain component mRNAs on mitochondria outer membrane. *Cell Metabol*. 2015;21:95–108.

20. Taymans JM, Nkiliza A, Chartier-Harlin MC. Deregulation of protein translation control, a potential game-changing hypothesis for Parkinson's disease pathogenesis. *Trends Mol Med*. 2015;21(8):466–472.

21. Rhinn H, Qiang L, Yamashita T, et al. Alternative α-synuclein transcript usage as a convergent mechanism in Parkinson's disease pathology. *Nat Commun*. 2012;3:1084.

22. Dhungel N, Eleuteri S, Li L-b, et al. Parkinson's disease genes VPS35 and EIF4G1 interact genetically and converge on α-synuclein. *Neuron*. 2015;85:76–87.

23. Ottone C, Galasso A, Gemei M, et al. Diminution of eIF4E activity suppresses parkin mutant phenotypes. *Gene*. 2011;470(1-2):12–19.

24. Lee Y, Karuppagounder SS, Shin J-H, et al. Parthanatos mediates AIMP2-activated age-dependent dopaminergic neuronal loss. *Nat Neurosci*. 2013;16:1392–1400.

25. Kthiri F, Gautier V, Le HT, et al. Translational defects in a mutant deficient in YajL, the bacterial homolog of the parkinsonism-associated protein DJ-1. *J Bacteriol*. 2010;192(23):6302–6306.

26. Lin W, Wadlington NL, Chen L, Zhuang X, Brorson JR, Kang UJ. Loss of PINK1 attenuates HIF-1α induction by preventing 4E-BP1-dependent switch in protein translation under hypoxia. *J Neurosci*. 2014;34:3079–3089.

27. Imai Y, Gehrke S, Wang H-Q, et al. Phosphorylation of 4E-BP by LRRK2 affects the maintenance of dopaminergic neurons in Drosophila. *EMBO J*. 2008;27:2432–2443.

28. Martin I, Kim JW, Lee BD, et al. Ribosomal protein s15 phosphorylation mediates LRRK2 neurodegeneration in Parkinson's disease. *Cell*. 2014;157(2):472–485.

29. Gehrke S, Imai Y, Sokol N, Lu B. Pathogenic LRRK2 negatively regulates microRNA-mediated translational repression. *Nature*. 2010;466(7306):637–641.

30. Gillardon F. Interaction of elongation factor 1-alpha with leucine-rich repeat kinase 2 impairs kinase activity and microtubule bundling in vitro. *Neuroscience*. 2009;163:533–539.

31. Macleod DA, Rhinn H, Kuwahara T, et al. RAB7L1 interacts with LRRK2 to modify intraneuronal protein sorting and Parkinson's disease risk. *Neuron*. 2013;77:425–439.

32. Funayama M, Hasegawa K, Kowa H, Saito M, Tsuji S, Obata F. A new locus for Parkinson's disease (PARK8) maps to chromosome 12p11. 2-q13.1. *Ann Neurol*. 2002;51(3):296–301.

33. Moore DJ, West AB, Dawson VL, Dawson TM. Molecular pathophysiology of Parkinson's disease. *Annu Rev Neurosci*. 2005;28:57–87.

34. Martin I, Kim JW, Dawson VL, Dawson TM. LRRK2 pathobiology in Parkinson's disease. *J Neurochem.* 2014;131(5):554–565.

35. West AB, Moore DJ, Biskup S, et al. Parkinson's disease-associated mutations in leucine-rich repeat kinase 2 augment kinase activity. *Proc Natl Acad Sci USA.* 2005;102:16842–16847.

36. West AB, Moore DJ, Choi C, et al. Parkinson's disease-associated mutations in LRRK2 link enhanced GTP-binding and kinase activities to neuronal toxicity. *Hum Mol Genet.* 2007;16:223–232.

37. Smith WW, Pei Z, Jiang H, Dawson VL, Dawson TM, Ross CA. Kinase activity of mutant LRRK2 mediates neuronal toxicity. *Nat Neurosci.* 2006;9:1231–1233.

38. Cookson MR. Cellular effects of LRRK2 mutations. *Biochem Soc Trans.* 2012;40:1070–1073.

39. Cookson MR. LRRK2 pathways leading to neurodegeneration. *Curr Neurol Neurosci Rep.* 2015;15(7):42.

40. Manzoni C, Mamais A, Dihanich S, et al. Pathogenic Parkinson's disease mutations across the functional domains of LRRK2 alter the autophagic/lysosomal response to starvation. *Biochem Biophys Res Commun.* 2013;441(4):862–866.

41. Kett LR, Boassa D, Ho CC, et al. LRRK2 Parkinson disease mutations enhance its microtubule association. *Hum Mol Genet.* 2012;21(4):890–899.

42. Ramonet D, Daher JP, Lin BM, et al. Dopaminergic neuronal loss, reduced neurite complexity and autophagic abnormalities in transgenic mice expressing G2019S mutant LRRK2. *PLoS One.* 2011;6(4):e18568.

43. Dodson MW, Zhang T, Jiang C, Chen S, Guo M. Roles of the Drosophila LRRK2 homolog in Rab7-dependent lysosomal positioning. *Hum Mol Genet.* 2012;21(6):1350–1363.

44. Nikonova EV, Xiong Y, Tanis KQ, et al. Transcriptional responses to loss or gain of function of the leucine-rich repeat kinase 2 (LRRK2) gene uncover biological processes modulated by LRRK2 activity. *Hum Mol Genet.* 2012;21(1):163–174.

45. Leger-Silvestre I, Milkereit P, Ferreira-Cerca S, et al. The ribosomal protein Rps15p is required for nuclear exit of the 40S subunit precursors in yeast. *EMBO J.* 2004;23(12):2336–2347.

46. Rouquette J, Choesmel V, Gleizes PE. Nuclear export and cytoplasmic processing of precursors to the 40S ribosomal subunits in mammalian cells. *EMBO J.* 2005;24(16):2862–2872.

47. Hansen M, Taubert S, Crawford D, Libina N, Lee SJ, Kenyon C. Lifespan extension by conditions that inhibit translation in Caenorhabditis elegans. *Aging Cell.* 2007;6(1):95–110.

48. Herzig MC, Kolly C, Persohn E, et al. LRRK2 protein levels are determined by kinase function and are crucial for kidney and lung homeostasis in mice. *Hum Mol Genet.* 2011;20(21):4209–4223.

49. Miklavc P, Ehinger K, Thompson KE, Hobi N, Shimshek DR, Frick M. Surfactant secretion in LRRK2 knock-out rats: changes in lamellar body morphology and rate of exocytosis. *PLoS One.* 2014;9(1):e84926.

50. Ali IK, McKendrick L, Morley SJ, Jackson RJ. Truncated initiation factor eIF4G lacking an eIF4E binding site can support capped mRNA translation. *EMBO J.* 2001;20(15):4233–4242.

51. Lomakin IB, Hellen CU, Pestova TV. Physical association of eukaryotic initiation factor 4G (eIF4G) with eIF4A strongly enhances binding of eIF4G to the internal ribosomal entry site of encephalomyocarditis virus and is required for internal initiation of translation. *Mol Cell Biol.* 2000;20(16):6019–6029.

52. Svitkin YV, Herdy B, Costa-Mattioli M, Gingras AC, Raught B, Sonenberg N. Eukaryotic translation initiation factor 4E availability controls the switch between cap-dependent and internal ribosomal entry site-mediated translation. *Mol Cell Biol.* 2005;25(23):10556–10565.

53. Fujioka S, Sundal C, Strongosky AJ, et al. Sequence variants in eukaryotic translation initiation factor 4-gamma (eIF4G1) are associated with Lewy body dementia. *Acta Neuropathol.* 2013;125(3):425–438.

54. Tucci A, Charlesworth G, Sheerin UM, Plagnol V, Wood NW, Hardy J. Study of the genetic variability in a Parkinson's Disease gene: EIF4G1. *Neurosci Lett.* 2012;518(1):19–22.

55. Schulte EC, Mollenhauer B, Zimprich A, et al. Variants in eukaryotic translation initiation factor 4G1 in sporadic Parkinson's disease. *Neurogenetics.* 2012;13(3):281–285.

56. Lesage S, Condroyer C, Klebe S, et al. EIF4G1 in familial Parkinson's disease: pathogenic mutations or rare benign variants? *Neurobiol Aging.* 2012;33(9):2233 e2231-2233 e2235.

57. Nuytemans K, Bademci G, Inchausti V, et al. Whole exome sequencing of rare variants in EIF4G1 and VPS35 in Parkinson disease. *Neurology.* 2013;80(11):982–989.

58. Kalinderi K, Bostantjopoulou S, Katsarou Z, Dimikiotou M, Fidani L. D620N mutation in the VPS35 gene and R1205H mutation in the EIF4G1 gene are uncommon in the Greek population. *Neurosci Lett.* 2015;606:113–116.

59. Blanckenberg J, Ntsapi C, Carr JA, Bardien S. EIF4G1 R1205H and VPS35 D620N mutations are rare in Parkinson's disease from South Africa. *Neurobiol Aging.* 2014;35(2):e441–443:445.

60. Gagliardi M, Annesi G, Tarantino P, Nicoletti G, Quattrone A. Frequency of the ASP620ASN mutation in VPS35 and Arg1205His mutation in EIF4G1 in familial Parkinson's disease from South Italy. *Neurobiol Aging.* 2014;35(10):e2421–2422:2422.

61. Chen Y, Chen K, Song W, et al. VPS35 Asp620Asn and EIF4G1 Arg1205His mutations are rare in Parkinson disease from southwest China. *Neurobiol Aging.* 2013;34(6):e1707–1708:1709.

62. Li K, Tang BS, Guo JF, et al. Analysis of EIF4G1 in ethnic Chinese. *BMC Neurol.* 2013;13:38.

63. Zhao Y, Ho P, Prakash KM, et al. Analysis of EIF4G1 in Parkinson's disease among Asians. *Neurobiol Aging.* 2013;34(4):e1315–1316:1311.

64. Nishioka K, Funayama M, Vilarino-Guell C, et al. EIF4G1 gene mutations are not a common cause of Parkinson's disease in the Japanese population. *Parkinsonism Relat Disord.* 2014;20(6):659–661.

65. Sudhaman S, Behari M, Govindappa ST, Muthane UB, Juyal RC, Thelma BK. VPS35 and EIF4G1 mutations are rare in Parkinson's disease among Indians. *Neurobiol Aging.* 2013;34(10):e2441–2443:2442.

66. Deng H, Wu Y, Jankovic J. The EIF4G1 gene and Parkinson's disease. *Acta Neurol Scand.* 2015;132(2):73–78.

67. Villa N, Do A, Hershey JW, Fraser CS. Human eukaryotic initiation factor 4G (eIF4G) protein binds to eIF3c, -d, and -e to promote mRNA recruitment to the ribosome. *J Biol Chem.* 2013;288(46):32932–32940.

68. Vosler PS, Gao Y, Brennan CS, et al. Ischemia-induced calpain activation causes eukaryotic (translation) initiation factor 4G1 (eIF4GI) degradation, protein synthesis inhibition, and neuronal death. *Proc Natl Acad Sci USA.* 2011;108(44):18102–18107.

69. Martin I, Dawson VL, Dawson TM. Recent advances in the genetics of Parkinson's disease. *Annu Rev Genom Hum Genet.* 2011;12:301–325.

70. Minones-Moyano E, Porta S, Escaramis G, et al. MicroRNA profiling of Parkinson's disease brains identifies early downregulation of miR-34b/c which modulate mitochondrial function. *Hum Mol Genet.* 2011;20(15):3067–3078.

71. Kabaria S, Choi DC, Chaudhuri AD, Mouradian MM, Junn E. Inhibition of miR-34b and miR-34c enhances alpha-synuclein expression in Parkinson's disease. *FEBS Lett.* 2015;589(3):319–325.

72. Yang Y, Gehrke S, Haque ME, et al. Inactivation of Drosophila DJ-1 leads to impairments of oxidative stress response and phosphatidylinositol 3-kinase/Akt signaling. *Proc Natl Acad Sci USA.* 2005;102(38):13670–13675.

73. Nagakubo D, Taira T, Kitaura H, et al. DJ-1, a novel oncogene which transforms mouse NIH3T3 cells in cooperation with ras. *Biochem Biophys Res Commun.* 1997;231(2):509–513.

74. Singh Y, Chen H, Zhou Y, et al. Differential effect of DJ-1/PARK7 on development of natural and induced regulatory T cells. *Sci Rep.* 2015;5:17723.

75. Scarffe LA, Stevens DA, Dawson VL, Dawson TM. Parkin and PINK1: much more than mitophagy. *Trends Neurosci.* 2014;37(6):315–324.

76. Tain LS, Mortiboys H, Tao RN, Ziviani E, Bandmann O, Whitworth AJ. Rapamycin activation of 4E-BP prevents parkinsonian dopaminergic neuron loss. *Nat Neurosci.* 2009;12:1129–1135.

77. Liu S, Lu B. Reduction of protein translation and activation of autophagy protect against PINK1 pathogenesis in Drosophila melanogaster. *PLoS Genet.* 2010;6(12):e1001237.

78. Corti O, Hampe C, Koutnikova H, et al. The p38 subunit of the aminoacyl-tRNA synthetase complex is a Parkin substrate: linking protein biosynthesis and neurodegeneration. *Hum Mol Genet.* 2003;12(12):1427–1437.

79. Zimprich A, Benet-Pages A, Struhal W, et al. A mutation in VPS35, encoding a subunit of the retromer complex, causes late-onset Parkinson disease. *Am J Hum Genet.* 2011;89(1):168–175.

80. Vilarino-Guell C, Wider C, Ross OA, et al. VPS35 mutations in Parkinson disease. *Am J Hum Genet.* 2011;89(1):162–167.

81. Gupta A, Dawson VL, Dawson TM. What causes cell death in Parkinson's disease? *Ann Neurol.* 2008;64(Suppl 2):S3–15.

82. Guzman JN, Sanchez-Padilla J, Wokosin D, et al. Oxidant stress evoked by pacemaking in dopaminergic neurons is attenuated by DJ-1. *Nature.* 2010;468:696–700.

83. Sulzer D, Surmeier DJ. Neuronal vulnerability, pathogenesis, and Parkinson's disease. *Mov Disord.* 2013;28(1):41–50.

84. Hetz C, Mollereau B. Disturbance of endoplasmic reticulum proteostasis in neurodegenerative diseases. *Nat Rev Neurosci.* 2014;15(4):233–249.

85. Buffington SA, Huang W, Costa-Mattioli M. Translational control in synaptic plasticity and cognitive dysfunction. *Annu Rev Neurosci.* 2014;37:17–38.

86. Gitler AD, Bevis BJ, Shorter J, et al. The Parkinson's disease protein alpha-synuclein disrupts cellular Rab homeostasis. *Proc Natl Acad Sci USA.* 2008;105(1):145–150.

87. Gómez-Suaga P, Rivero-Ríos P, Fdez E, et al. LRRK2 delays degradative receptor trafficking by impeding late endosomal budding through decreasing Rab7 activity. *Hum Mol Genet.* 2014;23:6779-6796 ddu395.

88. Su YC, Qi X. Inhibition of excessive mitochondrial fission reduced aberrant autophagy and neuronal damage caused by LRRK2 G2019S mutation. *Hum Mol Genet.* 2013;22(22):4545–4561.

89. Winslow AR, Chen CW, Corrochano S, et al. alpha-Synuclein impairs macroautophagy: implications for Parkinson's disease. *J Cell Biol.* 2010;190(6):1023–1037.

90. Xue S, Barna M. Specialized ribosomes: a new frontier in gene regulation and organismal biology. *Nat Rev Mol Cell Biol.* 2012;13(6):355–369.

91. Spence J, Gali RR, Dittmar G, Sherman F, Karin M, Finley D. Cell cycle-regulated modification of the ribosome by a variant multiubiquitin chain. *Cell.* 2000;102(1):67–76.

92. Iadevaia V, Huo Y, Zhang Z, Foster LJ, Proud CG. Roles of the mammalian target of rapamycin, mTOR, in controlling ribosome biogenesis and protein synthesis. *Biochem Soc Trans.* 2012;40:168–172.

93. Laplante M, Sabatini DM. mTOR signaling in growth control and disease. *Cell.* 2012;149(2):274–293.

94. Chantranupong L, Wolfson RL, Sabatini DM. Nutrient-sensing mechanisms across evolution. *Cell.* 2015;161(1):67–83.

95. Efeyan A, Zoncu R, Sabatini DM. Amino acids and mTORC1: from lysosomes to disease. *Trends Mol Med.* 2012;18(9):524–533.

96. Gao X, Zhang Y, Arrazola P, et al. Tsc tumour suppressor proteins antagonize aminoacid-TOR signalling. *Nat Cell Biol.* 2002;4(9):699–704.

97. Tee AR, Fingar DC, Manning BD, Kwiatkowski DJ, Cantley LC, Blenis J. Tuberous sclerosis complex-1 and -2 gene products function together to inhibit mammalian target of rapamycin (mTOR)-mediated downstream signaling. *Proc Natl Acad Sci USA.* 2002;99(21):13571–13576.

98. Cybulski N, Hall MN. TOR complex 2: a signaling pathway of its own. *Trends Biochem Sci.* 2009;34(12):620–627.

99. Alayev A, Holz MK. mTOR signaling for biological control and cancer. *J Cell Physiol.* 2013;228(8):1658–1664.

100. Hung CM, Garcia-Haro L, Sparks CA, Guertin DA. mTOR-dependent cell survival mechanisms. *Cold Spring Harb Perspect Biol.* 2012;4(12).
101. Sarbassov DD, Ali SM, Sengupta S, et al. Prolonged rapamycin treatment inhibits mTORC2 assembly and Akt/PKB. *Mol Cell.* 2006;22(2):159–168.
102. Ma XM, Blenis J. Molecular mechanisms of mTOR-mediated translational control. *Nat Rev Mol Cell Biol.* 2009;10:307–318.
103. Ganley IG, Lam du H, Wang J, Ding X, Chen S, Jiang X. ULK1. ATG13. FIP200 complex mediates mTOR signaling and is essential for autophagy. *J Biol Chem.* 2009;284(18):12297–12305.
104. Wazir U, Wazir A, Khanzada ZS, Jiang WG, Sharma AK, Mokbel K. Current state of mTOR targeting in human breast cancer. *Cancer Genomics Proteomics.* 2014;11(4):167–174.
105. Kim YC, Guan KL. mTOR: a pharmacologic target for autophagy regulation. *J Clin Invest.* 2015;125(1):25–32.
106. Bove J, Martinez-Vicente M, Vila M. Fighting neurodegeneration with rapamycin: mechanistic insights. *Nat Rev Neurosci.* 2011;12(8):437–452.
107. Lipton JO, Sahin M. The neurology of mTOR. *Neuron.* 2014;84(2):275–291.
108. Heras-Sandoval D, Perez-Rojas JM, Hernandez-Damian J, Pedraza-Chaverri J. The role of PI3K/AKT/mTOR pathway in the modulation of autophagy and the clearance of protein aggregates in neurodegeneration. *Cell Signal.* 2014;26(12):2694–2701.
109. Ingolia NT, Ghaemmaghami S, Newman JRS, Weissman JS. Genome-wide analysis in vivo of translation with nucleotide resolution using ribosome profiling. *Science.* 2009;324:218–223.
110. Kitchen RR, Rozowsky JS, Gerstein MB, Nairn AC. Decoding neuroproteomics: integrating the genome, translatome and functional anatomy. *Nat Neurosci.* 2014;17(11):1491–1499.
111. Heiman M, Schaefer A, Gong S, et al. A translational profiling approach for the molecular characterization of CNS cell types. *Cell.* 2008;135(4):738–748.
112. Brichta L, Shin W, Jackson-Lewis V, et al. Identification of neurodegenerative factors using translatome-regulatory network analysis. *Nat Neurosci.* 2015;18(9):1325–1333.
113. Thoreen CC, Chantranupong L, Keys HR, Wang T, Gray NS, Sabatini DM. A unifying model for mTORC1-mediated regulation of mRNA translation. *Nature.* 2012;486:109–113.

Index

Printed in the United States
By Bookmasters